Construction
Accounting
Manual

RICHARD S. HICKOK, CPA, is Chairman Emeritus of the international accounting firm of KMG Main Hurdman. A certified public accountant in New York and several other states, Mr. Hickok has been involved in construction accounting throughout his career. He served as Chairman of the AICPA Construction Contractor Guide Committee that authored *Construction Contractors Audit and Accounting Guide*. He has also served as Chairman of the Future Issues Committee of the AICPA and as a trustee of the Financial Accounting Foundation.

Construction Accounting Manual

Editor

RICHARD S. HICKOK, CPA

Chairman Emeritus
KMG Main Hurdman

Simonoff Accounting Series

WARREN, GORHAM & LAMONT
Boston • New York

This publication is designed to provide accurate and authoritative information in regard to the subject matter covered. It is sold with the understanding that neither the author nor the publisher is engaged in rendering legal, accounting, or other professional service. If legal advice or other expert assistance is required, the services of a competent professional should be sought.

Contributing Authors

Partners and management group personnel of KMG Main Hurdman:

Thomas E. Brightbill
(Chapter 5)

Charles L. Jacobson
(Chapter 2)

Francis J. Callahan
(Chapter 13)

Larry D. Jaynes
(Chapter 8)

John L. Callan
(Chapter 10)

John E. McEwen
(Chapter 5)

Bernard D. Dusenberry
(Chapter 3)

Joseph F. Potter
(Chapter 18)

Richard S. Hickok
(Chapters 1 and 11)

Roger B. Shlonsky
(Chapter 17)

Other contributors:

Eugene S. Abernathy
President, Eugene Abernathy, CPA, PC
(Chapter 4)

Kenneth R. Alderman
Vice-President—Finance, Jones Group
(Chapter 9)

Vincent J. Borelli
Senior Vice-President, Marsh & McLennan, Incorporated
(Chapter 14)

Norman G. Fornella
Vice-President—Finance and Computer Services, Gibbs & Hill, Inc.
(Chapter 7)

Robert Gardella
Senior Vice-President, The Bank of New York
(Chapter 12)

Jack P. Gibson
  Vice-President and Managing Editor, International Risk
  Management Institute, Inc.
  (Chapter 15)

Edward B. Lozowicki
  Partner, Pettit & Martin
  (Chapter 16)

William S. McIntyre
  President, ARTEX Insurance Agency, Inc.; Chairman, American
  Risk Transfer Insurance Company, Ltd.
  (Chapter 15)

Thomas W. McRae
  Vice-President, American Institute of Certified Public Accountants
  (Chapter 3)

Elliott C. Robbins
  Vice-President—Finance, the L.E. Myers Company Group
  (Chapter 11)

G. Barry Wilkinson
  CPA; former partner, KMG Main Hurdman
  (Chapter 6)

Preface

CONSTRUCTION ACCOUNTING MANUAL has been prepared with the purpose of covering virtually every significant accounting and financial area of concern to contractors, their financial personnel, CPAs serving the industry, and users of contractors' financial statements. These groups will find that the *Manual* constitutes an extremely valuable and practical reference guide for addressing most accounting and financial issues related to the construction industry. Furthermore, the *Manual* will be updated at least annually, reflecting emerging accounting and auditing standards that have an impact on contractors, changes in law and regulations, and other relevant matters that affect the financial operations of contractors. The intent has been to create a one-volume reference book to assist contractors, financial and accounting personnel, and users of contractors' financial statements in performing their responsibilities more effectively by providing comprehensive and current coverage of topics of financial relevance to the construction industry.

When Joe Palazzolo and Gene Simonoff first discussed the possibility of compiling a book on financial matters pertinent to the construction industry, I was, of course, flattered but less than enthusiastic. However, after several meetings, I agreed to undertake the project after a plan was jointly developed that allowed me to continue my other activities and responsibilities concurrently. One of the major factors in the "go" decision to accept this challenge was my long professional involvement and interest in construction accounting issues, which started early in my professional career and has continued over the ensuing years, culminating with my service as Chairman of the AICPA Construction Contractor Guide Committee that authored the *Construction Contractors Audit and Accounting Guide*, published in 1981.

For many years, there was a dearth of literature on contractor accounting and financial subjects; at the same time, there was an almost complete absence of authoritative accounting pronouncements on the unique accounting and financial input on aspects of the industry. In addition, there was no existing reference book that dealt comprehensively with a broad scope of financially related topics unique to the construction industry. Finally, any reference text published prior to 1981 had to be considered out of date because it could not reflect many of the recent developments that have had a significant impact on the construction industry. For these reasons and many more, *Construction Accounting Manual* seemed to be a necessary and welcome addition to the existing literature.

One further comment is necessary on the contributing authors. All contributors are experienced, knowledgeable professionals in their respective areas of expertise. They blend a strong theoretical and technical understanding with a practical "hands-on" approach to the real world. In reading and reviewing their contributions, I was impressed! It is a privilege for me as editor to be associated with this group of talented professionals from so many diverse backgrounds, all of whom have an overriding interest in the construction industry.

ACKNOWLEDGMENTS

I hope that I have appropriately noted the significant contributions of the contributing authors, for they made it happen and they deserve the credit. (I would like to particularly mention Vince Borelli, Senior Vice-President, Marsh & McLennan, Inc., an old friend and a professional in every sense of that word, who died on May 8, 1985.)

In addition, I would like to acknowledge the contributions of others. First, I had the advice and guidance of a talented Editorial Advisory Committee composed of KMG Main Hurdman partners knowledgeable and experienced in the construction industry. Their suggestions as to the scope of the *Manual*, as well as their assistance in the selection of the contributing authors, was invaluable:

> Bernard D. Dusenberry, *Retired Partner, KMG Main Hurdman, Pittsburgh*
>
> V.C. Lokey, *Partner, KMG Main Hurdman, Charlotte*
>
> Abraham J. Nassar, *Partner, KMG Main Hurdman, Boston*

I also cannot overlook the advice, guidance, and assistance of Joseph DiBenedetto, C.M.A. Professor of Accounting, Pace University (Westchester). Joe's diligent help in reviewing, editing, and often rewriting the submitted drafts of the contributing authors was invaluable, especially when the material often required considerable effort to reflect properly their respective knowledge and expertise.

There were others whose names are not included as contributing authors but who, nevertheless, helped to make it all happen through their participation. At the risk of omission, there are several who cannot be overlooked: Mervyn L. Koster, a retired partner of KMG Main Hurdman, Los Angeles—mentor and friend—whose background and knowledge in the construction industry is awesome and who worked with many of the giants of the industry. In addition, the contributions of Robert E. Decker, Senior Tax Partner, KMG Main Hurdman, San Francisco; Jack V. Hoffman, retired Managing Partner, KMG Main Hurdman, San Francisco; and Murray B. Schwartzberg, Tax Manager, KMG Main Hurdman, New York, were invaluable.

I cannot fail to acknowledge the contribution of Lynne Farber, my editor at Warren, Gorham & Lamont, Inc. Without her enthusiasm, drive, unfailing good humor, and competence, it would not have happened. I also want to acknowledge my secretary, Adele M. Thaler. We have worked together for over thirty years, and her loyalty and support made it all a lot easier.

Lastly and most importantly, I want to thank Boots. It takes something special to be married to a CPA and she is something special!

RICHARD S. HICKOK

New York, New York
June 1985

Summary of Contents

Table of Contents

1 The Construction Industry

2 Accounting for Joint Ventures

3 Income Recognition

4 *Contract Costs*

5 Management Information Systems

6 Internal Controls

7 Internal Auditing

8 External Auditing

9 *Relationship With a CPA Firm*

10 External Reporting

11 Cash Management

12 Financing

13 Contract Change Management

14 *Surety Bonds/Company Agent*

15 *Risk Management*

16 Legal Matters

17 Litigation Support

18 Taxation and Tax Planning

Appendix

Glossary

Index

The Construction Industry

Richard S. Hickok

The author wishes to acknowledge the contributions of Mervyn L. Koster, retired partner of Main Hurdman, to the contents of this chapter.

INTRODUCTION

The faces of cities are constantly being renewed. As the economy expands, new replaces old; old is done over. The economic growth of our society is reflected in its construction: new corporate headquarters and office buildings stretch toward the sky; low, sleek facilities are developed for production; roads are built to take one there and back. No two structures are alike, yet all share a common purpose: to promote efficiency through change.

Construction is the process of organizing materials, labor, and capital resources in order to build edifices, roads, bridges, and the like. The rise and fall of construction activity signals the growth and renewal or, conversely, the decline of business. With creativity and energy, thousands of construction entities continually remodel the appearance of the cityscape and the countryside. Those involved in the construction business range from the local contractor for whom thousands of dollars in sales is significant to the giant company with billions of dollars in sales; each contributes to the industry and creates its own special part of the overall change.

Construction is an industry that is not only different from all others, but is diverse from within as well. Tall buildings, superhighways, bridges, hydroelectric dams, and even entire cities are the subject matter of this industry. Endeavors as simple as repairing a corporate parking lot, rehabilitating an old warehouse, or just installing a piece of equipment are also within its domain.

The construction industry is comprised of many parts that together make up a very large piece of the economy. In the United States, the construction industry accounts for 5 percent of the gross national product (GNP).

The purpose of this reference work is to provide as complete an understanding of the construction business from a financial viewpoint as possible. Some basic concepts and definitions are developed in this chapter, so that the reader who is not expert in this field can gather basic information for the in-depth material that follows.

The 1984 edition of *Statistical Abstract of the United States* sampled 181,000 construction establishments and separated those entities into three major groups: general building contractors, heavy construction contractors, and special trade contractors. These groups accounted for $98.1 billion of the total $98.9 billion value added to the GNP by construction. Of this total, $32.3 billion was in the residential housing industry. The residential building industry is not explored in this book because the overlap existing between the construction and the real estate industries would make detailed development of the contract construction business unnecessarily complicated. Furthermore, the nature of the residential building industry is real estate development from an owner's viewpoint, whereas the focus of this book is on contract building.

BASIC TERMS

"Contractor"

A contractor is an entity, be it a proprietorship, partnership, corporation, joint venture, or other legal form, that engages in the activity of building (or demolishing). The contractor may function in many different spheres, or participate in a number of different roles for any given construction job.

"General Contractor"

The general contractor is the prime contractor; he deals directly with the owner of a project and assumes the full responsibility for the ultimate completion of the job. The general contractor may subcontract for the performance of the general contract with other firms in the industry that are special trade contractors.

"Subcontractor"

The subcontractor is usually a specialist within the industry. A contractual relationship exists between the subcontractor and the contractor. The subcontractor must convince the general contractor that the part of the total contract that had been delegated to him has been brought to a satisfactory completion. Among the specialty contractors recognized in the 1984 *Statistical Abstract* are plumbers, heating contractors, painters, electricians, carpenters, concrete contractors, and excavators.

"Construction Manager"

The nature of a particular job, if it includes characteristics such as large size or remote location, may dispose an owner to hire a contractor as a construction manager. This position is an agency arrangement wherein the contractor agrees to supervise and coordinate the construction activity on the project—a function that may include the negotiation of contracts for the owner by the construction manager with other contractors. This differs from the situation where a contractor negotiates for its own interest with subcontractors, because of the agency relationship between the owner and the contract manager.

"Turnkey"

A builder may provide an owner with a complete package of services, from site selection and project design, to financing, to the eventual construction. Through this so-called turnkey operation, the builder has more complete control of the entire construction cycle than it would with the more traditional role of a building contractor. This extra control comes about through the additional services—especially those non-building services—provided by the turnkey operator.

INDUSTRY CHARACTERISTICS

The construction industry is diverse by its nature. The size divergence encompasses the local and limited as well as the multinational and multibillion. The types of work done range from the specialty contractor with a single focus, such as concrete work, to the general contractor who deals in every phase of a project. In spite of this diversity, certain characteristics, such as the fixed contract price, are common to most contractors. The items explained in the following sections are some of the characteristics that are common to the construction business.

Contract

The performance of work under a contract with a customer is the basic thread that runs throughout the industry. The agreement between the owner and the contractor details both the type of work to be done on tangible property and the design specifications of that work.

Performance

The usual contract specifies the work to be completed, the payments to be made, and the modes for determining when partial and total performance are satisfied. Normally, the contractor will remain contractually liable to the owner until the entire project is finished. Partial performance levels are stated in contracts to determine items such as payment dates and amounts and construction schedules. Total performance is met upon the ultimate completion of all material contract terms.

Job-Site

The activities of the contractor are on the job-site of the owner, and the product of the contractor tends to be unique as opposed to the uniform product of a manufacturer. Because the contractor's activities are on the owner's property, the actual control of the contracted-for structure is often out of the hands of the contractor.

Common Characteristics

Significant common characteristics of the contract and the contractor include:

- The generation of sales through a bidding and estimating process

- The estimation of costs needed to finish a project long before it has even begun

- A risk of nonperformance or loss, especially in fixed-price contracts

- An insurance process (bonding) at all stages of construction, even preconstruction

- The use of cost and revenue accumulation techniques by project, rather than by job function

Each of these areas is extensively developed in subsequent chapters of this book.

AREAS OF SPECIALIZATION

This section defines some of the different types of general contractor specialties and identifies the unique problems associated with each of them.

Building Construction

The largest and most familiar segment of the industry is building construction, including office buildings, manufacturing, and process plants. Usually one general contractor deals with the owner, while employing subcontractor trade specialists. The building contractor is labor- and material-intensive. Although this contractor may be a giant multinational entity, more often it is a local company and operates within a limited geographic area with which it is familiar.

Heavy Construction

Heavy construction is a second major type of specialization. Road construction, usually for government units, is one of the primary focuses of heavy construction. Labor costs, material costs, and large capital requirements in the form of equipment are major factors on projects of this nature. The locations of the work performed, which are often remote and away from labor centers, and the necessity of a skilled labor force make these contracts very different in their execution than the more common commercial building specialty. Where building construction is normally a fixed-price contract, road construction is often unit-priced (e.g., cost per mile). Factors over which the builder has little control, such as weather and remote location, add to the higher risk inherent in this type of construction. Idle equipment and its cost are often non-controllable factors, which do not exist to the same degree in the building construction specialty.

Subcontract Specialties

Subcontract specialties are almost limitless in variety. The specialty contractor tends to be localized geographically, and expert in a single portion of the construction project. The marketing of its services is done through those general contractors most familiar with that specialist's work. One problem area common to subcontractors involves the coordination of their job function with that of other subcontractors; this area is often inadequately planned by the general contractor.

TYPES OF CONSTRUCTION CONTRACTS

The American Institute of Certified Public Accountants (AICPA) Audit and Accounting Guide, "Construction Contractors," (1981) known as the *Guide* identifies four basic types of contracts. They are classified by pricing arrangements: (1) fixed-price or lump-sum contracts, (2) time-and-material contracts, (3) cost-type (including cost-plus) contracts, and (4) unit-price contracts. The following sections detail the composition of these contracts and describe some of the variations contained in them.

Fixed-Price or Lump-Sum

The fixed-price, or lump-sum, contract provides for a single price for the total amount of work on a particular project. Normally, no adjustment to the price is made, regardless of the actual cost experience or level of performance of the contractor. One obvious risk is that in bidding for the job, the contractor may improperly overestimate the contingencies inherent in accepting this type of contract and, therefore, could lose the opportunity to do the work by failing to be the lowest acceptable bidder. Conversely, the contractor may fail to foresee all of the problems that potentially arise and could underbid, thus creating a position where little profit—if not substantial loss—could result. The long-term nature of some construction contracts also creates the possibility that a failure to recognize future economic changes, in such areas as interest rates and rates of inflation, could result in further problems of job profitability.

The fixed-price contract has several variations specifically enumerated in the *Guide*. These variations include specific provisions that modify the fixed-price contract. The most common of these provisions are:

- *Economic price adjustment* – provision allowing adjustments to be based on specified contingencies, such as changes in material price or labor rate. The adjustments can either increase or decrease the contract price. Contracts with a cost-of-living labor index included are an example of this type of provision.

- *Prospective periodic redetermination of price* – revision of the contract price after the passage of an initial period of performance, a delivery, or a defined base number of units. This type of pricing is used in governmental contracting.

- *Retroactive redetermination of price* – provision for a ceiling price (the fixed price) that, after completion of the project, is retroactively redetermined as being still within the ceiling, and in which the effectiveness and efficiency of management, as well as the actual costs incurred, are key components in the redetermination of price.

- *Firm target cost incentives* – establishment of target profits and costs at the outset of the contract, along with a price ceiling. A formula is then devised, based on the final negotiated cost-to-target-cost ratio, for establishing the final profit and the total contract price.

- *Successive target cost incentives* – provision for initial target costs and profits, a price ceiling, and a formula by which to fix the target profit. This variation further provides a point in production for the application of the formula.

- *Performance incentives* – provisions made in the contract whereby the contractor is rewarded for surpassing, or penalized for failing to meet, specific performance targets.

- *Level-of-effort term contract* – provision applied, usually to research and development projects, wherein the contractor must devote a specified level of effort for a stated period of time to earn a fixed amount of revenues.

The revenue and cost recognition problems created by the variations of the basic firm fixed-price contract are further developed in Chapters 3 and 4.

Time-and-Material

Contractor payments are computed at a basic price above cost. The base for costs is more often some standard plus a markup. The payment may be computed using direct labor hours (actual or standard) multiplied by standard hourly rates, with a separate cost accumulation for materials and indirect expenses. Variations of the time-and-material contract include:

- Time at a marked-up rate
- Time at a marked-up rate plus materials at cost
- Time and materials at marked-up rates and a guaranteed maximum cost
- Labor only
- Labor and materials only

It is common for contracts of this nature to contain ceiling or "upset" prices, which allow the owner to stop costly overruns. These contracts may also have provisions that call for any cost savings (standard vs. actual) to be divided among the parties.

The contractor's primary risk is in the poor planning of the standards used for computing the revenue base. A failure to stay within the upset price can result in a marked decrease in the profitability of a project. Thus, the reasonableness of the upset amount is the key factor to be considered when entering into a time-and-material contract.

Cost-Type

This contract category provides for the reimbursement of clearly defined allowable costs plus a fee that represents the contractor's profit. The best efforts of the contractor are the prime requirement in the cost-type contract. The term "best efforts" means that the contractor is required to put forth a standard of effort in performance of the contract based on the commercial standards of the industry. The work is contracted to be accomplished within a standard time and for a stated upper-dollar limitation. The outside ranges of completion dates and the maximum total costs are stated in the contract.

The cost-type contract has some common variations, including:

- *Cost-sharing-fee* – provision not included and the contractor is reimbursed for only a part of the costs incurred.

- *Cost-without-fee* – similar to the prior type of contract except that all costs are reimbursed.

- *Cost-plus-fixed-fee* – reimbursement of the contractor for costs, and receipt by the contractor of a predetermined fixed amount as the contract profit.

- *Cost-plus-award-fee* – the contractor's fee composed of a fixed part and a variable part. The fixed part is negotiated in advance and does not vary with performance. The variable part is an award based on performance in such areas as quality, timeliness, and cost-efficiency. This variable portion award is usually subjectively measured based on predetermined contractual criteria, which may include an award for early delivery; this happens most often in government contracting.

- *Incentive based on cost* – determination of the contractor's fee by use of a formula, which is a type of flexible revenue budget containing target, minimum, and maximum fees, and the method for changing the application of the formula.

- *Incentive based on performance* – computation of increases and decreases to a stated fee relative to the contractor's performance measured against specific targets provided in the contract.

Unit-Price

This contract variation is most often used where the quantity of certain components is not readily determinable prior to performance. The total contract price becomes the sum of the units multiplied by a fixed unit price. Variations in unit-price contracts are similar to those enumerated in fixed-price contracts. These unit-price contracts include:

- Cost-sharing
- Cost-without-fee
- Cost-plus-fixed-fee
- Cost-plus-award-fee
- Incentive based on cost
- Incentive based on performance

In some contracts of this type, unbalanced unit prices arise because different fixed prices are assigned to units delivered at different times (e.g., a higher price awarded for units delivered by a certain date).

CONTRACT MODIFICATIONS

Changes in Specifications

In the construction industry, modifications of a project's original contract is the norm. The modification may be as simple as the substitution of a similar quality material for one specified in the contract, or it may be as complex as the addition of a substantial amount of new construction. The control of change orders, claims, extras, and back charges by construction management is a key to the success of any construction undertaking.

Change Orders. A "change order" is defined as a modification that effectively alters a contract term or terms without adding new provisions. Either the owner or the contractor may initiate the change order. It may be specific, such as a change in the type of materials to be used, or general, such as a change in the period for completion of work. Since change orders are an innovation of sorts, both the work to be performed and the price must be negotiated. Often, the scope of the work and the price, or just the price, is not completely settled at the point of implementation, and accounting and legal problems consequently arise.

Claims. Claims are amounts in excess of an agreed-upon price, which a contractor attempts to collect from the customer because of customer-caused delays or for some other reason that is not the contractor's fault.

The basis for a claim often takes the form of a contractor seeking some monetary settlement for contractor-initiated or unapproved change orders.

Extras. Extras are additional work requested by the owner and are billed and accounted for separately from the original contract amount. Since the source of the extras is the owner, accounting and collection difficulties tend to be less severe than those encountered in either change orders or claims.

Back Charges. Back charges are reallocations of contract revenues between the contracting parties. Back charges are caused by one of the parties performing a task, or incurring a cost, that by contract was to have been borne by another party. An owner may bill a general contractor, or a general contractor may bill a subcontractor, for back charges. An example of a back charge is a charge for a subcontractor's use of a contractor's personnel.

The critical importance of the costs involved in all of the areas of specification changes is conducive to the eruption of disputes between the parties and, thus, costs must be properly controlled so that the ultimate goal of successful job completion can be peaceably attained.

Force Majeure

A construction contract, like any contract, may be modified or even discharged because of events that lead to the "impossibility" of performance. "Force majeure" refers to acts of God, such as floods, earthquakes, other types of natural disasters, or any other unanticipated events that are excuses for delay. The excusing of contract performance or the modifying of the duties of a promisor have major impact on a construction project. The legal rules of force majeure are that the excuse will be allowed if the event: (1) makes the agreed-upon contractual provision impractical (an extension of impossible), and (2) results from facts that the promisor had no reason to anticipate and for which he is not responsible.

The modification created by a force majeure can be so significant as to stop a job completely—as where construction of a nuclear power plant is interrupted by protesters—or so minor as to cause a short delay—as in the case of a snowstorm that requires a three-week change in construction schedules.

Force majeure modifications are not allowed merely because a project has become unprofitable. In fact, in several cases increased costs of over 100 percent have been found not to be excessive or unreasonable and, therefore, were not the cause of excused delay.

Delay

The amount of time allotted for performance of the contract is a provision of most contracts. The decision to commit to an entire project often hinges on the timing. Financing, tax planning, use of the facility, and profitability are all structured around a construction time frame. Failure to meet time requirements is a breach of contract, and the language of the agreement must therefore deal with this possibility.

- *Excusable delays* – those delays fitting into several categories, with the common bond that the delays are caused by a factor that is not the fault of the party seeking an extension of the time requirements.

- *Compensable delays* – those delays resulting, through the fault of the owner or his agents, in an increase in the contractor's risk or cost. The contractor accrues a cause of action upon which monetary damages can be assessed.

- *No-damage-for delays* – those delays an owner is liable for, caused by those of his actions or omissions that are not forgiven by a specific contract provision. To reduce or, perhaps, to define the specific contract risks, many nonfederal contracts are including no-damage clauses. The typical clause will exculpate the owner, while compensating the contractor by allowing additional time on the contract.

Subcontractors. The relationship of general contractor to owner is parallel to that of subcontractor to general contractor. The rules applicable to changes in specifications, force majeure, and delay are similar. The subcontractor may be at a more material risk than the contractor, since the consequential result of his actions or failures to act may have an impact on an entire project, rather than merely on the subcontracted portion thereof.

A time-of-essence provision is frequently imposed on subcontractors and results in specific liquidated damage amounts. The financial and accounting issues generated by this legality should be considered in an overall understanding of these delay items.

Penalty and Incentive Provisions

Construction contracts frequently contain provisions that modify the total revenues expected to be earned. These clauses allow for penalties or incentives to be earned by the contractor for failing to meet or for meeting certain predetermined targets of performance. Although these provisions are not technically contract modifications, they do, in fact, revise the gross revenue to be earned or create other modifications to the total job profitability. One example is an incentive award given to a general contractor for completing a phase of a project by a stated date prior to the contracted-for completion date.

Penalty and incentive provisions increase accounting risk because the estimated profits and costs to complete can be materially affected by their implementation, thus creating a potentially misleading financial disclosure. The user of construction company financial statements must be aware of the use of estimates, the potential impact of uncertainties such as the penalty and incentive provisions, and the fact that actual profits (or losses) on construction projects are often not accurately measurable until the project is finished.

BONDING

A unique feature of the construction industry is demonstrated by the bonding procedures that exist along each step of the construction job. The general purpose of bonding is to insure the owner's interest in the project by assuring that only qualified, responsible contractors are participants in any undertaking. The surety process is a three-party involvement between the bonding company, the owner, and the contractor. (This same surety relationship can exist between the bonding company, the general contractor, and the subcontractor.) Suretyship and bonding are discussed in detail in Chapter 14. The types of bonding usually existing in a construction project are bid security bonds, performance bonds, and payment bonds.

Bid Bond

The bid, or bid security, bond frequently takes the form of a bank-guaranteed or certified check, and is equal to a percentage of the expected total estimated cost of a project. This bid bond protects the owner from the unqualified bidder. It further provides that if the contractor who is awarded the contract refuses to sign, the owner will receive from the surety the difference between the contractor's bid and the next lowest acceptable bid, subject to the maximum bid bond liability. A bid bond or deposit is also used to reimburse an owner for the costs of rebidding a job.

Performance Bond

The performance bond, which is often required of the contractor awarded the contract, provides the owner with protection against the contractor's failure to perform in accordance with the provisions of the contract. This bond is paid for by the contractor, but is included in the expenses of construction and, thus, is covered by revenue generated by the contract.

Payment Bond

The payment, or labor-and-materials, bond provides protection for suppliers of labor, materials, and supplies. If the general contractor fails to make required payments to his own labor, his suppliers, or his subcontractors, the surety will make payment. This bond relieves the owner of the prospect of liens or other legal action against the site. Furthermore, this bond is also part of the "cost" of construction and is effectively included in the construction contract bid.

ACCOUNTING

Accounting Methods

The contract-construction business differs from other more common businesses, such as manufacturing, in that the accounting cycle—the period of time for the conversion of cash into inventories, inventories into receivables, and receivables into cash—is protracted. Construction projects may take many years to complete. The length of this cycle is part of the reason the accounting profession has long recognized that unique methods of accounting are required for the proper measurement of income.

Percentage-of-Completion. The percentage-of-completion method has been developed in accounting as an outgrowth of the pervasive accounting principles. This method attempts to match the profit earned in a particular accounting period with the effort expended during that period.

The percentage-of-completion method requires the accountant to measure and make judgments concerning the reasonableness of estimates provided by the contractor and the owner. Because of this critical reliance on estimates, the method can be used only where the dependability of the estimates is measurable with some precision. Even so, the percentage-of-completion method is preferred to the completed-contract method.

In addition to the threshold requirement of reasonably dependable estimates, other criteria must be satisfied before the percentage-of-completion method can be adopted. The rights and duties of the parties, clearly defined in contract form—including the work to be done, the consideration for the contract, and the terms and manner of settlement—are further conditions to the use of this method. The probability of performance of contract terms by both the buyer-owner and the seller-contractor are also requirements.

In spite of the ostensibly restrictive rules that govern its use, the percentage-of-completion method is appropriate wherever a reasonable measure of contract performance can be made during the life of a construction project.

Completed-Contract. The recognition of contract revenues and costs at the completion of a project, as in the completed-contract method, is consistent with the theory that the revenue is generated from sales, not production. In construction accounting, the use of this method will likely result in the income statement realization of revenues in a period in which perhaps only a slight amount of productive effort is expended. This can lead to material distortion of the periodic nature of income. For these reasons, among others, the completed-contract method is not preferred and should be used only in very limited circumstances.

The completed-contract accounting method in the construction industry is restricted to short-term contracts, which are usually of less than one accounting year in duration, and to those contracts for which estimates are not reasonably dependable. Any other use of the completed-contract method may be deemed not in accordance with generally accepted accounting principles (GAAP). A detailed discussion of revenue recognition methods appears in Chapter 3.

Profit Center

In accounting for construction contractors the base element for all accounting measurement is the profit center. In its Statement of Position (SOP) 81-1 the AICPA defines the "profit center" as a single contract, which, under certain specified circumstances, may be a combination of two or more contracts, a segment of a contract, or a group of combined contracts. The SOP provides further guidelines for the selection of profit centers and their appropriateness in alternative circumstances. (See also Chapter 3, "Income Recognition.")

FINANCIAL NEEDS

Although the financial needs of the contractor are explained in greater depth in Chapters 10, 11, and 12, a general overview of reporting and financing requirements is appropriate here.

Internal Reporting

Management control over every part of a construction project requires the proper and timely reporting of information. So that the data relevant to a specific job can be generated in a form usable for control purposes, an adequate system of internal accounting control must be maintained.

   Budgets. A contractor develops cost budgets through the bidding process. The contractor uses the bid as a starting point from which to work, since, in the great preponderance of contract-construction work, the production of a tangible asset results from the precise designs and specifications of architects and engineers, and since the bids are designed from these specifications. A well-designed budget will be able to measure the actual performance on a project against the anticipated performance as stated in the bid at given intervals. The success or failure of a company is measured by its ability to meet the contracted-for production requirements. The information for the step-by-step analysis of budget variances is brought out through the internal accounting system, but the variances from budgeted costs must be reviewed and understood through an ongoing management system, which uses and acts upon the data provided by the analyses.

Record-Keeping. Whether a handwritten or a computerized system of accounts is used, the contractor's records must be kept on a disaggregated basis. Job-by-job analyses of accounts must be maintained. A simplified system should include, as a minimum, work-in-process and billings on work-in-progress for each active project. The details of the work-in-progress account must be maintained by some form of subsidiary ledger and are ordinarily detailed by the cost categories outlined in the original bid. Whatever the system used, it should lend itself to being a management tool rather than a mere accounting document.

External Reporting

Annual Financial Statements. The basic external reporting time period for measurement of a contractor's accounting information is one year. The objective in these statements is to give potential users, creditors, owners, and current or prospective customers some insight into the fiscal condition of the company. The annual time period is long enough to allow measurement comparisons to be realistic, while not being so long (e.g., over the actual cycle) as to become untimely.

Since the nature of the long-term construction contract is that any one project will often extend over several accounting years, the basis of income and asset measurement can become complex. The determination of an exact percentage of completion on a particular job at a particular date is fraught with the inaccuracies inherent in making any accounting estimates. The precision, or lack thereof, on each estimate applied to each job will have an impact on the usefulness of the annual statements. A performance disclosure on a job-by-job or profit-center basis is a qualitatively useful concept that can be an aid to outside users in understanding the operating effectiveness of the company.

Interim Financial Statements. The compilation of financial statements on a more frequent than annual basis for use by those outside the firm is often helpful although not, in fact, mandatory. Monthly or quarterly statements that give information to lenders or creditors on the ongoing operations of the enterprise and provide assurance of the contractor's ability to meet contractual or borrowing requirements are salient. In the preparation of these interim statements, the same accounting principles used for annual reports should be applied.

Credit Granting

Owner-Generated Construction Loan. The planning of any construction project by an owner requires, as preliminary steps, the arranging of both temporary construction loan financing and permanent financing. The ability of the owner to generate the loans needed and to procure them at favorable rates of interest is a function of both the creditworthiness of the owner and the reputation of the contractor. Although the raising of capital for any type of construction venture is placed first on the ability of owner, the soundness and integrity of the contractor also has a substantial impact. In any analysis of the ability to complete a project, the prime construction lender needs information on the functioning of the contractor.

Job Profitability Analysis. A historical retrospective of a contractor's prior work goes a long way toward measuring the contractor's specific ability to complete a particular project. The credit grantor's investigation should include: a measurement of the estimating abilities of the contractor's staff, an evaluation of the profitability of previously completed work, a review of contracts in progress and the strain imposed on economic resources thereby, and a broad-based determination of contract backlogs. The preceding items indicate the contractor's ability to perform and the likelihood of success in each particular matter.

Analytical Review. A valuable method of testing the creditworthiness of a contractor is to review key elements in the financial statements. The areas that should be evaluated by the credit grantor include basic ratio analysis and trend projections.

In applying ratio analysis to contractors, working capital, net worth, debt-to-equity, and industry-wide comparisons are frequently used. If the credit grantor can conclude from a summary of the analysis that the ratios are a strong indicia of borrowing ability, then a valid credit relationship can be developed.

Trend projections indicate to the lender the place where the contractor is likely to be when the debt comes due and usually predict the contractor's ability to take on larger or more leveraged contracts. If the trends are true, the lender is able to determine with some certainty whether the job is within the realistic purview of the contractor. A measurement of the real asset growth over time is mandatory in evaluating the loan decision, since this reflects the longer-term pattern of a contractor's operations.

Warning Signs. The prospective lender must recognize the warning signs put out by contractors, so that the lender makes appropriate loans when they should be made and is able to save loans already made before they fail. The following are among the most frequently recognized warning signs used by lenders to monitor and make construction loans.

The Contractor's Historical Performance. A contractor who has frequent loss jobs should become suspect. The contractor's inability to bring a job in at bid prices or, alternatively, his being so weak in estimating that the bid, even if properly executed, creates a loss, is valid cause for concern.

Changes in Volume. Increases or decreases in the dollar volume of projects indicate possible credit problems. A contractor who has inadequate volume to absorb company overhead or who, by accepting too much work, has stretched company internal resources to breaking, may be in a jeopardy position. The impact of the current project on the internal functioning of the company must be ascertained.

Other Areas. Credit grantors should review other features of the contractor's financial composition such as:

- Qualitative aspects of assets such as receivables
- Investments in outside activities
- Joint ventures with unfamiliar partners
- New fields of construction
- Geographical changes

OPERATIONS

Generating the Contract

The first step in the process of doing business for the construction contractor is getting the work. In the field of subcontract specialty, often the best way to have a thriving business is to develop a reputation with one or a few general contractors who will seek out the specialty contractor when he is needed. This system works as long as the subcontractor's work retains its reputed quality, the price asked for that work is within reasonable parameters, and the general contractors remain faithful. In a somewhat similar manner, the general contracting company can develop a reputation within an industry such that owners will seek him out, requesting that he submit a bid on a particular contract. The general contractor maintains his reputation by performance and by coming up with a "right" price. However, this is not how the majority of the work is generated. More commonly the contractor—general or specialty—must go through the bidding process. To do this effectively, a properly functioning system of estimating and bidding must be developed.

Cost Estimating. The first step in the process begins when an owner decides to start a project. The owner hires an architect or engineer to draw plans and prepare expected cost estimates. If, after the initial process is completed, the project remains feasible (an estimated-cost-to-benefit decision), then detailed specifications are prepared so that prospective bidders can produce cost estimates. In preparing the cost estimate, the contractor analyzes the specifications to determine quantities of materials, necessary equipment, and hours of labor for every phase of the work to be performed. "Takeoffs," or quantity surveys of the materials required for a job, prepared by the architect/engineer or by independent firms are often available to and used by the contractor's estimating department. The sum of all of the inputs becomes the contractor's basis for the total cost estimate.

Bidding. Simply stated, the bid is the estimated cost of construction, computed by the contractor, plus a markup or profit margin. The factors that enter into the markup include:

- Contractor's previous experience in dealing with this owner
- Prospect of profitable change orders
- Competition and the economy
- Financing and cash flow requirements
- Weather
- Location

Entering Into the Contract

Signing. The successful bidder at some point actually signs a document that lists the rights and privileges of each of the parties. Unlike more traditional contract execution, however, the construction contract is not yet fully defined at that time. In the normal sales contract, the selling/negotiating process ends with the signed paper, and the performance begins thereafter. A construction contract is not that clear-cut. Negotiation is carried on during the entire period of performance. The first signing is merely the basic understanding; layered upon it are the change orders, the back charges, and the like, which are integral to every assignment. The owner, the contractor, and the advisors (CPAs, lawyers, and sureties) all must be aware of the true nature of the signed construction contract so that they may execute their roles appropriately.

Leaving Money on the Table. A contractor whose low bid is accepted is always in the situation of second guessing himself. The question of How low was I? must enter into the thought process. On many public sector and some private sector projects the submitted bids become a matter of record. A low bidder who finds himself materially lower than the next lowest bidder might be wise to question the efficacy and realism of the bid process he employs. Although nothing may be wrong with his bidding process, the low bidder could find that leaving a lot of "money on the table" may indeed be a dangerous warning sign.

Planning the Work

The job-site is the location where most of the work is done on a construction project. The contractor must prepare the work to be done at this location; the movement of equipment, material, and workers to the site must all be orchestrated by the contractor. The project manager must attend to such matters as preparing the field office, installing the utilities, analyzing the delivery schedule for purchased materials, and organizing the necessary equipment and labor. If a job is major in scope, the field office then becomes the administrative and accounting office as well as the operating control.

All the factors necessary for meeting contract deadlines must be planned well in advance. A production plan, with each cost broken down in detail, allows the project manager and the contractor to compare costs, by categories, against the standards set in the estimating and bidding processes.

Project Management

An organization lives or dies by its management. The goals in the construction business are to get work by submitting reasonable and profitable bids and then to bring that work to completion within the parameters set by the estimates. Results affect profits, and each event that occurs in the course of construction affects results. The basics of management call for planning, organizing, supervising, controlling, and implementing the philosophy of the business. The *Guide* lists 13 functions involved in project management:

1　Resource planning
2　Project start-up
3　Estimating
4　Scheduling
5　Project administration
6　Technical performance
7　Procurement and material planning
8　Labor planning and control
9　Subcontractor management
10　Support equipment and facilities
11　Project accounting
12　Project management reporting
13　Operations analysis

Each of these functions has its own set of details that must be planned by construction management. The list is provided merely as a starting point so that a well-managed (and therefore more likely profitable) construction project will result.

FORMS OF ORGANIZATION

The founder-entrepeneur of any organization is required to make several substantial decisions before the entity ever opens its doors for business. For the construction contractor, these initial decisions may be even more critical than they are for more usual businesses, such as retailing or manufacturing.

Construction is a high-risk venture; the dollar amounts of each transaction tend to be significant and are often enormous. The possibility of making very large profits or incurring very large losses exists in each venture. The immense capital equipment investments needed to perform one project may lose their utility because a series of related projects, which are expected to materialize, never do.

In instituting a form of organization, the contractor should base his decision on those factors that have the most significant impact. These factors include:

- Accounting method to be employed
- Income or loss expected
- Financing required
- Capital available vs. capital needed
- Liabilities resulting from the venture
- Personal wealth of the owners
- Tax consequences

The business of construction contractors is most often contracted through one of three basic forms of business organization. In order to select the best form of organization, contractors need to have some insight into:

1 Proprietorship

2 Partnership

3 Corporation

Within each of the second and third categories there is an almost limitless subset of business forms, each with its own reason for existing and its particular advantages and disadvantages.

Proprietorship

The construction industry has over one million entities; approximately 80 percent of them fall into the category of proprietorships. They are generally small, with gross revenues of less than $100,000, and usually operate in a limited local area. The owner is an all-purpose manager who arranges the necessary finances and does the bidding, purchasing, and "riding herd" on the project to completion. For ease of formation, simplicity of operation, and potentially favorable income tax treatment, this form of organization is often used.

- *Ease of formation.* No formal legal documents, no special filings, and no separate business bank accounts are required. An individual can merely decide to start a business as a contractor and then do so.

- *Simplicity of operation.* The owner is the manager; no one requires the owner to report; and the owner does not expect anyone in turn to report back. The individual proprietor is on site substantially all of the time.

- *Potentially favorable income tax treatment.* Tax rates for individuals are graduated. They start at a very low percentage (11 percent) and rise quickly within the first $100,000 of taxable income to 50 percent. If the proprietor expects income to be low, or if there are other losses that he can deduct against the contracting income, the net tax impact of the proprietorship may well be more favorable than the tax impact created by another business form (see Figure 1-1).

Based upon the 1984 tax rates, an individual with taxable income of $18,000 would pay a tax of $2,181, whereas a corporation with the same taxable income would be subject to a tax of $4,200, a difference of $2,019. However, if taxable income reached $100,000, then the individual would pay a tax of $32,935, while the corporation would be liable for only $25,750, a savings of $7,185.

Partnership

The owners of a partnership may be individuals, or they may be another type of entity. A partner's interest most often is based on his capital contribution, whether in cash or other assets, and is generally defined in a written agreement. (Many other methods of determining partnership interest are possible.) All income or loss is credited to the partners in accordance with the partnership agreement. Two basic types of partnerships exist—the general partnership and the limited partnership.

- *The general partnership.* A general partnership is an association of persons organized for the purpose of carrying on, for profit, a business as co-owners. The Uniform Partnership Act (1914) defines a "person" as either an individual, a business corporation, or any other entity having the same rights, privileges, and responsibilities as an individual.
  General partners share in the management of the partnership, divide profits and losses, and have unlimited liability for any debts of the organization.

- *The limited partnership.* A limited partnership is an association made up of both general (management) partners and limited (investment) partners. The limited partnership differs from the general partnership in that the limited partner has no voice in management, and his liability for partnership debts is limited to his capital investment.

FIG. 1-1 Comparison of Individual and Corporate Tax Rates
(for 1984 tax years)

SCHEDULE X
For Use by Single Individuals

Taxable Minimum	Taxable Maximum	Pay	+ Percentage on Excess	CORPORATION RATES At Level of Taxable Income	
$ 0 – $	2,300	$ 0	0%	$ 0 – $ 25,000	15%
2,300 –	3,400	0	11		
3,400 –	4,400	121	12		
4,400 –	6,500	241	14		
6,500 –	8,500	535	15		
8,500 –	10,800	835	16		
10,800 –	12,900	1,203	18		
12,900 –	15,000	1,581	20		
15,000 –	18,200	2,001	23		
18,200 –	23,500	2,737	26		
23,500 –	28,800	4,115	30	25,000 – 50,000	18
28,800 –	34,100	5,705	34		
34,100 –	41,500	7,507	38		
41,500 –	55,300	10,319	42	50,000 – 75,000	30
55,300 –	60,000	16,115	48		
60,000 –	70,000	18,371	48		
70,000 –	81,800	23,171	48	75,000 – 100,000	40
81,800 –	90,000	28,835	50		
90,000 –	100,000	32,935	50		
100,000 and over		37,935	50	100,000 and over	46

The Partnership Agreement. The agreement is a comprehensive document that details the purposes and functions of the organization, and lists the partners, the percentages of the ownership, the amount of each capital contribution, and the methods of determining value where it consists of assets other than cash.

The courts have long been filled with lawsuits arising from misunderstandings of certain facets of the partnership agreement's intent. Because of the almost infinite number of combinations possible in determining the amounts and distribution of profits, the content of the type and the value assigned to assets constituting each capital contribution, the lines of management, the procedure for changes in this authority, and the purpose and length of time needed to construct the entity, it is vital that the agreement clearly outline at least each of these areas and that each partner has a complete understanding as to how the agreement applies to the individual partner's interest.

Specific provisions should be made in the partnership agreement to deal with items as varied as:

- Profit- and loss-sharing
- Nonmonetary capital contributions
- Partners' compensation for direct services
- Payments to partners for administrative duties
- Partnership use of individual partner's equipment
- Sales of capital assets
- Liquidation and buy-out of partner's interests
- Admission of new partners

These items can become problems to the existence of the partnership entity unless adequate planning for each prospective problem is made. A basic set of rules established at the inception of the organization should provide appropriate commentary on items such as:

- The name of the operating entity that will be the contracting organization
- An outline of the nature of the work that will be undertaken
- The intended life of the entity and the steps that will be taken in the event of liquidation or termination, whether caused by legal action or by agreement between the partners
- An agreement as to the capital necessary to finance the intended activity, which bank accounts the funds will be kept in, and, if additional funds are required, what the proportional amount that the partners will contribute should be
- The nature of concessions to any partner for extraordinary services contributed
- An outline of the duties and related benefits of partners who assume managerial duties
- A designation of which parties are allowed to sign contract documents, bank loans, and bonds, and which parties can negotiate on matters concerning suppliers and subcontractors

Advantages. The principal advantage of a partnership is the joining of two or more entities with the necessary expertise to conduct operations that could not be done separately. Other advantages of the partnership form include the ability, under most circumstances, to raise more capital than a single individual could; the absence, with minor exceptions, of any tax at the partnership level; and the gathering of an appropriate mix of talents, which allows the contractor to perform new types of work and thus enables the firm to grow.

Disadvantages. Partnerships are not independent legal entities; therefore, not only is the capital already invested in the partnership at risk, but the total resources of the individual partners are subject to attachment by creditors as well. This at-risk provision is an influence that has led to the creation of limited partnerships, which, in effect, allow the limited partners to be subject to a liability equal only to that of the capital investment.

The tax law recognizes that the partnership is not an independent legal entity and therefore treats most partnership transactions as mere pass-throughs. This makes the income and expense of the partnership proportional income and expenses of the partners. If the individual partners are in high tax brackets, the net income of the partnership passes through and is subject to the high individual rates.

"Corpnerships"

These are defined as partnerships in which all of the members are corporations, thus avoiding the problem of a potentially higher tax created by the earnings of the partnership being attributed to the partners. This form of partnership also avoids the problem of partnership termination caused by the death of one of the partners, since corporations are not mortal. However, the corpnership does have disadvantages of its own, similar to those of corporations but with the further inherent disadvantage of having to operate with partners.

Many construction entities are partnerships. Some of these are long-lasting, well-operated, profitable, and have partners that cooperate effectively. The more common partnerships, however, are likely to be of short duration, created for a single project (the joint venture situation), and operated by a committee of the partners. Usually, the success of this sort of partnership depends solely on the quality of the specific job entered into by the partners.

Corporation

Definition. A corporation is a separate legal entity that comes into existence through the issuance of a state charter. This charter gives the corporation certain rights, powers, and obligations. One of the most significant legal attributes of a corporation is the limited liability of its stockholders. Corporate creditors can make claims only against corporate assets and not against the assets of the shareholders.

Advantages. In addition to the protection afforded shareholders through corporate limited liability, the corporate form of organization yields the following benefits:

- *Ease of capital formation.* A corporation can obtain financing through the sale of its stock to persons not part of management.

- *Unlimited life.* A corporation will continue to exist even after the death of its owners, whereas a partnership would have to be reorganized upon the death of a partner.

- *Tax benefits.* The corporate form is subject to a maximum federal tax of 46 percent of taxable income and can avail itself of tax benefits such as pension, profit-sharing, and other fringe benefit programs.

Disadvantages. The corporate form of organization has the following negative characteristics:

- *Tax on distribution.* Corporations that make distributions to shareholders in the form of dividends create a double tax on such distributions. The earnings are taxed both upon receipt by the corporation and, again, upon distribution to the shareholders. Corporations are also subject to a potential tax on undistributed earnings that are deemed excessive.

- *Limits on borrowing.* Most lending institutions subject the small corporation to very stringent lending requirements. Banks very often lend to these small corporations only at high rates of interest and with the personal guarantees of the corporate owners. The impact is that the limited liability advantage of the corporate form is effectively negated.

S Corporations. The Internal Revenue Code allows electing corporations (tax option or Subchapter S) to be treated as partnerships for tax purposes. This eliminates the corporate tax without eliminating the inherent liability protections of the corporation.

CHAPTER 2

Accounting for Joint Ventures

Charles L. Jacobson

INTRODUCTION

Growth is survival for the construction contractor. The primary limitations on the ability to grow are a lack of capital, a shortage of expertise, and an inability to obtain bonding. Construction organizations that suffer from any of these limitations often find that their difficulties can be overcome by entering into a different form of organization: the joint venture. By merging resources, experience, and bonding capacity with another complementary organization, the contractor makes it possible for his company to bid, accept, and execute construction projects that will eventually, if not immediately, lead to growth of the firm. Each of the members of the joint venture shares in the risks and ultimate rewards of taking on larger and more prestigious construction projects.

Frequently, the joint venture has a short life-span, since it is designed and created for the sole purpose of bidding on, negotiating for, and completing one specific contract. This single-purpose joint venture will last only as long as the one project lasts, which can be less than one year or as long as seven or eight years on major projects.

Some joint ventures are designed to be permanent in nature. In joint ventures of this type, the parties involved usually recognize a good matching of talents among the participants and decide that for all contracts of a specific type the parties will pool their resources and share in the cycle of business. The venturers generally never go beyond the joint venture form of organization – to a merger, for example – because each of them has his own separate business quite different from the business of the joint enterprise.

For accounting purposes, the joint venture is generally an independent entity, which keeps separate records and issues separate financial reports for each venture. The joint venture, undertaken to complete a specific project, is often liquidated when that project is completed. There are some joint ventures that do not maintain separate accounting records. In such cases, the members submit a joint bid, although they perform various aspects of the job separately. Progress payments are made directly to the venturer who performs the work; thus, there is no problem associated with the allocation of cash receipts.

In reporting investments in joint ventures, the consolidation, partial or proportionate consolidaton, equity, expanded equity, and cost methods are each acceptable, depending on the circumstances. All have been used, to a greater or lesser extent in the industry, both singly and in combination with each other.

This chapter first reviews the principal reasons for entering joint ventures and then examines in detail several accounting considerations from the investors' point of view. However, the methods of accounting for joint ventures are as varied as the types of joint ventures, and there is little authoritative literature on the subject. As is so often the case, there may be more than one solution to the same problem. The Financial Accounting Standards Board (FASB) is currently in the process of reexamining the accounting principles that relate to consolidations and the equity method. Recent developments in this area should be considered, since the rules for both consolidated financial statements and for equity accounting are the basis for the preponderance of theory used to develop the reporting methods employed by joint ventures.

FORMS OF JOINT VENTURES

The majority of joint ventures are formed for one specific purpose; nevertheless, there is a great deal of creativity involved in the organizational form of the joint venture entity. Partnerships that have corporations as partners, limited partnerships with a key general partner and mere investors as the limited partners, traditional corporations with the venturers as shareholders, and subsidiary corporations created to function as either stockholders or partners are all common forms within the joint venture framework. Each of these different types of joint ventures has a rationale for its use. While all generate similar types of accounting issues and problems, each has its own form for reporting financial information. The tax implications of a particular joint venture form very often will influence, if not dictate, the way in which the organization is designed. The tax implications are discussed later in this chapter and also in Chapter 18.

Partnerships

A partnership is an association of two or more persons (or other entities) designed to carry on, as co-owners, a business for profit. The partnership is usually, and in theory should always be, evidenced in writing. The form of the agreement is most often a contract made between the parties for the mutual participation in the profits that may accrue from property, credit, skill, or industry; and that will be furnished in predetermined proportions by the parties.

General Partnerships. A general partnership is one in which the parties carry on all of their trade and business, whatever it may be, for the joint benefit and profit of all the parties concerned. The total capital may or may not be limited; the contributions may be of equal or unequal amounts; and the profits and losses may be shared equally, in proportion to the capital contributed or the efforts put forth, or in any other way to which the partners agree. The common thread in the general partnership is that all of the general partners share in the management of the partnership. The general partnership form is most often used where the venturers perceive that any risk of loss is limited to the capital contributed to date or that the possibility of a loss greater than the amount of contributed capital is remote, and that alternative forms of organization have a greater economic cost.

Limited Partnerships. A limited partnership is comprised of one or more general partners who manage the business and who are personally liable for partnership debts, and one or more limited partners who contribute capital and share in profits and losses, but who take no part in the management of the business and who incur no liability with respect to partnership obligations beyond their agreed-upon capital contribution. The liability of the general partner is sometimes further mitigated by forming a limited partnership with a corporation serving as the general partner. In order to take advantage of the limited partnership form, the provisions of Section 1 of the Uniform Limited Partnership Act (1976) must be complied with in those states where the Act has been adopted.

Corporations

A joint venture structured as a corporation has been organized using the corporate form for one or several specific undertakings. Where the joint venture is so organized, it provides the venturers, be they other corporations or individuals, with the unlimited flexibility that the corporate form allows. The parties to the venture are able to avoid the direct liability inherent in a partnership while still being able to share in the management of the organization, something that the limited partner is ordinarily unable to do.

The corporation is an independent legal entity and is recognized as such by taxation authorities. It is because of the impact that tax authorities have on decision-making that the planning of a joint venture in the corporate form must include careful study of the tax implications. A review of the taxation of construction contractors, including joint ventures, is contained in Chapter 18. However, certain tax effects should be mentioned here for the joint venture corporation. These include:

- Dividends-received deduction
- Subsidiary corporation as a stockholder
- Subchapter S election corporations
- Double taxation
- Joint ventures structured as tax shelters

Dividends-Received Deduction. If the joint venture is structured as a corporation, where one or all of its shareholders are corporations, any distribution made by the joint venture to a corporate shareholder that qualifies as a "dividend" is subject to the dividends-received deduction. As a result, the corporate tax on this transaction at the shareholder level is substantially mitigated and the effective tax rate is reduced by as much as 85 to 100 percent. As a planning device, the dividends-received-deduction benefit is used as an important profit management and cash-flow-generating tool.

Subsidiary Corporation as a Partner/Stockholder. A joint venturer may decide that the proper vehicle for its investment in a corporate form joint venture is its own subsidiary. The subsidiary can be a going concern or a newly created entity. The subsidiary insulates the parent from difficulties such as law suits or losses arising from the joint venture and, at the same time, allows the parent to use available tax benefits.

If the subsidiary corporation is a partner in the joint venture, any losses of the venture may become deductions on the parent's tax return. The net income of the subsidiary corporation attributable to its partnership share in the joint venture can be used as an offset against other ordinary income of the subsidiary or, in the case of a consolidated return, possibly, against the parent's other income.

If the subsidiary corporation is a shareholder in the joint venture organized as a corporation, then the distributions made to it will probably be subject to either of two levels of dividends-received deductions (85 or 100 percent). Losses of the subsidiary corporation attributable to the corporate form joint venture will not be subject to offset against other subsidiary (or parent) income, but will be subject to carry-back or carry-over against previous or future tax liabilities of the joint venture.

S Corporations. Corporations that are eligible may, by unanimous consent of the shareholders, elect the treatment of S corporation status. This election by shareholders can now be done on an annual basis. The election converts the taxable entity into a pass-through entity. The character of items of income, deduction, gain, loss, and tax credit become, in the hands of the shareholders, the same as they were in the hands of the electing corporation. The S corporation is not subject to the tax imposed on most other business corporations, except for a possible tax on investment income and/or capital gains.

The benefits to the investors, who may not be corporations, are that the double taxation of corporate profits is avoided and any losses of the S corporation may be offset against the other income of the stockholders.

Double Taxation. The corporate form of joint venture creates an automatic double tax for its shareholders. The double tax consists of a corporate tax at the joint venture level and a second tax, individual or corporate, at the shareholder level when the profits are distributed, or distributable, in the form of dividends (or as return of capital or as capital gains). As discussed under S Corporations above, various simple techniques can be used to avoid this double tax. Although not common, there are situations where the avoidance of the double taxation may not be economically justified (as where a larger single tax results). In these cases, careful study of the fact pattern is prudent and recommended.

Other Forms of Joint Ventures

The purpose of any form of business organization is to provide the means through which the activities of a business can be channeled most efficiently. The joint venture is most frequently formed as either a partnership (limited or general) or a corporation. However, these are not the only forms that are available and used in practice.

The construction joint venture, when not using the more traditional partnership and corporate structure, may use a form that provides each of the venturers with "undivided interests in ventures." The meaning of this term is that each investor-venturer owns an undivided pro rata share of each of the assets and is responsible for an undivided pro rata share of each of the liabilities; thus, each investor-venturer has an undivided interest in the net equities of the venture. Because of the form and the related legal substance of this organization, the accounting by the investor-venturer on its financial statements is that the pro rata, or percentage, share of the assets, liabilities, revenues, and expenses is recognized on the financial statements as separate items rather than in a single investment account.

Determining Venturers' Percentage Ownership

Most contracts between joint venturers set forth the exact percentage ownership of each of the joint venturers. Some agreements set different and variable allocations among the venturers. The areas most frequently affected by such agreement are:

- Profits and losses
- Identified cost allocations
- Revenue allocations
- Expense sharing
- Cash distributions
- Liquidating distributions

Each of these areas can be provided for in joint venture agreements in ways that change allocations at specified future dates or on the occurrence of designated future events. The accounting problem of the venturer becomes one of determining the appropriate percentage of ownership that should be used for financial-statement purposes. The general rule to be applied for the purpose of recognizing income or loss is that the percentage of ownership interest should be based on the percentage by which the profits and losses will ultimately be shared. There is a provision for an exception to this rule in the American Institute of Certified Public Accountants (AICPA) Audit and Accounting Guide, "Construction Contractors," (1981), known as the *Guide*, that states "If changes in the percentages are scheduled or expected to occur so far in the future that they become meaningless for current reporting purposes . . . [then] . . . the percentage interest specified in the joint venture agreement should be used with appropriate disclosures."

ADVANTAGES AND DISADVANTAGES OF JOINT VENTURES

Many joint ventures fail. If they do not fail, it often takes several years to reach the point where the return on investment (cash and time) makes them worthwhile. Why, then, do investors turn in increasing numbers to the joint venture form of operation? The answer can be found by reviewing some of the major advantages and disadvantages of the joint venture. Figure 2-1 summarizes the advantages and disadvantages of joint ventures versus other forms of operations.

FIG. 2-1   Advantages and Disadvantages of Joint Ventures

ADVANTAGES

Financing:
- Raises needed capital
- Permits specific project financing
- Creates the potential of off-balance-sheet financing
- Expands lending capacity

Spreading of risk through diversification

Flexibility:
- Combination of different managament styles
- Various accounting and reporting methods
- Increased selection and capacity of projects

Foreign investment requirements:
- Ability to meet foreign investment requirements
- Increased foreign opportunities

DISADVANTAGES

Loss of control

Lack of compatibility of investors

Harder to establish and to administer

Advantages

The needs that lead to the formation of a joint venture are financing considerations, spreading of risks, flexibility, and foreign-investment requirements.

Financing Considerations. Typically, a construction project is a large undertaking and requires considerable capital to satisfy both operating needs and bonding requirements. Often, this capital requirement is beyond the resources of one investor. The joint venture offers a means of pooling the resources of two or more investors in order to reduce the requirements of any one investor by itself. For example, a company with excess cash may form a venture with a developer, architect, and contractor, each of whom offers professional expertise.

A joint venture also offers the opportunity to arrange specific project financing, including the potential for "off-balance-sheet" financing, a type of financing that has become increasingly popular in recent years. Simply stated, the term means financing a project without increasing the debt on the investors' balance sheet. This is accomplished by having the debt on the books of the joint venture and accounting for the joint venture by a method other than the consolidation method (see discussion of consolidation method later in this chapter). The ability of the joint venture to obtain debt financing, however, is usually based on the projected cash flow of noncancellable supply contracts between the joint venture and one or more of the investors or a third party, or is guaranteed by the investors. In such cases, the joint venture's debt usually becomes a disclosure item in the footnotes of the investor's balance sheet rather than a liability on the balance sheet. Thus, while "off-balance-sheet" financing is achieved, "off-footnote" financing is not. Accounting Principles Board (APB) Opinion No. 18, "The Equity Method of Accounting for Investments in Common Stock" and Statement of Financial Accounting Standards (SFAS) No. 47, "Disclosures of Long-Term Obligations" should be referred to for accounting rules relating to disclosure requirements for equity method investees and debt guarantees.

Spreading of Risks. Because construction projects are often large and financially risky, investors strive to diversify their activities so they are not overly dependent on any one project. The joint venture offers an opportunity for investors, or venturers, to share the risk with other venturers. By doing this, more diversification can be accomplished and, thus, a project that would be beyond the business ability of a single venturer becomes viable.

Flexibility. The joint venture may offer the investor more flexibility in both the financial reporting and the day-to-day operations of the venture than would be possible if the project was totally assumed by one investor. This is especially true if the venturer is a large construction company where well-defined operating and financial policies and regulations may otherwise put extra burdens on the joint-project operations. By operating as a separate entity, apart from the investor's operations, management of the joint venture is better able to attend to the project, thus substantially reducing expensive corporate overhead requirements.

Having the accounting done by the investor for the joint venture operations offers considerable flexibility, as is discussed later in this chapter. However, the method of accounting must be selected before the venture begins operations. The different methods of accounting for investees include:

- Cost
- Consolidation
- Equity method
- Expanded equity method
- Partial or pro rata consolidation

Specific goals and policies of the investors, along with the economics of the project, determine the manner in which the venture is structured.

Foreign Investment Requirements. Many foreign countries have strict investment rules that require various percentages (often in excess of 50 percent) of a company to be locally owned before those investors will be allowed to operate in that country. In fact, many Arab countries mandate that a controlling interest in any company operating in their country be in the hands of local citizens or nationally owned business entities. Joint ventures are a convenient means to meet these investment requirements.

Disadvantages

The disadvantages of foreign joint ventures arise from the same basic characteristics that result in the advantages. The joint venture can result in the investor losing actual control and does, in every situation, require the venturers to work closely with an outside party.

Loss of Control. Companies are often wary of joint ventures because they can no longer maintain complete control. This is especially true in international joint ventures where the added degree of uncertainty, the investor's lack of international experience, or the unfamiliarity with the local environment add complications. Companies must be willing to give up some control if a joint venture is to be attempted. The company must protect itself against loss of control by using clear, well-defined contract terms.

Although bonding, contracts with owners, clearly defined construction plans, and detailed agreements among the venturers offer genuine business protection in domestic dealings, the international venturer remains at risk. The additional complications an international venturer must consider include:

- Political climate of the foreign country
- Currency restrictions

- Currency translation fluctuations
- Prospects for nationalization or expropriation of industries
- Tax treaties and tariffs
- Local labor force

Compatibility of Investors. Investors in a joint venture must be compatible – they must be willing to compromise. Lack of compatibility is a major reason for joint venture breakups, restructurings, and failures. This aspect of a joint venture demands that a well-written and complete joint venture agreement be drafted at the beginning of the venture, so that the rules of operation and management are clearly set forth.

ACCOUNTING CONSIDERATIONS WHERE ESTABLISHING JOINT VENTURES

This section discusses the usual methods of establishing joint ventures and the associated accounting considerations. No differentiation is made between a corporate joint venture and a partnership joint venture because the form of organization does not affect the accounting.

Form of Contribution

Joint ventures are established by investor contributions of varying amounts and proportions of new assets. These may include cash, tangibles, and intangibles. Tangible assets are such items as buildings and equipment, while intangible assets are such items as know-how, trade name, construction permits, customer base, and reputation. Generally, the total value of each investor's contribution to the joint venture becomes the proportion of its ownership percentage in the venture. The following are simple illustrations of contributions in a joint venture situation involving two investors, *A* and *B*, each owning 50 percent of the venture.

A contributes cash of $1,000, and

B contributes equipment with a fair market value of $1,000;

or

A contributes cash of $1,000, and

B contributes know-how and reputation, which for purposes of establishing ownership in the joint venture is valued at $1,000.

Note that in these situations each investor contributes an equal amount to the joint venture and each investor receives a 50 percent interest, even though the form of the contribution (cash vs. other assets) is different.

Most joint venture situations are not this simple. The remainder of this section discusses the accounting considerations where contributions consist of the following:

- Cash only
- Cash and equipment (no cash withdrawals)
- Cash and equipment (cash withdrawal by one investor)
- Cash and intangibles (no cash withdrawals)
- Cash and intangibles (cash withdrawal by one investor)

Some basic principles for investors with respect to recording investments in joint ventures should be established. A significant question that arises where non-cash contributions are involved is whether investors recognize a gain (or loss) at the time when the joint venture is established. This question is addressed in the following discussion of accounting issues.

Initial Recording Principles

The following principles can be used by investors as a guide in recording initial investments in joint ventures under each of the alternatives presented above.

Cash Only. Each investor should record its investment at the cash value contributed. No gains or losses are recognized.

Cash and Other Considerations. An example of other consideration is the contribution of future services, which should be recorded as an investment only when performed, and then only to the extent of the fair market value of the service performed. The problem here is the valuation of the services, which should be set by agreement of the parties before the services are performed. For example, the parties would have to agree to the dollar value to be assigned to the know-how a venturer will contribute.

Loss Recognition. Losses should be recognized when they occur. There should be no deferral of indicated losses. Losses are measured as

---

**FIG. 2-2 Accounting for Investors and the Joint Venture —
Initial Contributions Are Cash Only**

GIVEN

- *A* contributes $100,000.
- *B* contributes $100,000.
- There are no cash withdrawals. No gain or loss is recognized.

JOURNAL ENTRIES

Investors A & B

Dr	Investment in joint venture	100,000	
Cr	Cash		100,000

Joint Venture

Dr	Cash	200,000	
Cr	Equity – *A*		100,000
Cr	Equity – *B*		100,000

---

the difference between the book value of the asset contributed and the fair market value assigned by the joint venture entity when book exceeds fair value.

Difference Between Cost and Underlying Equity. The joint venture's net income should be adjusted for the difference between the cost (to the investor) and the underlying equity (in the joint venture) before the investor records its share of earnings from the joint venture. This is demonstrated in Figures 2-2 through 2-4.

Profit Recognition

Although the exhibits and commentary described above are only some of the numerous situations that are encountered in practice, they provide a framework for determining the appropriate accounting in all such situations. The primary issue to be considered whenever accounting of this type is needed is profit recognition. While important, the impact on the balance sheet caused by this type of transaction is secondary to the income statement consequence.

**FIG. 2-3 Accounting for Investors and the Joint Venture —
Initial Contributions Are Cash by One Venturer
and Appreciated Equipment by Another**

GIVEN

- A contributes $100,000.
- B contributes equipment with a fair value of $100,000 and a basis (book value) to B of $60,000.
- There are no cash withdrawals

JOURNAL ENTRIES

Investor A

Dr	Investment in joint venture	100,000	
Cr	Cash		100,000

Investor B

Dr	Investment in joint venture	60,000	
Cr	Net book value of equipment		60,000

The investment is recorded at the book value of the equipment.

Joint Venture

Dr	Cash	100,000	
Dr	Equipment	100,000	
Cr	Equity — A		100,000
Cr	Equity — B		100,000

The difference between B's equity in the joint venture and B's investment account ($40,000) will be recognized as income (reduction of losses) by B over the life of the equipment.

This gain recognition can be illustrated as follows. Assume that in Year 1, the joint venture recognizes a net income from operations of $30,000. B's share (50 percent) of this profit is $15,000. Assuming that the useful life of the equipment is 5 years, and that both B and the joint venture use straight-line depreciation, the equity method will yield the following entries on B's records. (To simplify the presentation, all entries below ignore tax accounting considerations.)

Investor B

Dr	Investment in joint venture	15,000	
Cr	Earnings attributable to joint venture		15,000
	To record the earnings of the joint venture in Year 1.		

Dr Investment in joint venture   8,000
Cr Earnings attributable to joint
       venture   8,000
To record the difference in net income
attributable to the excess depreciation
taken in Year 1 by the joint venture on
equipment contributed by *B*.

This difference is the $100,000 joint venture equipment carrying amount less the $60,000 carrying value that had appeared on *B*'s books before transfer divided by the useful life: ($100,000 − $60,000) ÷ 5 = $8,000.

The second entry will be repeated annually for the life of the equipment as an adjustment to the net income recognized by *B*.

Tax Considerations

Tax considerations are, of course, very important in the formation of joint ventures. The initial decisions made with respect to types of contributions and the legal form of the joint venture can often spell success or failure in the future. Either the venturer should be very current on tax matters or he should seek counsel before proceeding.

The main areas of consideration are:

- Initial contributions

- Joint venture earnings

- Dissolution (even if not planned)

Initial Contributions. The main factor to consider here is whether contributions made by the investors will be treated as taxable events. Once this has been determined, the nature of the gain or loss (ordinary or capital) must be established. The starting point for making this decision should be the planning of the transaction.

Joint Venture Earnings. The legal form of the joint venture determines the taxation method applicable. Corporations have a two-tier (or double) taxation, whereas partnerships are taxed only at the investor level. Tax option (Subchapter S) corporations and special partnership

**FIG. 2-4   Accounting for Investors and the Joint Venture —
Initial Contributions Are Cash by One Venturer
and Equipment by Another, and Where Cash Is
Withdrawn**

GIVEN

- *A* contributes $100,000 cash.
- *B* contributes equipment with a fair value of $160,000 and a basis (book value) to *B* of $80,000. *B* receives $60,000 in cash as part of the transaction

JOURNAL ENTRIES

Investor A

Dr	Investment in joint venture	100,000
Cr	Cash		100,000

Investor B

Dr	Cash	60,000
Dr	Investment in joint venture	50,000
Cr	Equipment		80,000
Cr	Gain		30,000

The entry is calculated as follows:

1. *B*'s total realized gain is $80,000. The fair value of the assets received is $160,000 ($60,000 cash plus a 50 percent share in the joint venture, which, after the transaction, has total assets of $200,000), less *B*'s basis in the asset given of $80,000. (The joint venture's assets are $40,000 in cash and a piece of equipment with a fair value of $160,000.)

2. *B*'s recognized gain is $30,000. The recognized gain computed in accordance with APB Opinion 29, "Accounting for Nonmonetary Transactions," is computed by the following formula:

$$\frac{\text{Monetary assets received}}{\text{Total assets received}} \times \text{Realized gain} = \text{Recognized gain}$$

or

$$\frac{\$60,000}{\$60,000 + \$100,000} \times \$80,000 = \$30,000$$

3. *B*'s "Investment in joint venture" account is the original basis of the asset given ($80,000), less the gain recognized ($30,000), or $50,000.

forms are available to alter these effects if the situation is appropriate. Some factors to consider include:

- Partnership income is generally taxed on a current basis

- Corporate income can be deferred

- Capital gains treatment may be available to the joint venture upon sale or termination and should be considered

Dissolution. Joint ventures usually are designed to have a limited life. In the situation where the joint venture was established for a specific purpose (e.g., to construct a particular building), the time of dissolution is predictable. Once the purpose for which the venture was formed is completed, the need for the joint venture ceases and it is terminated, as planned by the venturers. At other times, dissolution is not part of the plan and comes about for a variety of reasons (e.g., one of the venturers is bankrupt or the venturers can no longer agree on management). In these cases the joint venture can be terminated, but the termination process is much less efficient. The tax consequences of an eventual dissolution should be studied at the inception of the organization not at its demise — because the potential for serious tax consequences for early dissolution of the joint venture can cut into any profits earned during the operations of the venture.

Accounting for Differences Between Cost and Underlying Equity

As illustrated in Figures 2-3 through 2-5, there is often a difference between the investment shown on the investor's financial statements and the investor's proportionate share of equity in the net assets of the joint venture. In Figure 2-3, B's initial investment is $60,000, and its underlying equity in the net assets of the joint venture is $100,000. Such differences are generally a result of two factors:

1 Initial establishment of the joint venture, that is, the joint venture records assets contributed at fair market value, while the investor records its investment at a carry-over cost.

2 Different accounting methods used by the joint venture and the investor.

The general rule is that assets contributed by the investor should be accounted for in a manner that produces the result that would have been

FIG. 2-5   Methods of Accounting Generally Used for Joint Ventures

Type of Venture Ownership	Cost	Consolidation	Equity (a), (b)	Partial Consolidation
Corporate				
Over 50%		X		
20–50%			X	
Less than 20%	X			
General partnership				
Over 50%		X	X	
20–50%		X	X	
Less than 20%			X	
Limited partnership				
Over 50%		X	X	
20–50%			X	
Less than 20%	X		X	
Undivided interest				
Over 50%		X		X
20–50%				X
Less than 20%	X			X

(a) The expanded equity method also may be appropriate where the contractor's share of the joint venture's assets are shown on a one-line basis on the contractor's balance sheet, and the contractor's share of the joint venture's liabilities are shown separately on a one-line basis.

(b) The equity method is appropriate for less than 50 percent owned joint ventures where the contractor has the ability to exercise significant influence on the joint venture. That ability is presumed to exist for ownership interest of 20 percent or more. The presumption may be overcome by evidence to the contrary.

Source: Reprinted with permission from *Construction Contractors – A Summary of the 1981 Audit and Accounting Guide and Statement of Position 81-1* (New York: Main Hurdman, 1981).

produced had the assets not been in the hands of the joint venture, but, rather, were still a part of the investor's operations. For example, in the situation discussed in Figure 2-3, there was a difference of $40,000 between the investment of $60,000 and the underlying equity of $100,000. This difference resulted from a step-up of the asset values to fair market value on the joint venture's records. In that same example, it was suggested that the $40,000 difference be amortized over the expected period of benefit of the contributed assets. As a result, B is affected as if the assets had not been contributed, or as if they had been recorded by the joint venture at $60,000.

ACCOUNTING BY INVESTORS

The construction industry has been innovative in the various methods that it has used to account for joint venture operations. Several alternative methods, which are all acceptable, exist within a general framework. They are:

1 Cost method

2 Full consolidation method

3 Equity method (expanded equity methods)

4 Partial or proportionate consolidation method

The best one to use in any given venture is that which most aptly reflects the transactions of the entity. The important issues in determining the appropriate method of accounting for construction joint ventures relate to extent of control and meaningful reporting. Figure 2-5 summarizes the situations in which each method may be appropriate.

Cost Method

The cost method is appropriate where the ownership percentage in the joint venture is less than 20 percent, or where a greater than 20 percent ownership is deemed to be inappropriate for the use of the equity method. Under the cost method, the initial investment remains unchanged until it is sold or permanent impairment is indicated. Earnings of the joint venture are recorded only when distributions are received or dividends are declared. The appropriate accounting is illustrated as follows:

Minority Corp. has a 15 percent interest in Ultra Venture, Inc., a joint venture. The original investment was $100,000 cash. During its first year, Ultra Venture earned $200,000. Minority's 15 percent share of the $200,000 net income is $30,000. No dividends were declared in the first year. Minority records no accounting entry in Year 1, since no actual distribution was made.

On May 1, 19X2, Ultra Venture, Inc. declares and pays cash dividends totalling $50,000. Minority would record the following entry:

Dr   Cash $7,500
Cr   Dividend income from joint venture $7,500

This same pattern would continue through all subsequent dividend distributions; no income or loss is recorded as a direct result of the operations of the joint venture, and distributions are recognized only where accruable or received.

Equity Method

The equity method, the most common method of accounting for investments in joint ventures, is used where the ownership percentage of the venture is between 20 and 50 percent. In these situations, the investor is presumed to be able to exercise significant influence over the venture. If there is evidence indicating that the investor does not have significant influence, the cost method is used.

Accounting for a joint venture under the equity method involves the initial recording of the investment at cost and the subsequent adjustment of the carrying amount of the investment to recognize the investor's share of the earnings or losses of the venture. The amount of the adjustment is included in the determination of net income by the investor and reflects adjustments similar to those made in preparing consolidated financial statements (e.g., the elimination of intercompany transactions). Dividends received from the joint venture reduce the carrying amount of the investment.

Earnings of the joint venture are recorded on the books of the venturers when the income of the venture is reported. Distributions to the venturers are treated as mere asset exchanges. The appropriate journal entries are recorded as follows:

Equity Corp. has a 40 percent interest in Plusultra Venture, Ltd., a joint venture. The original investment was $100,000 cash. During its first year, Plusultra Venture earned $200,000. Equity's 40 percent share in the $200,000 net income is $80,000. No dividends were declared in the first year. Equity records the following entry, even though no distribution was made:

Dr   Investment in Plusultra Venture, Ltd. $80,000
Cr   Equity in earnings of joint venture $80,000

On May 1, 19X2, Plusultra Venture, Ltd. declares and pays cash dividends totalling $50,000. Equity would record the following entry:

Dr   Cash ($50,000 × 40%) $20,000
Cr   Investment in Plusultra Venture, Ltd. $20,000

Under the equity method, the investment in the joint venture is shown as a separate line item on the investor's balance sheet. The investor's share of the net income or loss of the venture is shown separately on the investor's income statement.

In applying the equity method, losses in excess of an investor's investment, loans, and advances should not be recorded by the investor unless the investor is committed to providing additional financial support to the joint venture. Such circumstances could result from:

- Legal obligations as a guarantor or general partner.
- Commitment based on such considerations as business reputation, intercompany relationships, and credit standing. Such a commitment might be evidenced by previous support by the venturer indicating that it would guarantee joint venture obligations, or public statements by the venturer of its intention to provide support.

Expanded Equity Method

The expanded equity method has the advantage of conveying to users of financial statements the size and scope of the investor's interest in the joint venture in situations where the simple equity method may be misleading. Although this method differs in balance sheet presentation from the equity method, the income statement presentation is the same. Under this method, the investor's share of the joint venture's assets is shown in a single line on the balance sheet, and the investor's share of the joint venture's liabilities is shown separately on another line. The net income is shown in the same manner as it is shown in the equity method.

A common use of the expanded equity method is to show the proportionate share of the joint venture's assets on a single line captioned "Company's share of assets of joint venture." In a similar manner, the proportionate share of the joint venture's liabilities would be shown as "Company's share of liabilities of joint venture."

Full Consolidation Method

The full consolidation method should be used where more than 50 percent of the joint venture is owned by the investor and there are no reasons for not consolidating. A reason for not consolidating a construction venture might be that the operations of the joint venture are not compatible with those of the venturer. For example, the venturer might be an insurance company required to use statutory accounting methods, while the construction joint venture might be using generally accepted accounting principles (GAAP). Under the full consolidation method, the complete amount of each of the assets, liabilities, revenues, and expenses of the joint venture is combined with the corresponding amounts of the investor. The interest of the minority investors in the net assets of the joint venture is shown in the consolidated balance sheet equity section and captioned "Minority interest." The minority investor's interest in the net income of the joint venture appears on the income statement as an adjustment to the consolidated net income, and may be captioned "Minority interest in net income."

Partial or Proportionate Consolidation Method

The partial consolidation method is not in common use. This method should only be used where an investor has an undivided interest in the assets and liabilities of a joint venture. Under the proportionate consolidation method, the investor's proportionate interest in each of the assets, liabilities, revenues, and expenses of the joint venture is combined with the corresponding amounts of the investor's own assets, without a distinction made between them. Since the proportionate rather than the full amounts of assets and liabilities are combined, there is no minority interest, as there would be in the full consolidation example. Disclosure of the use of this method and the details of its application to the financial statements is appropriate because most readers of accounting statements are unfamiliar with either partial or proportionate consolidation and, therefore, may be misled without the requisite detailed information.

Combination of Methods

An investor may also decide that the best method of presentation for a particular set of joint venture investments is to use a mixture of methods, that is, to mix features of both the equity method and the complete or partial consolidation method, or to mix the proportionate and complete consolidation methods with variations of other methods.

The most common combination involves the use of the equity method in the balance sheet and the proportionate consolidation method in the income statement. This combination gives the users of the financial information more details about the operations of the entity without cluttering the balance sheet. The use of this hybrid method requires complete disclosure in the financial statements.

DISCLOSURES BY INVESTORS

Disclosures about joint venture operations in the investor's financial statements should include:

- Name and ownership percentage of each significant joint venture.

- Important provisions of the joint venture agreement, including any resulting liabilities or contingent liabilities.

- Joint venture financial information in summary form, if significant.

- Intercompany transactions, pricing, and other related-party arrangements.

SALES TO A JOINT VENTURE

As a general rule, intercompany profit resulting from sales of materials, supplies, or services to a joint venture by an investor that controls the joint venture, either through majority voting interest or by another means (effective although not actual control), should not be recognized by that investor unless the following three conditions are met:

1 The transaction is entered into at a price determined on an arm's-length basis, that is, fair market value can be measured by comparable sales at normal selling prices to independent third parties or by competitive bids.

2 There are no substantial uncertainties regarding the venturer's ability to perform – such as those that may be present if the

venturer lacks experience in the business of the venture – or regarding the total cost of the services to be rendered.

3 The venture is creditworthy and has independent financial substance.

If the investor does not control the joint venture, it should recognize the intercompany profit to the extent of third-party interest in the joint venture. For nonmonetary transactions, profit should be recognized in accordance with APB Opinion 29, "Accounting for Nonmonetary Transactions."

DIFFERING ACCOUNTING PRINCIPLES

The accounting principles adopted by the joint venture may differ from GAAP. For example, cash-basis financial statements may be prepared by the joint venture, whereas accrual-basis statements are required by the investors. Such joint venture financial statements would first have to be adjusted to conform to GAAP before consolidation with the investor or application of the equity method.

CHAPTER **3**

Income Recognition

Bernard D. Dusenberry
Thomas W. McRae

INTRODUCTION

Income recognition is the process of assigning the revenue and expenses of an enterprise to appropriate accounting periods in order to determine reported net income. The revenue of an enterprise is generally measured by the charges made to customers for the sale of goods or the rendering of services. An enterprise's revenue results from the activities and operations that are identified in accounting as the earning process. The general principles for recognition of revenue are well established in accounting and are based on conventional revenue-realization principles. However, specialized principles apply to the construction industry.

The specialized generally accepted accounting principles (GAAP) for recognizing contract revenue and costs are the central focus in accounting and financial reporting for construction contractors. The application of revenue-recognition rules in the construction-contracting industry requires a sound knowledge of the earning process in the industry. To account for the performance of contracts for financial-reporting purposes, contractors and their accountants must learn to apply the specialized accounting principles and procedures that apply to construction contractors to dynamic businesses and explain their underlying rationale.

CONTRACTS AND FINANCIAL REPORTING

Financial reporting in the construction-contracting industry usually centers on individual contracts for recognizing revenue, accumulating costs, and measuring income. A contractor earns revenue from work performed under contractual arrangement with a customer; these arrangements usually entail building or making improvements on tangible property due to customer's specifications. The contract specifies the work to be performed, the price to be paid, the timing of performance, and the accounting for the contract. Accountants must thoroughly understand the contracts on which financial reporting is based and the characteristics of those contracts that affect income recognition. The presentation of information in conformity with the specialized GAAP and the evaluation of profitability rests on such an understanding. The methods and bases of measurement require estimates of contract revenue, contract costs, extent of progress toward completion, and gross profit. These factors are essential for assessing the reasonableness of reported revenue, costs, and gross profit allocated to accounting periods.

Individual Contracts

The focus on individual contracts is a unique aspect of financial reporting in the construction-contracting industry. Thus, the method used in recognizing income on construction contracts differs from those methods used in other business environments. A manufacturing company, for example, produces a series of homogeneous units of a product and determines income for all units of products sold during a prescribed period. The revenue for the period is the proceeds from the units sold; the cost of earned revenue is the aggregate cost of the units determined by using inventory-costing techniques, usually without specific identification. In contrast, a contractor must usually account for each contract separately. The contract is generally presumed to be the profit center for accounting purposes, unless circumstances indicate that there is a need to combine a contract with similar contracts, or segment a contract into its components.

Contract Terms Governing Accounting

Contract terms are the basic determinants of revenue (prices paid by customers), timing of performance (contract milestones), specifications (performance requirements), and payment terms (billing arrangements).

Revenue. The revenue from a contract is the total amount that a contractor expects to collect from the customer for contract performance. The starting point is the basic contract or the formula by which the basic price can be determined. Under some circumstances the contract may also provide for incentive payments and penalties. The price is often adjusted by change orders and claims. The total expected revenue from a contract is the total amount that can reasonably be collected from the customer – the net amount derived from the basic contract price, contract options, change orders, claims, and amounts determined by contract provisions and incentives. The basic contract price may be relatively fixed or completely variable, depending on the type of contract. For accounting purposes, a contractor is often required to estimate the total amount of revenue earned or to be earned as performance under the contract proceeds. This amount is subject to the vagaries of changing circumstances and may not be determinable with certainty until the terms of the contract have been fully performed.

Some fixed-price contracts have escalation clauses relating to price redetermination, incentive, penalty, and other pricing provisions. Time-and-material contracts may have guaranteed maximums or assigned markups for labor and materials, each of which requires careful analysis in price determination. The various forms of cost-type contracts may have differing terms relating to reimbursable costs, overhead recovery percentage, and fees. These various provisions affect accounting choices in income recognition.

Timing of Performance. Contracts usually specify performance targets or milestones. The ability of a contractor to meet these targets must be considered in accounting for contracts, since failure to meet targets may result in reduced revenue or increased costs. For example, a contract may specify interim dates for completion of different stages of the work, as well as the date required for final completion of the project. Penalties are usually assessed for failure to meet target dates.

The required timing of performance may be indicated in the specifications for bids. The contractor must consider the required timing in preparing cost estimates and bids and in selecting a measure of progress toward completion for accounting purposes. The contractor's plans for initiation of the subcontractors are affected by the need to meet the dates for completion of each phase of the work. Without careful planning of these activities, the profitability of a contract can be seriously affected.

Specifications. Contract specifications usually describe the scope of the work to be performed and detail matters such as type and quality of materials, standards of performance, and separable phases or elements of the project. Other matters that may be covered in the specifications include the responsibility for acquiring materials, procedures for initiating change orders, and procedures for handling claims and back charges. Here, as in timing of performance, penalties are imposed for failure to perform in accordance with the specifications.

The type and scope of performance specifications are factors that must be considered in establishing procedures for income recognition. Under some circumstances, for example, customer-furnished material may need to be excluded from a contractor's revenue and cost. Contract revenue and contract costs should include all materials for which the contractor has an associated risk. If, for example, the specifications require the contractor to be responsible for the nature, type, characteristics, or specifications of materials furnished by the customer for the ultimate acceptability of performance of the project based on such materials, these materials should be included in both contract revenue and contract costs in the periodic reporting of operations.

Payment Terms. Contracts specify payment terms or billing arrangements, which are usually not correlated with income-recognition procedures in financial accounting. Billing arrangements in the construction-contracting industry differ from those used in other industries. A typical manufacturing company, for example, normally bills a customer on the shipment of its product and recognizes the amount billed as revenue. In contrast, billing arrangements in the construction-contracting industry vary widely among companies because different types of contracts require different ways of measuring performance for billing purposes.

The amount and timing of payment may depend upon:

- Completion of certain stages of the work
- The amount of cost incurred
- Estimates of percentage of completion
- Predetermined time schedules
- Quantity measures

A contractor's method of recognizing revenue is unrelated to the payment schedule specified in the contract. However, billing practices are tied very closely to the required payment schedules. Payment schedules and billing practices are negotiated to generate cash flow in order to help finance the progress of work on a contract. They may often be structured to achieve "front-end loading," which is the assignment of a disproportionate amount of revenue to the early work on a contract.

Most contracts have retention provisions that permit customers to withhold a designated percentage (e.g., 5 to 10 percent) of each billing until completion or until some other condition, such as a usability guarantee, is met. Retentions are important to cash flow, and their collection in accordance with the terms of the contract should be monitored by the contractor.

Risks

Contractors are exposed to significant risks in the business environment in which they operate. The exposure to risk is an important financial-reporting consideration that underlies the procedures established for recognizing income for contractors. The prices for work to be performed are established by bids or negotiations based on cost estimates and expected profit margins; thus, the total revenue on a contract is estimated long before reliable cost estimates can be generated. The prices established – particularly for fixed-price contracts – are usually not subject to modification solely because of changes in construction that arise from factors such as inflation and unexpected price increases for materials and labor. Thus, a contractor faces the risk of unexpected developments, such as changing work conditions and prices, over the life of a contract. Since a contractor usually works on the customer's premises and has only lien rights against property under construction, he must often make significant commitments of resources to a project without financial assurances of recovery.

   The fallibility of estimates is one substantial source of risk in accounting for contracts. A contractor's ability to make reasonably dependable estimates is a fundamental assumption of contract accounting. Yet contractors are exposed to all of the risks, which may vary from contract to contract, to which other business enterprises are exposed, as well as the hazards inherent in the contracting process. Thus, the actual results often differ from the original estimates simply because of the nature of the business.

Contract Modifications

Modifications in the work being performed and in accounting estimates over the life of a contract are inherent in the contracting process. The American Institute of Certified Public Accountants (AICPA) Audit and Accounting Guide, "Construction Contractors" (1981), known as the *Guide*, states on page 7 that:

> Management control orders, claims, extras, and back charges are of critical significance in construction activity. Modifications of original contract frequently result from change orders that may be initiated by either the customer or the contractor. The nature of the construction industry, particularly the complexity of some types of

projects, is conducive to disputes between the parties that may give rise to claims or back charges. Claims may also arise from unapproved change orders. In addition, customer representatives at a job site sometimes authorize a contractor to do work beyond contract specifications, and this will give rise to claim from "extras." The ultimate profitability of a contract often depends on control, documentation, and collection of amounts arising from such items.

The modifications inherent in contract performance are significant factors in determining procedures for recognizing income on contracts.

Length of Contract Term

Income recognition in the construction-contracting industry is a process that involves measuring the results of relatively long-term events and assigning those results to relatively short accounting periods. Contracts are often performed over extended periods that span more than one accounting period. The most recent accounting pronouncement on contract accounting, the AICPA Statement of Position (SOP) 81-1, "Accounting for Performance of Construction-Type and Certain Production-Type Contracts," does not treat the length of a contract as a defining criterion for determining the contracts to which its provisions apply, primarily because of the difficulty in designating a cutoff between long-term and short-term contracts. As a result, the same basic accounting rules apply to accounting for the performance of all contracts for which specifications are provided by customers for the construction of facilities or the provision of related services in financial statements prepared in conformity with GAAP. Nevertheless, a contractor whose contracts are relatively short-term and turn over rapidly faces considerably less risk and uncertainty than a contractor with long-term contracts.

BASIC ACCOUNTING ISSUES

Determining revenue, expenses, and net income is relatively more complicated for construction contractors than it is for most other types of business enterprises. The problems involved relate to the special realization rules for revenue recognition and for exceptions to the general rule, expense recognition principles, the use of estimates, and related uncertainties. SOP 81-1 states in paragraphs 109 to 110 that:

> The determination of the points at which revenue should be recognized as earned and costs should be recognized as expenses is a major accounting issue common to all business enterprises engaged

in the performance of contracts of the types covered by this statement. Accounting for such contracts is essentially a process of measuring the results of relatively long-term events and allocating those results to relatively short-term accounting periods. This involves considerable use of estimates in determining revenues, costs, and profits and in assigning the amounts to accounting periods. The process is complicated by the need to evaluate continually the uncertainties inherent in the performance of contracts and by the need to rely on estimates of revenues, costs, and the extent of progress towards completion.

Income recognition in a construction-contracting environment must cope with all of these factors in measuring and reporting periodic net income.

Realization of Revenue and Recognition of Costs as Expenses

Under GAAP, conventional realization rules determine the points at which revenue should be recognized as earned. These rules specify the events or points in time when an increase in assets from the revenue-generating activity of an enterprise is justified. Realization rules tell the accountant when a change in an asset has objectively become sufficiently definitive to warrant recognition in the accounts. Revenue is generally recognized when the earning process is complete, or substantially complete, and exchange has taken place. For most enterprises, this means that revenue is recognized when a sale is made. In some circumstances, however, realization may be deemed to occur earlier or later than the point of sale, such as at the time of production for special-order goods. In contract accounting, revenue is generally recognized as construction progresses; this represents a departure from the general realization rule. The exception is a hybrid realization rule based on the availability of evidence of the ultimate proceeds from a contract and the consensus that a better measure of periodic income results.

When revenue is recognized, closely related expenses (i.e., the cost of earned revenue) must be recorded and assigned to the same period as the related revenue. Under the realization rules, costs are related to revenue on the basis of cause and effect, systematic and rational allocation, or the period within which the expenses were incurred. The cause-and-effect rule is the most direct way of associating revenue and expenses. In contract accounting, costs are accumulated by contract and are recorded as expenses when revenue from the contract is recorded as realized. This procedure further suggests that costs are also recognized as expenses as construction progresses.

Justification for the Exception to the Realization Rule

The realization rule for contracts has been described as a modifying principle in accounting. One justification for the special realization rule is that recognizing revenue in strict conformity to the general realization rule produces results that are considered to be unreasonable. Another justification relates to the peculiar nature of the events that occur under a contract. SOP 81-1 draws an analogy between these factors and a sales transaction, and suggests that what is taking place is a continuous sale. The SOP states in paragraph 220 that:

> Under most contracts for construction of facilities, production of goods, or provision of related services to a buyer's specifications, both the buyer and the seller (contractor) obtain enforceable rights. The legal right of the buyer to require specific performance of the contract means that the contractor has, in effect, agreed to sell his rights to work-in-progress as the work progresses. This view is consistent with the contractor's legal rights; he typically has no ownership claim to the work-in-progress but has lien rights. Furthermore, the contractor has the right to require the buyer, under most financing arrangements, to make progress payments to support his ownership investment and to approve the facilities constructed (or goods produced or services performed) to date if they meet the contract requirements. The buyer's right to take over the work-in-progress at his options (usually with a penalty) provides additional evidence to support that view. Accordingly, the business activity taking place supports the concept that in an economic sense performance is, in effect, a continuous sale (transfer of ownership rights) that occurs as the work progresses.

Under this view, which has considerable merit, the special realization rule for contracts is only a refinement of the general rule to recognize the events that occur under a contract. It is not an exception to the general rule; rather, it is a special application required by the general rule.

Periodic Reporting—Interim Measures for Short Time Periods

Most of the difficult problems in accounting can be attributed to the need to allocate the results of the earning process to relatively short time periods. Theorists in accounting have often noted that measuring net income over the life of a business enterprise would be relatively simple because most of the allocation problems would disappear. To serve the informational needs of users, however, financial reporting requires the periodic reporting of results. George O. May characterized the difficulties in his book, *Financial Accounting: A Distillation of Experience* (1943):

> The problem of allocation of income to particular short periods obviously offers great difficulty – indeed, it is the point at which conventional treatment becomes indispensable, and it must be recognized that some conventions are scarcely in harmony with the facts. Manifestly, when a laborious process of manufacturing and sale culminates in the delivery of the product at a profit, that profit is not attributable, except conventionally, to the moment when the sale or delivery occurred. The accounting convention that makes such an attribution is justified only by its demonstrated practical utility.

The difficulty arises because most revenue is the joint result of the many profit-directed activities of an enterprise. The problem is compounded in contract accounting because the focus is individual contracts, yet most contracts cover extended periods. Complex rules must therefore be followed in assigning the results of contract performance to annual and interim periods.

Estimates and Related Uncertainties

Income recognition in contract accounting is complicated by the need (1) to rely on estimates of revenue, costs, and extent of progress toward completion; and (2) to evaluate continually the uncertainties inherent in the performance of contracts. The degree of the reliability of estimates of revenue, costs, and extent of progress will have an impact on both the accuracy of measurement of contract performance and the appropriateness of the accounting method selected.

Although accounting literature has given little attention to the difficulties in estimating contract revenue, the various elements of revenue make estimates of total contract revenue subject to a high degree of uncertainty. Estimating total contract cost is difficult because of uncertainties relating to future prices and conditions that may be encountered as the contract progresses. The extent of progress toward completion is an uncertain estimate, often determined by indirect factors that depend on other estimates, such as the amount of actual costs incurred as a ratio of the estimate of total costs.

BASIC ACCOUNTING AND REPORTING REQUIREMENTS

The basic contract accounting and reporting requirements specify the acceptable methods under GAAP of income recognition and the bases for making the choice between them. The specification of the requirements raises some controversial issues concerning whether the acceptable alternatives may be used under similar circumstances, and whether they may be used simultaneously by the same entity. Other issues relate to distinctions between (1) acceptable and unacceptable methods, and (2) between those methods acceptable for financial-reporting purposes and those acceptable for income-tax-accounting purposes.

Accounting Literature

The literature on contract accounting has been relatively sparse. The basic principles, which are still in effect, were enunciated in Accounting Research Bulletin (ARB) 45, "Long-Term Construction-Type Contracts," issued in 1955 by the Committee on Accounting Procedure (CAP) of the AICPA. An earlier AICPA pronouncement dealt with three narrow areas relating to government contracts. That pronouncement is included as Chapter 11, "Government Contracts," in ARB 43, "Restatement and Revision of Accounting Research Bulletins," issued in 1953 to revise and restate the 42 Accounting Research Bulletins previously issued. It contains provisions that deal with accounting problems arising under cost-plus-fixed-fee contracts, accounting for government contracts subject to renegotiation, and accounting for terminated war and defense contracts. These particular provisions are not relevant to most construction contractors. Until 1981, the only other significant accounting pronouncement dealing with accounting for construction contractors was the AICPA Audit and Accounting Guide, "Audits of Construction Contractors," issued in 1965, which combined into one document the provisions of two earlier guides – an audit guide for contractors and an accounting guide for contractors, issued by the AICPA in 1959.

In 1981, the AICPA issued a new Audit and Accounting Guide, "Construction Contractors," to update and revise the 1965 *Guide*. At the same time, the AICPA, through its Accounting Standards Division, issued an SOP on contract accounting, SOP 81-1, "Accounting for Performance of Construction-Type and Certain Production-Type Contracts," to provide comprehensive guidance on the application of GAAP to, among other areas, accounting for construction contractors with primary emphasis on income-recognition methods. SOPs had been used by the AICPA Accounting Standards Division to provide guidance on the application of specialized accounting principles, until the Financial Accounting Standards Board (FASB), the body now authorized to issue authoritative pronouncements on accounting and financial reporting, announced its intention to incorporate the guidance found in AICPA guides and SOPs into its pronouncements and to take direct responsibility for providing future guidance. SOP 81-1 was issued separately, and also included an appendix to the 1981 *Guide* in order to provide a frame of reference for accounting guidance.

The issuance of the *Guide* and SOP 81-1 marked the completion of a long, complex project, which had started in 1975. The process of preparing and issuing the two documents took more than five years, and covered the full range of due process procedures for issuing an accounting pronouncement. When the project started, over 15 years had already elapsed since serious attention had been given to contract accounting by professional bodies. Robert Morris Associates (RMA), a professional organization of bank-lending officers, was the catalyst for the project. The Board of Directors of the RMA adopted a resolution urging the AICPA to update and revise the 1965 *Guide* to recognize and deal with changes that had taken place in the construction-contracting industry since 1965. The RMA's request was based on a survey that showed numerous deficiencies in the financial-reporting practices of construction contractors. The RMA identified several areas of accounting and financial reporting – including, in particular, the area of income recognition – as needing attention. A task force appointed by the AICPA to consider the RMA's request agreed with the RMA's assessment of the situation, and the AICPA appointed a 17-member committee of its members, all of whom had wide experience in accounting and financial reporting for the construction-contracting industry, to undertake the project.

As the project progressed, it became clear to the committee that (1) some substantial changes in emphasis were necessary in contract accounting; (2) contract accounting was not the exclusive preserve of construction contractors; and (3) many unrelated industries had used and sometimes abused contract-accounting principles. For these reasons, the project was broadened to cover the preparation of the SOP on contract accounting, including both construction-type and production-type contracts.

The provisions of the *Guide* and the SOP represent a significant expansion of the literature on contract accounting and should improve financial reporting in the construction-contracting industry. The *Guide* and the SOP provide contractors and accountants with comprehensive guidance for preparing and auditing the financial statements of construction contractors and significantly affect all parties interested in contractor financial statements – such as bankers, sureties, and investors – by affording them more understandable financial statements.

Following the issuance of the new *Guide* and SOP 81-1, the FASB, as the first step in its plan to incorporate specialized industry-accounting principles into its pronouncements, issued Statement of Financial

Accounting Standards (SFAS) No. 56 with the tongue-twisting title, "Designation of AICPA Guide and Statement of Position (SOP) 81-1 on Contractor Accounting and SOP 81-2 Concerning Hospital-Related Organizations as Preferable for Purposes of Applying APB Opinion 20." By making the provisions of the *Guide* and the SOP preferable for changing accounting policies and procedures in accordance with APB Opinion 20, "Accounting Changes," SFAS 56 enhanced their authority. However, SFAS 56 does not require contractors to change their existing policies (which may differ from those required by the *Guide* and the SOP) to conform to those provisions. The FASB has announced plans to reconsider the provisions of the *Guide* and the SOP after approximately one year to decide whether it should issue a pronouncement incorporating these provisions into its standards, which are enforceable under Rule 203 of the AICPA's Code of Professional Ethics. This step will further enhance the authority of the specialized principles in the *Guide* and the SOP by making them GAAP. Financial statements purported to be in conformity with GAAP would then be required to conform to these provisions.

Two Acceptable Methods for Contract Accounting

Despite the issuance of the *Guide* and the SOP, which contains comprehensive guidance on contract accounting, ARB 45 provides the basic accounting requirements on income recognition for construction contractors. ARB 45 requires the use of either of two basic acceptable methods in accounting for contracts, the percentage-of-completion method or the completed-contract method. The percentage-of-completion method states that income from the performance of a contract is recognized as performance as the contract progresses, and that the recognition of revenue and profit on a contract is related to the costs incurred in providing the services required. The completed-contract method requires income from the performance of a contract to be recognized upon completion of the contract; during the production cycle, costs incurred in the performance of the contract and any revenues received, or billed, are reported as deferred items in the balance sheet.

ARB 45 generally describes the circumstances under which each of the two generally accepted methods of contract accounting are preferable and how those methods should be applied in practice. The percentage-of-completion method is preferable only where estimates of costs to complete and of the extent of progress toward completion are reasonably dependable. The completed-contract method is to be used where the "reasonably dependable" criteria cannot be met. The basis for determining "reasonably dependable" estimates is subjective, since the bulletin does not provide criteria for measuring whether those conditions exist.

Acceptable Methods Tailored to Circumstances

SOP 81-1, based on the same general concept as ARB 45, more carefully identifies the circumstances under which either of the two methods are to be applied. By providing extensive guidance on the application of the concepts in ARB 45, SOP 81-1 significantly expands the requirements for contract accounting. The basic thrust of the SOP is to change the manner in which the two basic acceptable methods of contract accounting are applied in practice. Although SOP 81-1 does not change the basic rule for the choice between the two acceptable methods, it does provide detailed objective criteria for determining where circumstances appropriate for the use of the percentage-of-completion method are present. The SOP views the choice between the two acceptable methods as a broad policy question relating to the nature of an entity's predominant type of contracts; it requires a contractor to adopt one method for substantially all contracts, instead of making a choice based on each individual contract.

In the application of ARB 45 before SOP 81-1 was issued, there had always been some dispute as to whether the percentage-of-completion method and the completed-contract method were intended to be acceptable alternatives under the same conditions. In practice, the two methods were widely used on that basis; contractors could use either method without regard as to whether they met the circumstances prescribed for the use of the percentage-of-completion method. The dispute over whether the two methods were acceptable alternatives under the same circumstances hinged on the interpretation of the term, "preferable," as used in ARB 45. Some contended that the bulletin merely designated the conditions under which the use of the percentage-of-completion method was deemed "preferable," but did not require the use of that method in those circumstances; thus, either method could be used. Others believed that by designating the circumstances in which the use of the percentage-of-completion method was deemed preferable, the bulletin required the use of the percentage-of-completion wherever the criteria were met. SOP 81-1, in paragraph 21, adopts what is believed to be the correct view: that the two methods should not be used as acceptable alternatives under the same circumstances and that if the criteria for the use of the percentage-of-completion method exist, this method must be used; the completed-contract method is therefore proscribed. Since under most conditions, construction contractors can make reasonably dependable estimates, the percentage-of-completion method of accounting becomes the method to use.

Adopting a Basic Policy Decision

As recommended by SOP 81-1, a contractor should adopt one of the acceptable methods of income recognition as a broad accounting policy decision for all contracts. The choice relates to the nature of the contractor's predominant type of contract and the contractor's ability to make reasonably dependable estimates of contract revenue, contract costs, and the extent of progress toward completion. As previously noted, the guidance on the choice of method in SOP 81-1 is based on the presumption that most contractors have the ability to make such estimates. The presumption is established if the contracts clearly set forth the enforceable rights of the parties and if all the parties have the ability, and thus can be expected, to perform their obligations under the contracts. SOP 81-1 contends that entities engaged on a continuing basis in the performance of contracts as defined, and for whom contracting represents a significant part of their operations, should have the ability to make reasonably dependable estimates to justify the use of the percentage-of-completion method; persuasive evidence to the contrary is necessary to overcome this presumption.

The ability to make such estimates is an essential element of the construction-contracting business. Moreover, a central rule of GAAP implies that a contractor must have such an ability to be able to apply GAAP correctly. For a contract on which a contractor anticipates a loss, GAAP requires recognition of the entire loss, as soon as it becomes evident, without regard to the income-recognition method used by the contractor. Without the ability to update and revise estimates continually with some degree of confidence, a contractor could not meet this essential requirement of GAAP.

Since the percentage-of-completion method should be employed under most circumstances, the discussion of income recognition in this chapter focuses on the application of that method, giving only limited attention to the completed-contract method.

Unacceptable Accounting Methods for GAAP

Other than the two acceptable methods under GAAP, income-recognition accounting methods are to be used only in those circumstances where GAAP financial statements are not required. Income is sometimes recognized, for example, on the cash basis, billing basis, or some hybrid basis, none of which is GAAP. On the cash basis, revenue is reported and income is measured on the basis of cash receipts and disbursements. On the billing basis, the amounts billed or billable under contract during a period are recognized as revenue, and costs incurred (paid and accrued) during the period are recognized as the cost of earned revenue. SOP 81-1 states that only by coincidence would these methods produce results that approximated the results under the two methods acceptable under GAAP. As stated in the preceding section, the discussion of income recognition in this chapter is concerned only with the methods acceptable for GAAP purposes.

However, methods of income recognition acceptable for income tax purposes are not limited to those acceptable for GAAP purposes and are covered in detail in Chapter 18.

COMPUTATION OF INCOME UNDER PERCENTAGE OF COMPLETION

In order to provide a framework for discussing the basic method of income recognition for contracts, a description and examples of the procedures for computing earned revenue, cost of earned revenue, and gross profit on a contract are presented. The relationships among all of the factors used in applying the method are shown in the discussion of the computation. These factors include:

- Estimated total contract revenue
- Estimated total contract cost
- Estimated extent of progress toward completion

Noting how these factors are used in financial reporting helps clarify the underlying accounting used for contracts.

The periodic revenues and expenses reported under the percentage-of-completion method are determined by any one of several alternatives. The alternatives are convenient ways for measuring the amount of work done and, therefore, the revenue earned, on a contract in any specific accounting period. Among the methods used for measuring revenue, cost, and profitability on a contract are:

- The revenue-cost approach
- The gross-profit approach

Both of these methods require that a measure of the extent of progress toward completion be selected, and that this measure be applied in determining periodic income. ARB 45, in paragraph 4, identifies as acceptable measures of progress the cost-to-cost, efforts-expended, and units-of-work-performed methods. The ARB further states that depending on the circumstances of use and the manner in which a method is applied, any particular method may or may not achieve the goal of measuring the extent of progress on a contract.

The cost-to-cost method of estimating the percentage of completion is computed by dividing the actual contract costs to date by the sum of the actual costs to date and the estimated costs to complete. This fraction, or percentage, can then be used with either the revenue-cost or gross-profit approaches to determine the revenues, costs, and profitability of a contract to be recognized in an accounting period.

The efforts-expended method is an input measure of percentage of completion. A significant measure of contract completion is chosen, such as labor-hours, machine-hours, or materials quantities. A ratio, or percentage, is then computed, for example, by dividing the actual labor-hours worked by the sum of the actual labor-hours worked and the estimated labor-hours needed to complete the contract. This percentage is then applied in a manner similar to the application of the cost-to-cost method. The assumption in the efforts-expended method is that revenues and, therefore, profits are derived from the actual work done and that the means selected for measuring that work, such as labor-hours, is the best guideline to efforts expended.

The units-of-work-performed method differs from the efforts-expended method in that it is an output measure. The ratio, or percentage, of completion is based upon actual units produced, such as linear feet of steel, compared with estimated total units expected to be produced. Once computed, the ratios are applied in a manner similar to those previously discussed.

Computational Methods

In practice, two equally acceptable computational methods are used to measure revenue and costs. One, the revenue-cost approach, multiplies the estimated percentage of completion, computed by one of the methods previously detailed, by the estimated total contract revenue and by the estimated total contract cost to determine both earned revenue and cost of earned revenue. The other, the gross-profit approach, multiplies the estimated percentage of completion by the estimated gross profit — the difference between the estimated total contract revenue and the estimated total contract cost — to determine the estimated gross profit earned to date.

Revenue-Cost Approach

Under the revenue-cost approach, earned revenue is determined by multiplying the estimated percentage of completion by the estimated total revenue. The cost of earned revenue is determined by multiplying the same percentage of completion by estimated total cost. The difference between these two amounts is the gross profit earned to date. The process is an inventory-like procedure. The excess of earned revenue and cost of earned revenue computed for the current period over the amount computed at the end of the preceding period are the amounts reported in the income statement.

To illustrate, let us assume that AG Company has a contract to build an apartment building for $9 million. The company expects to complete the project over a period of 24 months at an estimated cost of $7.2 million, with an estimated gross profit of $1.8 million. At the end of its current reporting period, AG Company showed the status of the contract as follows:

Contract Status Item	Current Period	Preceding Period
Total costs incurred	$5,500,000	$1,750,000
Extent of progress (engineers' estimate)	75%	25%
Billings to date	$7,650,000	$3,150,000
Cash collected	$6,885,000	$2,835,000

Based on the illustration, the computation of earned new issue for the current period would be:

	Current Period	Preceding Period
Earned revenue to date (75% of $9,000,000 and 25% of $9,000,000)	$6,750,000	$2,250,000
Less: Amount, preceding period	2,250,000	
Earned revenue, current period	$4,500,000	

In the illustration, the cost of earned revenue would be computed as follows:

Cost of earned revenue to date (75% × $7,200,000)	$5,400,000
Cost of earned revenue at the end of the preceding period (25% × $7,200,000)	1,800,000
Cost of earned revenue, current period	$3,600,000

Under the revenue-cost approach, the amount of gross profit earned is merely the difference between earned revenue and cost of earned revenue as computed. Based on the illustration, the income statements for the current and preceding periods would appear as follows:

Income Statement Item	Total	Current Period	Preceding Period
Earned revenue	$6,750,000	$4,500,000	$2,250,000
Cost of earned revenue	5,400,000	3,600,000	1,800,000
Gross profit	$1,350,000	$ 900,000	$ 450,000

Gross-Profit Approach

The other acceptable computational approach focuses on the estimated total gross profit. The estimated percentage of completion is multiplied by the estimated total gross profit (the difference between estimated total revenue and estimated total cost) to determine the amount of gross profit earned to date. The actual costs incurred during a period are always treated as the cost of earned revenue. The revenue reported is a residual, the sum of the estimated gross profit earned and the amount of costs incurred for the period.

In the illustration, the computations under the gross profit approach would yield the following amounts for gross profit, cost of earned revenue, and earned revenue:

Gross profit to date (75% of $1,800,000)	$1,350,000
Gross profit at end of preceding period (25% of $1,800,000)	450,000
Gross profit, current period	$ 900,000
Cost incurred to date	$5,500,000
Cost incurred at end of preceding period	1,750,000
Cost of earned revenue, current period	$3,750,000
Earned revenue:	
Cost of earned revenue	$3,750,000
Gross profit earned	900,000
Earned revenue, current period	$4,650,000

Based on the illustration, the income statement under the gross-profit approach would appear as follows:

Income Statement Item	Total	Current Period	Preceding Period
Earned revenue	$6,850,000	$4,650,000	$2,200,000
Cost of earned revenue	5,500,000	3,750,000	1,750,000
Gross profit	$1,350,000	$ 900,000	$ 450,000

The two approaches may produce different amounts for earned revenue and cost of earned revenue unless the estimated percentage of the extent of progress toward completion is based on the cost-to-cost approach.

The two approaches are equally acceptable. The use of the revenue-cost approach is based on the view that earned revenue and the cost of earned revenue should both be allocated to a period on the basis of the extent of progress toward completion in accordance with the matching concept. This approach produces a consistent gross-profit percentage from period to period, unless estimates of total revenue and total costs change between periods. Revenues are reported on a basis consistent with performance. The use of the gross-profit approach is based on the point of view that costs incurred on a contract (1) can be objectively determined; (2) do not depend on estimates; and (3) are the accounting determinants of the amount that is shown as revenue. This approach may cause varying gross-profit percentages.

BASIS FOR THE PERCENTAGE-OF-COMPLETION METHOD

The primary test of an income-recognition method is whether, through its use, the revenue and expenses of an enterprise are allocated to accounting periods in a manner that relates realistically to the enterprise's efforts during those periods. The FASB in the first pronouncement resulting from its conceptual framework project, Statement of Financial Accounting Concepts (SFAC) No. 1, "Objectives of Financial Reporting by Business Enterprises," stressed that the objectives of financial statements "stem primarily from the needs of external users who lack the authority to prescribe the information they want." The primary objective in this statement is to provide information to third-party users of financial statements necessary to them "in making rational investment, credit, and similar decisions" and to aid them "in assessing the amount, timing, and uncertainty of prospective cash receipts" from their interests in the enterprise. Measured in terms of these objectives, the percentage-of-completion method, where applied by construction contractors, is more responsive than any other method to these goals.

Rationale for Use of Percentage-of-Completion Method

The basic rationale for the use of the percentage-of-completion method rests on the belief that with its use, the financial statements of construction contractors report the performance of an enterprise on a timely basis and portray the basic economic and legal realities of the construction-contracting business. Using this method, the income statement reports contract activity in the period in which it is performed, and the balance sheet shows the current status of contracts. The financial statement also shows the volume, mix, and profitability of current contract activity. In highlighting these factors, SOP 81-1 states in paragraph 22 that:

> Financial statements based on the percentage-of-completion method present the economic substance of a company's transactions and events more clearly and more timely than financial statements based on the completed-contract method, and they present more accurately the relationships between gross profit from contracts and related period costs. The percentage-of-completion method informs the users of the general purpose financial statements of the volume of economic activity of a company.

As previously noted, a basic rationale for the percentage-of-completion method is that the financial reporting resulting from it recognizes the economic and legal relationships between the parties to the contract. The relationships between buyers and sellers under contracts of the types performed by construction contractors are governed by both contract terms and well-established rights and obligations under law. The use of the percentage-of-completion method provides information in the financial statements that reflects events and conditions – such as the continuous passage of title to the customer, the contractor's right to compensation on an interim basis for services performed, and the earnings realized by the contractor as a result of work performed. In SOP 81-1, paragraph 22, it is noted that the "percentage-of-completion method recognizes the legal and economic results of contract performance on a timely basis." This statement neatly sums up the rationale for the use of the percentage-of-completion method under most circumstances.

Conditions for Use of Percentage-of-Completion Method

The use of the percentage-of-completion method requires that certain underlying conditions be present. These conditions include:

- Dependable estimates
- Contract specifications and
- Performance by parties

Dependable Estimates. The primary requisite for the percentage-of-completion method is that reasonably dependable estimates can be made of the three elements used in determining the status of contracts – contract revenue, contract cost, and the extent of progress towards completion. Estimates of contract revenue include not only the basic contract price but also anticipated change orders, incentive adjustments, penalties, and under some circumstances, contract options and claims. Contract costs and cost estimates include direct material, labor, and overhead, as well as projections of these same cost components in determining estimated total contract costs. (Accounting for contract costs is discussed in Chapter 4.) The extent of progress toward completion is estimated in several ways discussed more fully at the end of this chapter. The methods used depend on the type of contract, the data compiled by the contractor, and the contractor's previous experience. The prudent contractor generally uses not only a basic method for measuring progress but also one or more alternative methods for confirmation or job analysis purposes.

Contract Specifications. SOP 81-1 suggests that the requisite for reasonably dependable estimates is the existence of a contract or an agreement between contractor and buyer that contains certain specifications that are susceptible to measurement and to which estimates can be applied. The requisite conditions as stated in SOP 81-1 are:

- Contracts executed by the parties normally include provisions that clearly specify the enforceable rights regarding goods or services to be provided and received by the parties, the consideration to be exchanged, and the manner and terms of settlement.
- The buyer can be expected to satisfy his obligations under the contract.
- The contractor can be expected to perform his contractual obligations.

The requisite for use of the method includes a specification of a product or services to be provided. The specification may be in terms of a described physical structure, a machine, a production facility, the number of units of a product or service to be provided, or a facility or service that achieves stated objectives. The contracts should contain a stipulation regarding compensation, which can be in the form of a lump sum, a cost-related sum, or a price-per-unit of product or services. Contracts may specify different terms for separate phases or parts of the contract, or may include wide variations on the three basic compensation methods mentioned. Regardless of the variations, both the buyer and the seller should understand (1) how compensation is earned by the contractor and (2) all of the terms and procedures for payment.

Performance by Parties. In using the percentage-of-completion method, contractors and accountants consider not only what has happened to date in a project, but also what is expected to occur up until the time of completion. It is presumed that the parties are able to complete their performances or obligations under the contract. Thus for estimates to be considered dependable, the contractor must be able and willing; it is therefore expected that he will complete his obligations under the contract. The customer, in turn, must be expected to fulfill his obligations as they arise under the agreement. Under circumstances in which contracts meet these conditions, the percentage-of-completion method should be the basic accounting policy used.

Justification of Conditions

The view that most contractors have the ability to make reasonably dependable estimates is based on the logical presumption that entities engaged in the performance of contracts on a continuing basis are able to make estimates reliable enough to be used in applying the percentage-of-completion method. The history and continuity of the enterprise indicate an ability to produce financial and operating data and other information necessary to meet successfully the enterprise's contractual obligations. The important requisite is the resulting estimate, even though the form, detail, and totality of the information may differ greatly from one entity to another. SOP 81-1 recognizes that a contractor's record-keeping procedures may affect the results without changing the basic presumption that applies to most contracts and contractors. The SOP states in paragraph 24 that:

> Many contractors have informal estimating procedures that may result in poorly documented estimates and marginal quality field reporting and job costing systems. Those conditions may influence the ability of an entity to produce reasonably dependable estimates. However, procedures and systems should not influence the

development of accounting principles and should be dealt with by management as internal control, financial reporting, and auditing concerns.

Another consideration is the requirement in ARB 45 that regardless of the accounting method used, a loss on a contract should be recognized in its entirety when the loss becomes evident. The presumption is that under typical conditions, the entity has information and related estimates that make a loss apparent at a time much earlier than the final completion of the contract. The lack of such information and the inability to determine contract status would, consequently, be an indication that the entity's records are inadequate and could not enable the contractor to meet this important requirement of GAAP.

Three Acceptable Levels of Reasonably Dependable Estimates

SOP 81-1 identifies the bases for, or the approaches to, making contract estimates that are acceptable for accounting purposes. Estimates may be made as range estimates, or break-even estimates, depending on the circumstances.

Point Estimates. In most situations, contract revenue, contract costs, and the extent of progress towards completion should be estimated as single amounts, or point estimates. The estimate of total revenue should include not only the original basic contract price, but also the amounts anticipated for change orders, incentive awards, penalties, and the like. Under the point-estimate approach, the contractor makes his best estimate of the anticipated outcome on the contract and of the status of the contract. To illustrate, a building contractor in the first year of a contract makes the following estimates:

Item	Amount (in thousands)
Estimated total revenue	$1,200
Estimated total cost	900
Cost incurred to date	300
Estimated extent of progress toward completion (cost-to-cost basis)	33⅓%

Under the point-estimate approach, the contractor would report earned revenue of $400,000 ($1,200,000 × ⅓) cost of earned revenue of

$300,000 ($900,000 × ⅓) and gross profit of $100,000 ($400,000 − $300,000).

Range Estimates. Under some circumstances, contract revenue, contract costs, and extent of progress toward completion can be reasonably estimated only in ranges, although some level of profit is assured throughout the range. These circumstances may arise as a result of, for example, change orders, unforeseen conditions or developments, or contract terms. In these situations, SOP 81-1, in paragraph 25(b), requires that:

> If, based on information arising in estimating the ranges of amounts and all other pertinent data, the contractor can determine the amounts in the ranges that are most likely to occur, those amounts should be used in accounting for the contract under the percentage-of-completion method. If the most likely amounts cannot be determined, the lowest probable level of profit in the range should be used in accounting for the contract until the results can be estimated more precisely.

Under these circumstances, if the best estimates or most likely amounts within the possible ranges can be determined, those amounts should be used as the basis for estimating contract revenue, contract costs, and the extent of progress toward completion. If, however, the most likely amounts are not apparent, then those amounts within the ranges that produce the least profit should be used. For example, the building contractor in the previous illustration may only be able to make the following estimates:

Item	Amount (in thousands)		
	Low	Best Estimate	High
Estimated total revenue	$900	$1,000	$1,200
Estimated total cost	875	900	925
Cost incurred to date	300	300	300
Estimated of progress toward completion (cost-to-cost method)	33.3%	33.3%	32.4%

Under these circumstances, the contractor would report earned revenue of $333,333, cost of earned revenue of $300,000, and gross profit of $33,333 (33.3 percent of $100,000). If, however, the range was as shown and the contractor was unable to determine a best estimate, the contractor would then report earned revenue of $308,700 (34.3 percent of $900,000), cost of earned revenue of $300,000, and gross profit of $8,700.

Break-Even Estimates. In some circumstances, it is not realistic or prudent for a contractor to estimate a profit on a contract, even though a loss is not expected. Where this happens, estimating the final outcome on a contract is deemed to be impracticable, except to assure that no loss will be incurred. However, the estimates are still deemed to be sufficiently dependable to justify the use of the percentage-of-completion method. SOP 81-1, in paragraph 25(c), recommends that "in those circumstances, a contractor should use a zero estimate of profit; equal amounts of revenue and costs should be recognized until results can be estimated more precisely." A contractor should use that basis only if point estimates or range estimates are clearly not appropriate. The SOP requires that a change from a zero estimate of profit to a point estimate or a range estimate should be accounted for as a change in accounting estimates (APB Opinion 20). That treatment is consistent with accounting for normal, recurring changes in contract estimates, which are discussed later in this chapter.

To illustrate break-even estimates, let us assume that the contractor in the previous illustration is only able to make the following estimates:

Item	Amount (in thousands)
Estimated total revenue	$900
Estimated maximum cost	900
Cost incurred to date	300
Estimated extent of progress toward completion (cost-to-cost basis)	33.3%

Based on this illustration the contractor would report earned revenue of $300,000, cost of earned revenue of $300,000, and a zero gross profit.

Deviations From Basic Policy. A contractor whose basic accounting policy is the percentage-of-completion method may use the completed-contract method for some contracts. If a contractor has a contract for which reasonably dependable estimates are not available or inherent hazards make the estimates doubtful, the completed-contract method should be used for the individual contract. The contractor should disclose this departure in the accounting policies section of the financial statements.

Estimating Considerations

The ability of a contractor to produce reasonably dependable estimates is best indicated by the reliability of previous projections made by the contractor, not by the complexity of the contractor's estimating system. Estimates may be based on simple, elementary procedures or systems, or may be developed from complex procedures and systems requiring detailed information. Estimating procedures usually vary with the complexity of a project. Whatever the level of complexity of the estimating procedure or system, the resulting estimates should reflect the contractor's consideration of not only financial and production information but, also, of management, supervision, and control features. However, in the final analysis, the reliability and acceptability of estimates depend on their accuracy; detailed and complex procedures are not necessary to produce reasonably dependable estimates. Changing phases of the work, unforeseen developments, and the availability of more recent and complete information – all of which require contractors to revise and update estimates continually – are not indications that estimates are unreliable for accounting purposes. Such revisions are inherent in the process of accounting for contracts. SOP 81-1 states in paragraph 27 that a "contractor's estimates of total contract revenue and total contract cost should be regarded as reasonably dependable if the minimum total revenue and the maximum total costs can be estimated with a sufficient degree of confidence to justify the contractor's bids on contracts."

Inherent Hazards

ARB 45 suggests that inherent hazards may cause forecasts to be doubtful; however, the bulletin does not discuss the nature of these hazards. SOP 81-1 suggests that inherent hazards are rarely sufficient to impugn the dependability of contract estimates and makes a clear distinction between inherent hazards and inherent business risks. Inherent hazards are unusual risks or conditions beyond those ordinarily encountered in the contractor's range of activity. Inherent risks are nonrecurring, unrelated to contract activity, and not susceptible to reasonable estimates of probability. For example, inherent risks involve legal or governmental actions, new or untried technology, unreliable parties on whom performance depends, or ill-defined contract terms. In contrast, inherent business risks are those normally encountered in the contractor's activity, although they may vary from contract to contract. These risks are considered in the development of estimates and in the preparation of financial statements based on those estimates. Accordingly, estimates – point estimates, range estimates, and break-even estimates – are sufficiently reliable for the use of the percentage-of-completion method, unless specific and significant inherent hazards, beyond normal inherent business risks, are demonstrably present and affect the ability of the contractor to make dependable estimates.

DETERMINATION OF THE PROFIT CENTER

The term "profit center" describes the level of aggregation or disaggregation acceptable for the application of income-recognition procedures. The level of aggregation can affect the allocation of revenue, costs, and profits among accounting periods. For this reason, the accounting literature carefully specifies the level of aggregation appropriate to an income-recognition method.

Individual Contracts

SOP 81-1 states the basic assumption that each contract is the profit center for revenue recognition, cost accumulation, and income measurement. The unique characteristics of each contract make the individual contract the logical and normal profit center for accounting purposes. Thus, in determining revenue, accumulating costs, and calculating profit or loss, individual contracts are ordinarily used as the profit center. However, this assumption may be overcome under those conditions described in the following sections on combining contracts and segmenting contracts.

Combining Contracts

Contracts may be combined for accounting purposes in cases in which they are closely interrelated and in which reporting the contracts on a combined basis provides a better measure of performance than would reporting them on an individual contract basis. Combining contracts is acceptable in cases where a single project is covered by separate contracts written for technical, jurisdictional, or regulatory reasons. Where interrelated contracts have different project margins, combining is also acceptable, so that the financial statements report a consistent profit margin for the overall project. In paragraph 37, SOP 81-1 recommends, but does not require, that a group of contracts be combined for accounting purposes if the contracts:

>   a. Are negotiated as a package in the same economic
>   environment with an overall profit margin objective.

Contracts not executed at the same time may be considered to have been negotiated as a package in the same economic environment only if the time period between the commitments of the parties to the individual contracts is reasonably short. The longer the period between the commitments of the parties to the contracts, the more likely it is that the economic circumstances affecting the negotiations have changed.

b. Constitute in essence an agreement to do a single project. A project for this purpose consists of construction, or related service activity with different elements, phases, or units of output that are closely interrelated or interdependent in terms of their design, technology, and function or their ultimate purpose or use.

c. Require closely interrelated construction activities with substantial common costs that cannot be separately identified with, or reasonably allocated to, the elements, phases, or units of output.

d. Are performed concurrently or in a continuous sequence under the same project management at the same location or different locations in the same general vicinity.

e. Constitute in substance an agreement with a single customer. In assessing whether the contracts meet this criterion, the facts and circumstances relating to the other criteria should be considered. In some circumstances different divisions of the same entity would not constitute a single customer if, for example, the negotiations were conducted independently with the different divisions. On the other hand, two or more parties may constitute a single customer if, for example, the negotiations were conducted jointly with the parties to do what in essence is a single project.

Contracts that meet all of these conditions may be combined for the purpose of income recognition and the determination of the need to provide for losses, if the practice is followed consistently.

Segmenting Contracts

Another level of aggregation that may be used for income-recognition purposes is described in accounting literature as "segmenting a contract," which is the process of breaking up a larger unit into smaller units for accounting purposes. SOP 81-1 states in paragraph 39 that "a single contract or a group of contracts that otherwise meet the test for combining may include several elements or phases, each of which the contractor negotiated separately with the same customer and agreed to perform without regard to the performance of the others." The SOP also suggests that "if those activities are accounted for as a single profit center, the reported income may differ from that contemplated in the negotiations for reasons other than differences in performance." However, segmenting the project allows the contractor to assign revenue to the different elements or phases in a manner that achieves different rates of profitability based on the relative value of each element or phase to the estimated total contract revenue.

In paragraph 40, the SOP provides two sets of criteria for a project that may consist of either a single contract or a group of contracts and may be segmented, if the segments or phases of the project have different rates of profitability.

The first set of criteria in the SOP depends on whether certain steps were taken in the process of entering into the contract and whether those steps are documented and verifiable. The steps are:

a. The contractor submitted bona fide proposals on the separate components of the project and on the entire project.

b. The customer had the right to accept the proposal on either basis.

c. The aggregate amount of the proposal on the separate components approximated the amount of the proposal on the entire project.

If the contractor takes these steps and they are documented and verifiable, the contractor may segment the contract, that is, he may treat each element as a separate contract for accounting purposes. The rationale here is that the contractor agreed to perform distinct functions for the customer, which were aggregated for the convenience of the parties.

There are circumstances in which segmenting is justified. These circumstances relate to situations in which there is a single contract, or a group of contracts covering a single project, to which the steps required in the first set of criteria clearly do not apply. SOP 81-1, paragraph 41, states the following criteria for these circumstances:

a. The terms and scope of the contract or project clearly call for separable phases or elements.

    b. The separable phases or elements of the project are often bid or negotiated separately.

    c. The market assigns different gross profit rates to the segments because of the factors such as different levels of risk or differences in the relationship of the supply and demand for the services provided in different segments.

    d. The contractor has a significant history of providing similar services to other customers under separate contracts for each significant segment to which a profit margin higher than the overall profit margin on the project is ascribed.

    e. The significant history with customers who have contracted for services separately is one that is relatively stable in terms of pricing policy rather than one unduly weighted by erratic price decisions (responding, for example, to extraordinary economic circumstances or to unique customer-contractor relationships).

    f. The excess of the sum of the prices of the separate elements over the price of the total project is clearly attributable to cost savings incident to combined performance of the contract obligations (for example, cost savings in supervision, overhead, or equipment mobilization). Unless this condition is met, segmenting a contract with a price substantially less than the sum of the prices of the separate phases or elements would be inappropriate even if the other conditions are met. Acceptable price variations should be allocated to the separate phases or elements in proportion to the prices ascribed to each. In all other situations a substantial difference in price (whether more or less) between the separate elements and the price of the total project is evidence that the contractor has accepted different profit margins. Accordingly, segmenting is not appropriate, and the contracts should be the profit centers.

    g. The similarity of services and prices in the contract segments and the services and the prices of such services to other customers contracted separately should be documented and verifiable.

The SOP, in paragraph 41, states that in applying criterion (d), values assignable to the segments should be on the basis of the contractor's normal historical prices in terms of those services to other customers. The concept of allowing a contractor to segment on the basis of prices charged by other contractors was considered and rejected, because it does not necessarily follow that those prices could have been obtained by a contractor who has no history in the market.

Thus a contractor may segment a contract under two sets of conditions – on the basis of the contractor's normal historical prices or on the basis of prices charged by other contractors. Again, as with the requirements for combining contracts, such a disaggregation for accounting purposes is not required, although some contractors may find it advantageous to do so. For example, a contractor who specializes in engineering, procurement of materials, and construction of plants may contract for all three elements on a particular project for a fixed, single fee. Since these phases – engineering, procurement, and construction – are clearly separable, and the profit margins applicable to each differ substantially as does the timing of performance, the contractor may well find it advantageous to segment the contract for accounting purposes. The practice of segmenting contracts may be used both for income recognition and for determining the need to provide for anticipated losses, if the practice is applied consistently in similar situations.

MEASURING EXTENT OF PROGRESS TOWARD COMPLETION

The ability to measure the extent of progress toward completion of a contract is an essential factor in determining interim measures of earned revenue, the cost of earned revenue, and the amount of earned gross profit. A contractor should select and use the best available method. In practice, one or more alternative measures, in addition to the contractor's basic measure of extent of progress toward completion, may be applied to a contract to provide a check and add a degree of validity to the determination.

Methods of Measurement

Several acceptable methods are available to a contractor for measuring the extent of progress toward completion under the percentage-of-completion method. Prominent among those methods is the cost-to-cost method, which uses the ratio of cost incurred to estimated total costs in determining contract performance under a contract. Other methods include variations of the cost-to-cost method, the efforts-expended method, the units-of-delivery method, and the units-of-work-performed method. Contractors and accountants often use experts such as engineers or architects to estimate the extent of progress toward completion.

Input Measures. SOP 81-1 paragraph 46 classifies the various acceptable methods as either input or output measures. Input measures are those that measure efforts and resources applied to a project. For example, the labor-hours method is an input measure. Input measures are indirect indicators of completion that rely on an assumed correlation between input and accomplishment, which may not always be obtainable. For this reason, input measures are subject to error to the extent that the assumed correlation is not, in fact, achieved. However, these measures are widely used because of the relative availability of input data.

The input method most widely used is the cost-to-cost method. A 1982 survey of the top 400 contractors, conducted by the accounting firm of Arthur Andersen & Co., shows that of the 86 percent of companies using the percentage-of-completion method, 71 percent use this input method. Although the cost-to-cost method has the advantage of using all cost factors in measuring progress, the specialized principles that apply to contract accounting require that the method be modified in cases where certain costs incurred may distort the percentage of completion achieved at a given point. For example, costs of standard materials accumulated at a job site on which no work has been performed should be excluded until they are integrated into the project. Specialized materials acquired for a specific job should generally be included in costs incurred to date. This is a rule that applies only to the computation of the extent of progress toward completion; it does not govern how such costs are accounted for in the accounting records of the construction contractor. The methods of accounting for costs are discussed in Chapter 4.

Other input methods include methods based on the ratio of efforts expended to total anticipated efforts, labor-hours, labor-dollars, machine-hours, or material quantities. These methods generally produce more realistic measures of performance than the cost-to-cost method and tend to reflect more specifically the contractor's input toward completion resulting from direct project efforts. In using methods based on efforts expended, such as labor-hours, it is desirable to use the efforts of both the contractor and any subcontractors to develop reasonably dependable estimates, particularly in cases where a significant part of the total effort is performed by subcontractors.

Output Measures. Output measures have the advantage of being a direct measure of progress. Under certain conditions that arise in many circumstances, however, such measures are difficult or impractical to obtain. The output measure most frequently and successfully used is units-produced, because on some projects, the number of units can be readily determined. In addition to specific items, units may be measured in terms of volume, weight, linear fee, and so forth. Output can sometimes be measured by stage of completion, which can be determined, for example, by the completion of providing steps in a sequence or of furnishing components of an integrated project. In the absence of reasonably reliable output measures, input measures must be used, despite their inherent shortcomings.

Measures of extent of progress toward completion by architects and engineers are used (1) as primary measures, in situations where information and expertise are available to assure acceptable results from using those measures; and (2) as secondary measures to confirm project status determined by other means. In using these measures, whether primary or secondary, the contractor and the accountant should review and understand the basis of the determinations made by architects and engineers.

PROBLEMS IN ESTIMATING CONTRACT REVENUE

The starting point in income recognition is the estimate of the amount of total contract revenue anticipated from a contract. As previously noted, the accounting literature has given little attention to the difficulties and uncertainties involved in that process. However, SOP 81-1, in paragraph 53, notes that estimating revenue is a complex process affected by a variety of uncertainties related to the outcome of future events, and points out that there is a need to revise the estimates periodically throughout the life of a contract as events occur and as uncertainties are resolved. The major factors to consider are: the basic contract price, which is subject to different levels of uncertainties depending on the type of contract; contract options; change orders; claims; and special contract provisions for possible incentive payments, termination, and penalties. Additional contract provisions, laws or governmental regulations, or other unusual factors may affect estimates of contract revenue and should also be considered.

Basic Contract Price

Estimated revenue begins with the determination of the basic contract price, which derives from the original terms of the contract. The process of estimating revenue varies depending on the type of contract, of which there are four general categories:

1 Fixed-price or lump-sum contracts

2 Unit-price contracts

3 Cost-type contracts

4 Time-and-material contracts

Regardless of the type of contract used, the total anticipated revenue may be reasonably stable or may be highly variable and, therefore, is subject to uncertainties and adjustments.

Fixed-Price or Lump-Sum Contracts. A fixed-price contract is one in which the price for the work to be performed is stated in terms of a fixed amount, which is not to be changed by reason of cost or performance. In practice, however, most fixed-price contracts are subject to price variations for economic price adjustments (escalations), incentives and penalties related to cost and performance targets, change orders, and price redetermination provisions.

Unit-Price Contracts. Unit-price contracts are those in which the contractor is paid a stated or fixed price for each unit produced or performed. In addition to the variations to which fixed-price contracts are subject, the estimated total revenue from a unit-price contract may vary according to the number of units produced.

Cost-Type Contracts. Cost-type contracts generally provide for reimbursement of defined contract costs plus a fee representing the contractor's profit. Fees may be earned in several ways, each of which affects the way in which estimated total revenue is calculated. A fixed fee is ordinarily allocated uniformly over the entire project. A percentage fee is generally recognized in proportion to cost incurred, taking into account upper and lower limits as fixed in the contract. A fee subject to incentive provisions is one in which there is a target fee related to a target cost or performance standard. The fee may then increase or decrease, usually within limits, depending on target costs or performance achievements. The estimated total revenue for a contract subject to incentive provisions is adjusted periodically to reflect anticipated changes in the amount of incentive payments that will be earned or the amount of penalties that will be incurred under the contract performance. In estimating the ultimate fee on complex projects where a significant amount of work remains to be performed, the experience of the contractor must be considered. The contractor and the accountant must necessarily exercise some judgment in making these assessments.

For cost-type contracts in which the contractor, in effect, acts as an agent for the customer, the estimate of total contract revenue should not include the related costs and cost reimbursements. Such a relationship occurs when a contractor orders material or arranges subcontracts for the account of the customer. The reasons these costs are excluded from estimated revenue are that (1) the contractor does not take title to the materials acquired, (2) there is little or no risk associated with the materials, and (3) the contractor is not the primary obligor for such materials. Under these conditions, the estimate of the contractor's revenue should be limited to the amount of the fee.

In other instances, materials may be purchased for the contractor's own account, and the contractor may therefore be obligated to subcontractors and suppliers. Where this happens, contractors act on their own account and at their own risk; their estimate of total revenue should include these anticipated costs and cost reimbursements. Similarly, if the customer purchases or arranges subcontracts or incurs other obligations that are for the account and responsibility of the contractor, the estimate of total revenue should include the cost of these items. SOP 81-1 states in paragraph 60 that a contractor's revenue and costs should generally include all items for which the contractor has an associated risk, including those items on which the contractor's fee is based.

Contract Options

Estimating total contract revenue may be affected by contract options. Contract options are provisions made in a contract for the possible performance of work beyond the basic contract price or for the addition of new items under the same contract. In estimating total contract revenue, it may be necessary to include amounts for the exercise of contract options. SOP 81-1, in paragraph 64, provides that contract options or additions should be considered extensions of the basic contract and treated as change orders, unless an option or an addition meets specified conditions for treatment as a separate contract. The circumstances specified for treating an option or an addition as a separate contract are:

   a. The product or service to be provided differs significantly from the product or service provided under the original contract.

b. The price of the new product or service is negotiated without regard to the original contract and involves different economic judgments.

c. The products or services to be provided under the exercise option are similar to those under the original contract, but the contract price and anticipated contract cost relationships are significantly different.

In addition, the conditions for combining contracts should be considered. Contract options and additions that meet the criteria previously cited in the section of this chapter, "Combining Contracts," may be combined with the original contract.

Change Orders

Change orders are modifications of one or more specifications or provisions of the original contract. Some change orders may have little or no effect on the project, while others, individually or collectively, may significantly change the anticipated revenue cost and performance characteristics of the contract. The accounting considerations differ for: (1) change orders approved by the customer and the contractor as to both scope and price, (2) change orders for which the work to be performed has been defined and agreed on but the adjustment to the contract price is to be negotiated, and (3) change orders for which both the scope and price are unapproved or in dispute.

Estimated contract revenue and contract cost should be adjusted in the normal ongoing estimating process to include change orders approved by the customer and the contractor as to both scope and price.

Change orders that are unpriced, that is, the work to be performed has been defined but the adjustment to the contract price has not been negotiated, are accounted for through careful consideration of the nature and circumstances of the change and some exercise of judgment by the contractor. Numerous, nominal change orders are evaluated as a group, based on the previous experience of the contractor with similar changes and customers. More significant unpriced change orders should be assessed based on their attributes as to the likelihood of recovery and the probable amounts. The likelihood of recovery is usually enhanced for change orders initiated by the customer in cases where the customer and the contractor have agreed on the merits and specifications of the change. The contractor's past experience in negotiating change orders may also be a favorable factor. SOP 81-1, in paragraph 62, provides guidelines on accounting for unpriced change orders, which require that:

a. Costs attributable to unpriced change orders should be treated as cost of contract performance in the period in which the costs are incurred if it is not probable that the cost will be recovered through a change in the contract price.

b. If it is probable that the cost will be recovered through a change in the contract price, the costs should be deferred (excluded from the cost of contract performance) until the parties have agreed on the change in contract price; or, alternatively, they should be treated as cost of contract performance in the period in which they are incurred, and contract revenue should be recognized to the extent of the cost incurred.

c. If it is probable that the contract price will be adjusted by an amount that exceeds the costs attributable to the change order and the amount of the excess can be reliably estimated, the original contract price should also be adjusted for the amount when the costs are recognized as cost of contract performance. However, since the substantiation of the amount of future revenue is difficult, revenue in excess of the costs attributable to unpriced change orders should only be recorded when realization is assured beyond a reasonable doubt, such as circumstances in which an entity's historical experience provides such assurance or in which an entity has received a bona fide pricing offer from a customer and records only the amount of the offer as revenue.

Thus, estimates of total revenue and the computation of earned revenue and the cost of earned revenue may or may not reflect amounts attributable to unpriced change orders, depending on the circumstances and the probability of recovery. It is possible that (1) the cost may be reflected in the income recognition computation without adjusting estimated revenue, (2) the amount of estimated total revenue and the amount of total estimated costs may be adjusted by the same amount, (3) the amount attributable to unpriced change orders may be deferred (not treated as a contract cost or included in the computation of earned revenue) without adjusting the estimated total revenue, or (4) the total estimated revenue may be adjusted to reflect the total adjustment anticipated because of the change order.

Change orders for which both scope and price are unapproved or in dispute should be treated as claims. The criteria for recognizing claims in the estimate of total revenue and in the computation of earned revenue are discussed in the next section.

Claims

Claims are amounts in addition to the basic contract price that a contractor seeks to recover from a customer or from others for: customer-caused delays, errors in specifications and design, contract terminations, change orders in dispute or unapproved as to scope and price, or other causes of anticipated additional costs. Amounts attributable to uncollected claims are included in estimated total revenue and in the computation of earned revenue only under very limited circumstances. These amounts are recognized only where their recovery is probable and they can be reliably estimated. Even under these conditions, the amount recognized should not exceed the relevant cost incurred. SOP 81-1, in paragraph 65, specifies conditions that must be satisfied for a claim to meet the two requirements of probability of recovery and reliability of the estimated amount:

a. The contract or other evidence provides a legal basis for the claim; or a legal opinion has been obtained, stating that under the circumstances there is a reasonable basis to support the claim.

b. Additional costs are caused by circumstances that are unforeseen at the contract date and are not the result of deficiencies in the contractor's performance.

c. Costs associated with the claim are identifiable or otherwise determinable and are reasonable in view of the work performed.

d. The evidence supporting the claim is objective and verifiable, not based on management's "feel" for the situation or on unsupported representations.

When these conditions are met, a contractor may adjust estimated total revenue to the extent of the costs incurred that are associated with the claim and may reflect the amount in the computation of earned revenue and the cost of earned revenue. However, some contractors follow the practice of adjusting revenue to recognize claims only when the amounts have been received or are awarded, without regard to whether the claims meet the conditions specified in SOP 81-1. This practice is acceptable under GAAP. In any event, the costs incurred that are attributable to claims are treated as a cost of contract performance and are reflected in the computation of earned revenue and the cost of earned revenue. The practice followed by the contractor in accounting for, and reporting, claims should be disclosed in the notes to the financial statements to afford the users of financial statements a basis for assessing their effect on earned revenue. In addition, amounts attributable to claims not recognized in the estimate of total revenue and in the computation of earned revenue are disclosed as contingent assets in the notes to the financial statement. This practice is required by SFAS 5, "Accounting for Contingencies."

ACCOUNTING FOR REVISIONS IN ESTIMATES

GAAP recognizes the necessity of using estimates and judgment in financial measurements as a significant characteristic and limitation of financial information. The complexity and uncertainty of economic activities seldom permit exact measurement. Thus, estimates and informed judgment must be used to assign amounts to the effects of transactions and other events that affect a business enterprise. Since the outcome of economic activity is uncertain at the time decisions are made, actual results often do not correspond to original expectations. The continuous nature of the economic activity of business enterprises requires that the measurement of relationships associated with relatively short intervals of time, such as a year or a quarter, be made on the basis of assumptions on conventional allocations. The jointness and complexity of economic activity make the computation of the precise effects of particular economic events impossible. These factors are recognized as a source of uncertainty in financial information and are the reasons why approximation and judgment are described as basic features of financial information. Estimates are a central factor in contract accounting, and the estimates used must be continually revised to reflect the latest available information. Understanding the effects of the required accounting for changes in the estimates and the rationale underlying the requirement are critical factors in assessing financial statements of construction contractors.

Critical Areas for Estimates – Revenue, Cost, Extent of Progress

In accounting for contracts under the percentage-of-completion method, the critical areas of estimates relate to revenue, cost, and extent of progress toward completion. Changes in the estimates of these elements are inevitable as the work on a project moves forward and as increasing amounts of data become available. The objective is to make the revisions in the estimates of those elements timely enough to be useful to management in monitoring progress and efficiency in the performance of contracts. Ordinarily, a contractor does not revise estimates every month on a five-year project, however, estimates may be revised monthly or more frequently on projects that last only a few months. As a general rule, estimates are revised frequently enough to assure that the financial statements reflect the best estimates available at the time of issuance. Unusual events that occur after the date of the financial statements are reflected in the estimates on which the financial statements are based, although they should be disclosed in the notes to the financial statements as subsequent events.

Method of Accounting for Revisions in Estimates

Revisions in estimates of contract revenue, contract cost, and extent of progress toward completion are accounted for as changes in accounting estimates. APB Opinion 20, "Accounting Changes," requires changes in estimates to be accounted for in the period of change if: (1) the change affects that period only, or (2) if the change affects both the period of change and future periods. APB Opinion 20 prohibits accounting for a change in estimates by a retroactive restatement of prior period amounts, that is, by a revision of previously issued financial statements to reflect information that became available only after the statements were issued. Retroactive restatement of previously issued financial statements is prohibited on the grounds that the statements reflected the best information available at the time when they were issued. Before the issuance of SOP 81-1, two alternative approaches to accounting for changes in estimates in the construction-contracting industry were followed. Under one approach, the change in estimate was accounted for in the period of change, so that the balance sheet at the end of the period of change and the accounting in subsequent periods were as they would have been if the revised estimates had been the original estimate. Under this approach, the financial statements were revised at the end of each reporting period to reflect fully the effects of estimates based on the latest available information. The effects of changes from previous estimates were reflected in the income statement for that period. Under the second approach, changes in estimates in the construction-contracting industry were accounted for in both the period of the change and subsequent periods. Under this approach, the effects of a change in the estimate were spread over the remaining term of the contract.

SOP 81-1 requires that the full effects of a change in the estimate be accounted for in the period of change, so that the balance sheet at the end of the period fully reflects the effects of the revised estimate, and the income statement for the period reflects the full effects of required adjustments to earned revenue, cost of earned revenue, and the extent of progress toward completion. The rationale for this position is that the estimates used in accounting for contracts should be based on the latest and best information available at the time when the estimates are made, and the financial statements should fully reflect the effects of the latest available estimates. Contractors who followed the approach of spreading the effects of changes over the remaining term of a contract were not precluded from continuing that practice, although SOP 81-1 provides the justification for a change to the preferred practice. Another requirement of APB Opinion 20 is that the notes to the financial statements disclose the effect of revisions in estimates if the effect is material. The requirement is stated somewhat ambiguously, and the financial statements of construction contractors may or may not disclose the effect of changes in estimates, because APB Opinion 20 states that the disclosure is not necessary for estimates made each period in the ordinary course of accounting for certain items. Some contractors and some accountants view contract estimates as estimates made each period in the ordinary course of accounting.

ACCOUNTING FOR PROVISIONS FOR ANTICIPATED LOSSES ON CONTRACTS

As previously noted, ARB 45 requires that the full amount of an anticipated loss on a contract be recognized in the period in which the loss becomes evident. Thus, contractors and their accountants review all contracts to determine the need for a provision for anticipated loss, and the financial statements reflect those amounts in the computation of earned revenue and the cost of earned revenue to date. The requirement to provide such provisions applies to both the percentage-of-completion method and the completed-contract method, which is discussed later in this chapter.

Provision for Loss

The necessity to provide a provision for a loss on a contract arises when the estimates of total contract costs exceed the estimates of total contract revenue. A contractor is required to make the provision for an anticipated loss in an amount equal to the total anticipated loss on the contract without regard for the state of completion at the time of recognition. For example, if the total estimated cost on a contract exceeds the total estimated revenue by $100,000, a provision for loss in that amount should be made at the time of the estimate, even though the contract may only be 50 percent completed at the time. On contracts that are combined or segmented for accounting purposes, the effective profit center (the combined contracts together or each segment separately) is used as an accounting unit for determining the need for a provision for loss. In this situation, an anticipated loss on one contract in a combined group of contracts is offset against an anticipated profit on other contracts in the group.

Measurement of an Anticipated Loss

All of the costs and revenues that will ultimately be recognized on the contract should be reflected in calculating the amount of the anticipated loss to be recognized. Thus, in measuring the loss, estimates of total contract revenue and costs include all elements based on the latest available information. For example, the estimate of total costs is determined by including all direct material and labor costs plus estimates of overhead costs that the contractor normally includes in the computation of the cost of earned revenue. The objective is to relieve future accounting periods of costs that must be charged to the contract and that will not be recovered from revenue. Also, in measuring the anticipated loss, estimates of total contract revenue include:

- Effects of penalty and incentive provisions
- Change orders
- Price redeterminations
- Escalations
- Cost sharing limitations

Expected revenue variations are also related to and correlated with current cost estimates.

Since the provision for an anticipated loss on a contract is the excess of the estimated total contract costs over the estimate of total contract revenue, the provision is recognized, as illustrated in Figure 3-1, as an additional cost in the income statement rather than as a reduction of revenue.

Subsequent Revisions in an Anticipated Loss

Subsequent revisions in a provision for an anticipated loss are treated as changes in accounting estimates. The financial statements report the entire effect of the change in the period of the change; the amount of the adjustment reflected in the income statement for the period of the change is the amount that would increase the cumulative loss to the amount of the current estimate. For example, if a contract that had shown a $50,000

FIG. 3-1   Illustration of Financial Statement Presentation for Contracts With an Anticipated Loss

Thayler Corp. entered into a contract to construct a building with an original contract price of $2 million and original estimated costs of $1 million. The company used percentage-of-completion accounting for this contract. At the end of the first year the contract was estimated to be 60 percent complete; however, the total costs on the contract have been revised to $2 million. The income statement for this contract can be summarized as follows:

Contract revenues (60% of $2,000,000)	$1,200,000
Contract costs ($200,000 expected loss plus 60% of $2,000,000 other contract costs)	1,400,000
Loss on contract	$ 200,000

Note that the entire anticipated loss ($200,000), and not merely 60 percent of the loss, is recognized in the current period.

profit at its 50 percent stage of completion shows a $30,000 loss at its 75 percent stage of completion, the effect of the income statement for the current period would be a loss of $80,000 bringing the cumulative results to the current anticipated loss of $30,000.

THE COMPLETED-CONTRACT METHOD

Thus far this chapter has focused on the percentage-of-completion method of income recognition for contracts, because most accounting issues unique to contracts relate to the application of that method. The other acceptable method under GAAP, the completed-contract method, is described and discussed briefly in this section. ARB 45 states that the use of the completed-contract method is preferable only where a lack of dependable estimates or the existence of inherent hazards cause forecasts to be doubtful.

In applying the completed-contract method, earned revenue, the cost of earned revenue, and the resulting gross profit on a contract are reported in the financial statements only when a contract has been completed or substantially completed. For interim periods during contract performance, contract costs and amounts billed and collected are reflected in the contractor's balance sheet; the income statements for those periods do not reflect any amounts of earned revenue, cost of earned revenue, or the estimated profit on the contract. As the work progresses, the contractor accumulates costs by contracts, usually in considerable detail. Interim billings to the customer, provided for in a contract, are accumulated in a balance sheet account; the resulting amounts of contract costs incurred and progress billings on the contract are offset against each other for balance sheet presentation purposes. For contracts on which the amount of cost incurred exceeds billings, the balance sheet shows the excess as an asset, somewhat in the nature of an inventory item; for contracts on which billings exceeded the amount of cost incurred, the balance sheet shows a liability — a nonmonetary item representing an advance on work to be performed. The resulting net balances shown on the balance sheet are relatively insignificant and provide little information on the extent of progress or on the profitability of a contract. For this reason, the financial statements under the completed-contract method are generally not considered indicative of current performance on the contract.

In many cases the exclusion of interim information from the financial statement produces a distortion of activities in the period during which the full revenue and cost of the contract are reported in the period of completion. In addition, there is ordinarily a mismatch of period cost — for example, interest costs on working capital and equipment loans — with the revenue with which those amounts should logically be associated when revenues are not reported until contract completion. This results in a deficiency in the financial statements because they do not provide a timely reflection of contract activity, earned revenue, cost of earned revenue, and estimated profit on contracts in progress. These factors make the completed-contract method inappropriate, except under particular conditions.

The completed-contract method, despite its inherent weaknesses in furnishing useful interim information, is the preferred method of contract accounting in some limited instances. In the past, the method has been widely used for a number of reasons. The most valuable and widely cited reason is that the risks and hazards faced by a contractor are so significant and pervasive that reliable profit estimates cannot be made until at or near the completion of a contract. A corollary to this reasoning is the conservative bias in accounting practice, which tends to discourage reporting of profits or favorable developments until they are virtually beyond question. Another important reason given for using the completed-contract method is for income-tax-reporting purposes, that is, to avoid paying taxes on a profitable contract until the contract has been completed. Using the same method for income-tax- and financial-reporting purposes is simpler. Reporting for income tax purposes is more fully discussed in Chapter 18.

Conditions for Use of the Method

The completed-contract method is acceptable where the reported financial position and results of operations would not be significantly different if the percentage-of-completion method were used. The correlation between the methods should encompass the balance sheet in terms of the net investment in contracts by the contractor and the net liability of the contractor to customers after consideration of the effective transfer of title and recognition of revenue earned from prior progress billings. More importantly, the income statement should reflect activity reasonably corresponding to that which would have been reported under the percentage-of-completion method. Earned revenue, cost of earned revenue, and gross profit earned are all significant factors in judging whether or not the income statements under the two methods are materially different. Under the requirements in SOP 81-1, the assessment of whether the results would be substantially the same is made on the basis of an overall evaluation of the nature of a contractor's operations, not by determining the results on both bases.

The conditions for the use of the completed-contract method are most commonly found in those circumstances in which a contractor has numerous contracts that are completed in a relatively short time, are individually small in relation to the contractor's total activity, and are similar in character to each other. Although some large companies may have contracts that meet these general criteria, contractors who more frequently qualify are apt to be specialty contractors. Specialization narrows the diversity of the work and often limits the time in which the work is done. To illustrate, contractors engaged in roofing, plumbing, painting, glazing, and the like often have a pattern of contracts and activity for which the completed-contract method is acceptable. In these instances, the cost and effort required to maintain percentage-of-completion accounting is greater than the benefits gained from the use of that method.

The use of the percentage-of-completion method depends on the existence of reasonably dependable estimates; if these estimates are not available, the completed-contract method becomes the method of choice. The chronic lack of dependable estimates is usually manifested in the inaccurate or unpredictable discrepancies in past estimates, or even worse, in the lack of data on which to make a rational estimate. However, the continuing lack of dependable estimates raises a more fundamental problem than choosing which method of contract accounting to employ.

As a matter of principle, ARB 45 requires that losses on contracts be recognized in their entirety when they are evident. An inability to produce dependable estimates for job status and outlook precludes meeting that requirement and raises the question of whether the accounting records can meet GAAP.

The completed-contract method is also the method of choice where inherent hazards are associated with the performance of the contract. As discussed under "Risks" at the beginning of the chapter, inherent hazards differ from inherent business risks and are not normally associated with typical contracts in the ordinary course of business. Inherent hazards are unusual, uncontrollable, and generally unpredictable dangers, which are not considered in the usual process of estimating and forecasting. Such hazards are rarely present and are limited to contracts such as those subject to political- or regulatory-action bodies, litigation, and ill-defined terms or conditions. An example of this sort of hazard is environmental litigation brought against the owner of a nuclear power facility.

Zero-Profit Margin as an Alternative

The zero-profit-margin basis of applying the percentage-of-completion method of accounting is a preferable alternative to the completed-contract method in some situations. These situations involve contracts in which uncertainties and lack of significant data early in the contract make meaningful profit estimates doubtful, but for which some level of profit is assured. When this occurs, the use of the zero-profit-margin basis of the percentage-of-completion method is preferable to the use of the completed-contract method. Using a zero-profit-margin basis results in recognizing earned revenue and cost of earned revenue in equal amounts until reasonably dependable estimates of the outcome can be made. Many contractors use this basis in the early stages of a contract, for example, up to the 15 or 20 percent level of completion of most or all of their contracts. They have found from experience that until that stage is reached they did not have enough data and experience with the problems and progress of a job to make informed and reliable estimates. As the job progresses and reliable estimates become possible, the normal percentage-of-completion method is resumed by reporting the difference as a change in an accounting estimate. To illustrate, if a zero-profit margin were used for the first 15 percent of a contract, but a reliable estimate of a $100,000 total contract profit could be made at the 25 percent completion level, the entire $25,000 profit to date would be recognized at that point, rather than prorating the total profit over 85 percent of the contract remaining after the zero-profit level. The principal advantage of using the zero-profit method is that the financial statements reflect the activity taking place during the reported period, even though the ultimate profit is still in doubt.

Determination of Completion Date

A contract is accounted for as completed when (1) all significant work on the contract has been done, (2) the specifications under the contract have been met, and (3) no significant risks remain. The point of completion may vary, depending on the type of project, the terms of the contract, and the contractor's prior experience. In practice, a contract is almost always deemed to be complete when 95 percent or more of the work is finished. The unfinished work typically consists of punch-list items, cosmetic and clean-up work, and demobilization activities. Completion is a matter of substance and reality over form. Indication of substantial completion may be acceptance by the customer, production or performance that meets contract specifications, or departure from the site. Whatever measures are applied, they should be applied consistently and disclosed in a note to the financial statement.

The prime guideline to be used in choosing the best method of contract accounting is the guide of best presentation. Since the percentage-of-completion method most clearly reflects the actual operations of most contracts and contractors, this is the method of choice. Alternative methods – completed-contract and zero-profit margin – should be used only if the percentage-of-completion method cannot be applied.

CHAPTER 4

Contract Costs

Eugene S. Abernathy

INTRODUCTION

A contractor obtains his work in a competitive environment, which curtails the potential for profit. To be successful, he must manage his costs well. Since he performs work on more than one contract at a time, it is not sufficient that he control all of the costs of his business as a single entity. Rather, he must account for each cost separately, by contract profit center. Moreover, he must account for these costs against the same tasks, or bid items, used in the buildup of his price quote. Only then will he be able to identify and control cost overruns as they occur, thereby preserving his opportunity to make a contract profit.

A contractor needs to file income tax returns and prepare financial statements; these responsibilities also require that he account for contract costs accurately. The contractor is confronted with an additional problem here. Not only must he assign his costs to different contracts, he must also assign them to different accounting periods. Due to the nature of his business, the fiscal-period cutoff is immensely more complex for the contractor than it is for someone in a nonconstruction-business enterprise.

HISTORICAL BACKGROUND

Until recently, accounting literature offered the contractor little direction in the management of his contract costs and even less direction regarding the accounting for such costs in financial statements. Suitable principles and methods were developed largely by individual contractors acting alone. In recent years, a great deal more has been published on the subject.

The American Institute of Certified Public Accountants

The American Institute of Certified Public Accountants (AICPA) deals with construction costs in its Accounting Research Bulletins (ARB) 43 and 45; its industry Audit Guides, *Audits of Construction Contractors* (now superseded), and *Audits of Government Contractors*; its industry Audit and Accounting Guide, *Construction Contractors* known as the *Guide*; and its Statement of Position (SOP) 81-1, *Accounting for Performance of Construction-Type and Certain Production-Type Contracts*.

The AICPA pronouncements are mostly concerned with financial reporting in accordance with generally accepted accounting principles (GAAP).

The Internal Revenue Service

Internal Revenue Service (IRS) regulations deal with income tax treatment of contract costs under both the percentage-of-completion and the completed-contract methods. Section 229 of the Tax Equity and Fiscal Responsibility Act (TEFRA) instructs the Treasury Department to amend its regulations relating to the completed-contract method of accounting for long-term contracts. The amended regulations require that contractors allocate certain costs to contracts that heretofore could have been deducted as period costs.

The Cost Accounting Standards Board

The Cost Accounting Standards Board (CASB), established by Congress in 1970, is a governmental board charged with the responsibility of promulgating cost accounting standards for large negotiated federal contracts. Nineteen such standards have been issued thus far. For contracts having Cost Accounting Standard (CAS) clauses, the CASB standards have the force of law.

SCOPE OF CHAPTER

The focus of this chapter is on the management of contract costs. Guidance on financial statement and income tax reporting requirements can be found in Chapters 11 and 18, respectively. Note that the cost principles set forth here substantially conform to GAAP and to income tax requirements. Further guidance can be found in the pronouncements of the AICPA and in applicable Treasury Department regulations. The cost principles set forth here do not conform in all respects to standards of the CASB. Reference should be made to Title 4, Subchapter G of the Code of Federal Regulations for an understanding of those principles.

A contractor cannot effectively control his costs without knowing which types of costs are contract costs and how to assign these costs to his contracts. Accordingly, the first objective of this chapter is to help the contractor improve his effectiveness in identifying contract costs and in assigning them to contract profit centers.

The second objective is to try to strengthen the contractor's cost accounting procedures for more effective managerial control. Discussions of procedures for comparing actual costs to estimates and standards, cost control through the use of an equipment profit center, and current issues in construction cost accounting follow.

GENERAL COST PRINCIPLES

The cost principles recommended herein classify contract costs into four possible cost types, subdivided as follows:

- Contract direct costs:
  a Direct materials
  b Direct labor
  c Subcontract costs
  d Other direct costs
- Equipment pools
- Overhead pools
- General and administrative cost pools (under limited circumstances)

Few problems are encountered in assigning direct costs to contracts, since they can be charged on an item-by-item basis. The assignment of equipment pools, overhead pools, and, if appropriate, general and administrative cost pools to contracts is more difficult. Therefore, considerable attention is directed here toward helping the contractor overcome these difficulties. In addition, the circumstances appropriate to the use of the general and administrative cost pool are described and evaluated.

CONTRACT DIRECT COSTS

Contract direct costs are those costs that can be specifically identified to a contract profit center. They include, but are not limited to, direct materials, direct labor, and subcontract costs.

Direct Materials

Direct materials costs are the costs of materials that become a part of the construction project, including freight costs. Accounting for customer-furnished materials depends on the nature of the contract. Paragraph 60 of SOP 81-1 provides the following guidance:

> If the contractor is responsible for the nature, type, characteristics, or specifications of material that the customer furnishes or that the contractor purchases as an agent of the customer, or if the contractor is responsible for the ultimate acceptability of performance of the project based on such material, the value of those items should be

included as contract price and reflected as revenues and costs in periodic reporting of operations.

Materials stored centrally for general construction usage are inventory, rather than contract, costs. Materials of nominal value, such as nails, are frequently accounted for as part of the overhead cost pool.

Direct Labor

Direct labor costs are those costs that can be specifically identified to a contract profit center, including costs of direct field supervision.

Direct labor costs include not only wages, but also all costs that are a function of wages (i.e., those that have been incurred because wages were paid). These include vacation and holiday pay, workers' compensation costs, payroll taxes, and retirement costs. (For federal income tax purposes, certain retirement costs are not required to be charged to contracts.)

Subcontract Costs

Costs incurred under subcontract agreements are direct contract costs. Contract profit centers should be charged with these costs, including the amounts of any retainage accrued but not due.

Subcontractors may be back charged for costs incurred or work performed by the contractor that should have been incurred or performed by the subcontractors. Amounts involved, if they are not in dispute and are considered collectible, should be applied to reduce contract costs.

Other Direct Costs

Costs that can be specifically identified to a contract profit center that are not direct materials, direct labors, and subcontract costs are categorized as other direct costs. Examples include costs of mobilization and demobilization, surety bonds, job-site utilities, and short-term rental of equipment.

Precontract costs, such as bid and proposal costs, can also be considered direct contract costs under certain circumstances. Paragraph 75 of SOP 81-1 states:

Costs that are incurred for a specific anticipated contract and that will result in no future benefits unless the contract is obtained should not be included in contract costs or inventory before receipt of the contract. However, such costs may be otherwise deferred, subject to evaluation of their probable recoverability, but only if the costs can be directly associated with a specific anticipated contract and if their recoverability from the contract is probable. . . . Costs appropriately deferred in anticipation of a contract should be included in contract costs on the receipt of the anticipated contract. . . . Costs related to anticipated contracts that are charged to expenses as incurred because their recovery is not considered probable should not be reinstated by a credit to income on the subsequent receipt of the contract.

EQUIPMENT POOLS

A contractor should rent major items of equipment to his contract profit centers, whether he owns or leases the equipment himself. The *Guide*, on page 26, explains how this can be done:

In establishing operating unit costs for construction equipment, contractors may apply rates arrived at under the so-called use rate theory. In applying this theory, the following factors should be considered: (1) the cost of the equipment, less estimates of its salvage value or rental if it is leased, (2) the probable life of the equipment, (3) the average idle time during the life or period of hire of the equipment, and (4) the costs of operating the equipment, such as repairs, storage, insurance, and taxes. A rate may be arrived at, which, based on the reported use of the equipment, will serve as a basis for charging the contracts on which the equipment is used. The cost of a contractor's equipment should be allocated to the particular contract on which it is used on a reasonable basis, such as time, hours of use or mileage.

This approach has the advantage of spreading the costs of extraordinary equipment repair over the contracts systematically rather than, for example, charging such costs to the contract for which the equipment was being used when a breakdown occurred. It is designed to recover all of the costs of the equipment by charging them to contract profit centers. Even idle equipment costs are assigned to contracts—the rationale being that independent lessors of equipment also attempt to recover their idle equipment costs through rental rates.

Since the rental rates used are based on estimates of costs, a profit or loss to the equipment pool will inevitably result. If material, the profit or loss should be eliminated by assignment to the contract profit centers.

The preceding concept is one of cost recovery. An alternative approach involves an equipment profit center. This concept has appeal to equipment-intensive contractors due to its emphasis on control of the equipment investment (for a more in-depth discussion, see the section in this chapter on managerial control of contract costs).

OVERHEAD POOLS

Overhead costs are those contract costs that cannot be specifically identified to contract profit centers. However, under limited circumstances, general and administrative costs not specifically identifiable to contracts can also be regarded as contract costs. These limited circumstances are explained more fully in the discussion of general and administrative cost pools that follows.

Examples of overhead costs are general supervision, supplies, quality control and inspection, and support services (such as procurement, payroll preparation, job billing and accounting).

Some accountants make a distinction between general overhead costs and job overhead costs. They consider job overhead to be inclusive of such costs as mobilization and demobilization, whereas job support services include costs such as field payroll and supervision, and costs of physical security. The cost principles recommended here classify these costs as other direct costs because, by being specifically identifiable to a contract profit center, they meet the criteria set forth here for contract direct costs.

The nomenclature used by contractors for these costs is not at issue. The notion of job overhead recognizes that some costs, while specifically identifiable to contracts, cannot be specifically identified to tasks or bid items within a contract. For example, costs such as employing security guards cannot be specifically identified to the masonry or erection tasks of a contract. This sort of situation exists regardless of what the pertinent costs are called (see the section, "Allocation to Tasks," for a more in-depth discussion).

Distinction Between General and Administrative Costs

The distinction between contract direct costs and overhead costs has already been discussed. This section examines the factors that distinguish contract overhead costs from general and administrative costs.

Accounting literature requires that if costs "cannot, as a practical matter, be associated with any other period . . . [because] they provide no discernible future benefits" (Accounting Principles Board (APB), Statement 4) and because they cannot be "clearly related to production" (ARB 43), they should be regarded as period costs. The *Guide*, on page 65, recognizes the applicability of this principle to construction contracts in stating that "[c]ontract costs are accumulated in the same manner as inventory costs."

From an analysis of the above references, it is obvious that costs can be classified as general and administrative if they fail to be classified as anything else. Their classification is thus not determined by the common characteristics that they may possess. It is also apparent from these references that the determinations are practical judgments of what is clear and discernible.

Allocation of Overhead Costs

Overhead allocations are inherently inaccurate. With this in mind, it is important that the following standards be met: (1) all significant costs that can be specifically identified to contracts should be charged as direct costs; (2) overhead allocation methods should be systematic, rational, and consistently applied (in fact, the contractor should have a written policy statement to support the consistency of overhead application); and (3) burden rates should be reviewed frequently and revised as necessary.

There are numerous ways that overhead can be allocated to contract profit centers. Some allocation bases are: direct materials costs, direct labor costs or hours, direct labor and subcontract costs or hours, total direct costs, machine hours, and output measures such as cubic yards of concrete poured. Obviously, these and other bases are not randomly appropriate, but should be selected only if rational for the type of overhead pool being allocated. For example, engineering occupancy costs could be allocated on the basis of engineering labor hours or labor dollars. The occupancy costs can be expected to benefit the contracts in approximately the same ratio that engineering labor hours or dollars benefit the contract.

A contractor should use one or more overhead pools and allocation bases as necessary to provide him with reasonable results. The nature of his contracting activities is the primary consideration in determining how sophisticated he must get. That is, if his contracting activities are homogeneous (involving the same or similar type projects and work force), even a rudimentary base may not produce serious distortions. The contractor may be able to assign overhead to a single pool and allocate it to contracts using a single allocation base, such as total direct costs.

Conversely, if contracting activities are heterogeneous (involving substantially different projects with widely differing labor requirements and approaches), the contractor's overhead allocations would be more complex. The reason for this is that his levels of support would be necessarily different for different types of contracting activity. He would therefore have several overhead pools, each with its own allocation base.

For example, assume a contractor's activities involve two different contract types. The first type requires that he purchase all the materials and provide all the labor necessary to construct the facilities. The second type is construction management, calling for the use of customer-furnished materials and subcontracted labor. Assume that the contractor wants to assign the central support costs of procurement and payroll to contracts. Using this example, if the contractor were to use total direct costs as the allocation base for procurement- and payroll-support costs, distortions would result. The construction management contracts would be burdened with a share of these costs, even though they benefit in little or no way from these support services.

A suitable allocation base for procurement-support costs might be contractor-furnished direct material dollars. However, if there is a substantial imbalance of high- and low-value materials requirements on the contracts, use of a dollar-value base may not spread the cost of the procurement effort equitably. In such a case, a better allocation base for procurement costs might be procurement labor hours. Suitable allocation bases for payroll-support costs might be direct labor dollars or direct labor hours.

In some instances, costs are assigned first to an intermediate cost pool using one allocation base and then reassigned, together with other costs in that pool, to their final base. For example, payroll-support costs might be partially assigned to the engineering department and then reassigned, together with other engineering costs, to contract profit centers, bid and proposal activities, and, possibly, research and development activities.

Overhead allocation procedures should be no more elaborate than those necessary to achieve reasonable results. Conceptual exactness must yield to practicality early on, or else the system will create more problems than it solves.

GENERAL AND ADMINISTRATIVE COST POOLS

As discussed above, general and administrative costs are, under most circumstances, considered period costs. As such, they are charged to an expense as incurred. However, there are two limited circumstances under which general and administrative costs can be assigned to contracts.

The first circumstance is discussed in ARB 45, which states:

When the completed contract method is used, it may be appropriate to allocate general and administrative expenses to contract costs rather than to periodic income. This may result in a better matching of costs and revenues than would result from treating such expenses as period cost, particularly in years when no contracts were completed.

Second, the *Guide* extends the application of this concept to government contracts accounted for by the percentage-of-completion method of accounting, using the rationale that "all costs under the contract are directly associated with contract revenue." The *Guide* states that:

> Practice varies among government contractors as to the extent to which costs are included in inventory. Some contractors include all direct costs and only certain indirect costs. . . . Other contractors record in inventory accounts all costs identified with the contract, including allocated general and administrative . . . expenses.

The CASB has promulgated its Standard 410 (4 Code of Federal Regulations (CFR) Part 410): "allocation of business unit general and administrative expenses to final cost objectives." Standard 410 provides guidance and imposes limitations on allocations of such expenses to contracts with CAS clauses.

The CASB defines general and administrative costs as:

> Any management, financial, and other expense which is incurred by or allocated to a business unit and which is for the general management and administration of the business unit as a whole. [General and administrative] expense does not include those management expenses whose beneficial or causal relationship to cost objectives can be more directly measured by a base other than a cost input base representing the total activity of a business unit during a cost accounting period.

The difference in the CASB definition and the criteria set forth herein may be superficial, for it is those costs that cannot be "clearly related to production" (the criteria set forth in this chapter) whose relationship to cost objectives is obscured sufficiently to preclude a "direct measurement base" (the criteria of the CASB).

ALLOCATION TO TASKS

Some contractors build up their bids by: (1) identifying direct costs to tasks or bid items; and (2) by then adding an overall factor at the end to provide compensation for overhead, general and administrative costs, and profit. Others allocate all contract costs (including overhead and, possibly, general and administrative costs) to tasks.

As previously mentioned, the notion of job overhead is predicated on the fact that some costs may be direct (specifically relatable) costs to the contracts, although they are indirect (not specifically relatable) costs to the tasks. Carrying this concept one step further, many contractors use the job overhead pool as a type of intermediate stop for general overhead that is not specifically relatable, but is allocable, to contracts. Once general overhead has been combined with job overhead, it, too, is allocated to tasks.

We have classified job overhead as other direct costs simply to emphasize the importance of direct charging costs where practicable, rather than allocating these costs through use of overhead or general and administrative cost pools. Although job overhead costs are direct costs to the contracts, they are usually indirect to the tasks. Thus, the problem of allocating job overhead costs to tasks has yet to be solved.

Costs specifically relatable to tasks should be direct charged as well. For example, equipment used specifically for site work would be charged to that task. Under different circumstances, such as in the case of pickup trucks for construction purposes general to all tasks, the equipment would not be direct charged to tasks. The equipment may also be used interchangeably between tasks to the point that specific identification or tasks is not practicable.

After direct charging all costs practicable to tasks, there still remains on every contract a pool of costs that cannot be direct charged. As stated, some contractors allocate these costs to tasks while others do not. Therefore, the questions that arise are: What purpose does such an allocation serve? and When should these allocations be made?

The answers lie in the nature of the contract and the ways in which the contract costs are controlled. In some cases, contract provisions stipulate unit prices by task. If such contracts are cost-plus, unit-priced, the need to assign all allowable costs to tasks is unquestionable. Since the contractor's compensation is cost-based, he could not be completely compensated without this accounting.

The need to assign all costs to tasks may exist even if the contract provisions do not base the contractor's compensation on cost. These contracts could be unit-priced or fixed-priced (lump-sum) contracts. The reason is that, while the contracts are not based on cost, disputes with owners and subcontractors frequently focus on costs and cost overruns. Some contractors use highly evolved systems of cost accounting to develop and support claims of cost overrun of tasks.

Individual contractors must weigh the expected benefits of full-absorption bid-item accounting against its expected costs. In making a decision, the contractor should assess the risk to which he is exposed in the environment in which he operates. The nature of his construction activities and the strength of his customer and supplier relationships are important factors that must be considered.

It is absolutely essential that costs be accounted for in the same terms that were used to estimate the contract prior to award. Otherwise, management control is lost since progress cannot be tracked against plan. However, if no useful purpose is served by assigning overhead and general and administrative costs to tasks, such assignments should not be made—either in estimating or in accounting. These costs can simply be provided for on an overall basis for both purposes.

Some accountants contend that control is lost if full-absorption task accounting is not performed. But, from the standpoint of the project manager, the opposite is more likely to be true. Loading tasks with costs over which he has no control obscures the project manager's effectiveness (or ineffectiveness) and may frustrate his purpose. His performance is best measured by separating those costs for which he is responsible from those for which he is not.

If the benefits of full absorption are determined to be not worth the cost, contract pricing will reflect an overall factor for overhead, general and administrative costs, and profits. Overhead (and possibly general and administrative costs) will be allocated to the contracts but not to the tasks. Most contractors decide to use this approach—a decision that may well be a sound one.

However, there is a real danger that must be addressed: the factor that is used needs to be reviewed and adjusted frequently. It is unfortunate that many contractors who control and account for direct contract costs nevertheless lose money due to the use of incomplete or obsolete overhead factors in their estimates. Many other contractors do not get work because the factors they use are excessive. A contractor should analyze his overhead accounts at least quarterly, so that he can ensure that his estimates are based on accurate rates.

Those contractors using full-absorption task accounting should subdivide each task of their contract accounting into the cost types recommended here: direct costs, including direct materials, direct labor, subcontracted costs, and other direct costs; and costs distributed from pools, including equipment pools, overhead pools, and, if used, general and administrative cost pools. In some instances, these distributed pools can be combined on the task records without losing important managerial

FIG. 4-1   **Distribution of Equipment Costs, Overhead Costs, and General and Administrative Costs to Tasks**

December 31, 19XX

Contract: ABC Chemical Plant
Task: Paving

Distributable costs:

Equipment pool	$414,162
Overhead cost pool	$241,720
General and administrative cost pool	$55,116

Direct costs:	This Task	All Tasks
Direct materials	$ 55,542	$5,647,488
Direct labor	114,526	746,291
Subcontract	542,416	2,461,412
Other direct costs	24,200	114,441
Total direct costs	$736,684	$8,969,632
Distributed costs:		
Equipment pool	$ 33,961	$ 414,162
Overhead pool	49,553	241,720
General and administrative cost pool	11,299	55,116
Total costs	$831,497	$9,680,630

control. This may be true even if the pools are assigned to tasks using different allocation bases.

A rational method should be used to distribute costs to tasks or the results may not be meaningful. But allocation procedures should be no more elaborate than necessary to achieve reasonable results. A single base of direct labor and subcontract hours or dollars is reasonable under many circumstances. Major pieces of equipment should be direct-charged to the tasks using rental rates and machine hours.

To illustrate, Figure 4-1 sets forth contract costs, including task direct costs, and distributable equipment, overhead, and general and administrative cost pools, together with their distribution to tasks. For this illustration, it is assumed that all contract costs have previously been distributed to the contract profit centers.

In Figure 4-1, $33,961 of equipment pool costs are distributed to this task, on the basis of total direct costs. This task's total direct costs of

$736,684 divided by the total direct costs of all tasks, $8,969,632, results in an 8.2 percent allocation rate. Total equipment pool costs assigned to this contract of $414,162, multiplied by 8.2 percent, equals $33,961.

The overhead and general and administrative cost pools are distributed on the basis of direct labor and subcontract dollars. The direct labor and subcontract dollars for this task total $656,942 ($114,526 + $542,416). Direct labor and subcontract dollars of all tasks total $3,207,703 ($746,291 + $2,461,412). $656,942 divided by $3,207,703 results in a 20.5 percent allocation rate. In this example, $49,553 of overhead costs are distributed to this task, determined by multiplying total overhead costs of all tasks, $241,720, by 20.5 percent. Similarly, $11,299 of general and administrative costs are distributed to this task, determined by multiplying total general and administrative costs of all tasks, $55,116, by 20.5 percent.

In this illustration, cost pools are distributed at month end, after all costs are known. If needed for managerial control purposes, these costs could have been applied to tasks earlier, using predetermined rates. If this procedure is followed, variances have to be reviewed and, if significant, assigned to the tasks.

FINANCIAL REPORTING

Cost-Plus Contracts

In some cases, contract provisions allow for costs that are not permitted by GAAP, and vice versa. Requirements, as explained in paragraph 57 of SOP 81-1, follow:

> Costs under cost-type contracts should be charged to contract costs in conformity with generally accepted accounting principles in the same manner as costs under other types of contracts because unrealistic profit margins may result in circumstances in which reimbursable cost accumulations omit substantial contract costs (with a resulting larger fee) or include substantial unallocable general and administrative expenses (with a resulting smaller fee).

Obviously, if these circumstances exist, the accounting system must have the capability to account for the contract both ways—in accordance with contract provisions and in accordance with GAAP. This might be accomplished by accumulating the costs allowed only one way—either as a separate cost component of the contract or of its individual tasks.

Comparing Alternatives *A* and *B*

SOP 81-1 sets forth two alternative approaches widely used in practice to account for contract revenues and contract costs under the percentage-of-completion method of accounting.

Under alternative *A*, the estimate of progress toward completion is used to recognize contract revenues and contract costs in the statement of income. Therefore, if a contract is 50 percent complete, 50 percent of the revenues and 50 percent of the costs would be recognized in income. To the extent costs actually incurred exceed or are less than 50 percent, an asset or liability would be recognized. A 50 percent gross profit would then be deduced from the use of the 50 percent factor in the measurement of both contract revenues and contract costs.

Under alternative *B*, the same contract would be accounted for by recognizing all contract costs (other than costs of certain stored materials) as costs of contract revenue. Revenue would then be recognized in whatever amount is necessary to achieve the amount of gross profit based on the 50 percent completion estimate.

Alternative *A* uses balance sheet deferrals to achieve uniform gross-profit rates throughout the contract (other than disparities resulting from changes in estimates). Alternative *B* avoids balance sheet deferrals, but does so at the expense of a uniform gross-profit rate. Under either approach, the absolute amount of recognized gross profit is the same. Moreover, if the cost-to-cost method is used to determine percentage of completion, the absolute amount of gross revenues, the cost of revenues, and the gross profit are the same under both methods. Where costs are the measure of progress toward completion, all costs would be assigned to the income statement under alternative *A*, just as they automatically are under alternative *B*.

A discussion of each of these two methods is included in paragraphs 79 through 81 of SOP 81-1 in the Appendix.

Financial Statement Disclosures

The *Guide*, on page 52, recommends the following financial statement disclosures for contract costs:

Contract costs.

1. The aggregate amount included in contract costs representing unapproved change orders, claims, or similar items subject to uncertainty concerning their determination or ultimate realization, plus a description of the nature and status of the principal items comprising

such aggregate amounts and the basis on which such items are recorded (for example, cost or realizable value).

2. The amount of progress payments netted against contract costs at the date of the balance sheet.

Deferred costs.

1. For costs deferred either in anticipation of future sales (precontract costs) or as a result of an unapproved change order, the policy of deferral and the amounts involved should be disclosed.

MANAGERIAL CONTROL OF CONTRACT COSTS

Standard Costing

With increasing frequency, contractors are using standard costing to analyze and control their costs. This technique permits them to identify and focus on the cause of cost overruns, with an accounting for variances of actual costs from standard and a subsequent analysis of each variance into its causes.

Direct materials variances are broken down into price and usage variances. Actual cost of materials is compared with what the cost would have been had the same quantities been purchased at standard cost. The difference between actual cost and standard cost is the price (or purchase-price) variance. With standard costing, both the actual and standard amounts of materials used are multiplied by the standard price of the materials. The difference between the two products is the materials-usage variance.

Direct labor variances are broken into rate and efficiency (usage) variances as well. This is accomplished by comparing the products of the actual labor hours at the actual rates with the same hours at the standard rates. The difference between the two is the rate variance. Both the actual and the standard amounts of labor hours are multiplied by the standard labor rates. The difference between the two products is the efficiency variance.

Some contractors account for variable overhead costs at standard and break down the variances into spending and efficiency variances. To do this, standards have to be expressed with reference to a common denominator, such as direct labor hours. The spending variance is simply the difference between actual variable overhead and the amount obtained by multiplying the base units (i.e., direct labor hours) by the standard rate. The efficiency variance is the difference between the products of the actual base units incurred and the standard base units expected at the standard rates.

For standards to be effective, they should be determined with great care. They must reflect realistic rather than optimum goals and should be reviewed frequently and revised as necessary.

The Equipment Profit Center

General Considerations. Equipment costs can be significant to a contractor whether he owns or leases his equipment. Generally, a contractor would lease most of his equipment if the following conditions were met:

1 He can meet his equipment needs through local, short-term rentals;

2 He can pass his rental costs through to the project owner;

3 He can lease and still be competitive; and

4 He can be assured of reliable equipment performance under such arrangements.

Unfortunately, it is becoming more difficult for the contractor to achieve these objectives through short-term rentals. Before he invests in his own equipment, however, he must understand equipment management from the perspective of the independent equipment lessor. To do this, he should view his equipment costs in the context of an equipment profit center.

Cost at Risk. In making a capital expenditure decision, the contractor should classify the components of the cost at risk in a way that permits him to deal in these terms. There are three costs at risk:

1 *Capital expenditure* – the initial investment minus the residual value;

2 *Investment maintenance costs* – repairs and maintenance, repair parts and depreciation (a type of maintenance cost, in that it measures the failure to keep the equipment in its original state); and

3 *Investment carrying costs* – insurance, taxes, storage, other equipment division costs, and interest, less equipment division income.

These are the costs at risk. Once a contractor acquires equipment, he attempts to recover the acquisition costs and to return a profit by using the equipment (and by realizing a residual value upon sale).

Return on the Equipment Investment. The contractor accomplishes this by "leasing" his equipment to his jobs, charging market rental rates. In this way, he views his equipment investment in the same way that the independent lessor of construction equipment views his. Since the lessor retains the risks of ownership in both cases, a return should be made on the investment that is equal to the risk.

Traditionally, the contractor has paid attention almost exclusively to the contract profit center, while almost totally ignoring the equipment profit center. Sometimes the contractor fails to appreciate the benefits of accounting for the equipment investment in this way; at other times, he overestimates the complexities of the operation. If the contractor could identify those items of equipment that are losing money and take corrective action, he could assess his equipment management in the critical areas of use and preventive maintenance. Consequently, he would be able to make better equipment buy-sell decisions.

The Capital Expenditure Decision. The capital expenditure request is the basis for the capital expenditure decision. It should be structured in terms of the classifications of cost at risk (the costs at risk in an equipment investment have been classified earlier in the chapter). That way, if the capital expenditure is made, the contractor can manage his costs against plan in these terms. For example, a capital expenditure request for a $100,000 excavator is illustrated in Figure 4-2.

For our purposes, only the first, second, and eighth (projected last) years of the capital expenditure request are illustrated.

Note that the after-tax return on investment is divided into two components: a cost of debt capital and a return on equity capital. The amount of interest becomes the cost of debt capital and the remainder is the projected return on equity capital. The rate is computed for all years and then discounted to derive the discounted rate of return.

The discounted rate of return and the assessment of investment risk become the principal determinants of the capital expenditure decision. Now, alternative investment opportunities can be weighed against each other, and the projected discounted rates of return and risks of each can be compared.

Managing the Equipment Investment. Assuming that the subject excavator is purchased, the contractor should manage his investment

FIG. 4-2 Capital Expenditure Request for $100,000 Excavator

PROJECTED RETURN ON INVESTMENT

	19X1	19X2	19X8
Hours used	800	1,400	700
Investment revenues at $25/hour	$20,000	$35,000	$ 17,500
Investment maintenance costs:			
Repairs and maintenance (includes repair parts)	$ 1,600	$ 2,800	$ 2,800
Depreciation at $10/hour	8,000	14,000	7,000
Other	800	1,400	700
	$10,400	$18,200	$ 10,500
Contribution to investment carrying costs and profits	$ 9,600	$16,800	$ 7,000
Contribution rate	9.6%	16.8%	7.0%
Investment carrying costs:			
Insurance	$ 200	$ 200	$ 200
Taxes	200	200	200
Storage	500	500	500
Other equipment division costs	800	800	800
Equipment income			(20,000)
	$ 1,700	$ 1,700	$(18,300)
Before tax return on investment	$ 7,900	$15,100	$ 25,300
Taxes at 50%, less investment credit	(6,050)	7,550	12,650
After-tax return on investment	$13,950	$ 7,550	$ 12,650
Cost of debt capital, less tax benefit	· 2,400	2,200	
Return on equity capital	$11,550	$ 5,350	$ 12,650
Average equity capital	$40,000	$45,000	$100,000
Rate of return	28.9%	11.9%	12.7%
Discounted annual rate of return (computation not shown)			12.39%

against the plan contained in his capital expenditure request. In doing so, however, he may find that the discounted rate of return upon which the purchase decision was based is not of much use. Rather, his attention may be directed to the contribution to investment carrying costs and profits projected in the capital expenditure request. This gauge is the critical measure of the effectiveness of the equipment-management function.

The life of a piece of construction equipment is determined primarily by use rather than time. Therefore, the return on one's investment is maximized by getting the greatest use from each piece of equipment. The primary control over an equipment investment should be its actual use

---

FIG. 4-3   Equipment Use Report for $100,000 Excavator

July 19XX

	Actual	Plan
Hours used	34	67
Investment revenues at $25/hour	$ 850	$1,675
Investment maintenance costs:		
Repairs and maintenance (includes repair parts) at $2/hour	$ 68	$ 134
Depreciation at $10/hour	340	670
Other at $1/hour	34	67
	$ 442	$ 871
Contribution to investment carrying costs and profits	$ 408	$ 804
Contribution rate, annualized	4.9%	9.6%
Cost (benefit) of equipment use variation:		
This month	$ 396	
Year to date	$2,642	
This month, annualized	$4,776	

---

against plan, accounting for the absolute cost of the usage variance. The usage variance is determined from the contribution to investment carrying costs and profits. Figure 4-3 illustrates an equipment usage report for this same excavator for July of the first year. In this example, the substantial unfavorable variance is the direct result of using the excavator for about half of the usage planned. Accounting for this variance is explained in the following discussion.

The starting point for managing an equipment investment is the segmenting of contract revenues and the assigning of appropriate portions to the equipment effort.

The rental rates should be sufficient to return a suitable profit over the life of the equipment and recover acquisition (less residual values), depreciation, maintenance and repairs, storage, insurance, taxes, and other costs. Normally, market rates would be used.

If contracts specify reimbursement of equipment costs at different rates, the reimbursement rates should not, in theory, be used in charging equipment use to contract profit centers. Nor should variances in actual rates of investment maintenance costs be reflected in charges to the equipment profit center. This approach preserves the purity of usage variance accounting in the equipment profit center.

Maintenance is charged to equipment profit centers at standard rates to achieve a meaningful matching of such costs with hours of use. For the same reason, depreciation is charged on a machine-hour basis.

A statement of income for the equipment division is set forth in Figure 4-4. This statement is a combination of the individual equipment profit centers. The accounting for the subject excavator is also shown.

The category of "unabsorbed equipment division costs" includes the effects of the rate variances not absorbed by the individual equipment items.

The combined return on equity capital in this example is negative because the equipment has not had sufficient use and because of the unfavorable absorption rates.

CURRENT ISSUES

Contractors use their cost accounting to expand their revenue base. Cost-plus contracts lend themselves to this concept, as do contracts with shared savings or other such incentive or penalty clauses. In addition, disputes arising on contracts that are not cost-based are frequently settled on a showing of cost.

For these reasons, contractors are understandably anxious to advance cost-accounting concepts to embrace even more costs than are traditionally characterized as contract costs. Areas fertile for development include: unsuccessful bid costs; interest and other financing costs; research and development costs; replacement cost accounting for construction plant and equipment; and opportunity or facilities capital costs.

Present cost-accounting concepts are in need of some revision to reflect the economic realities of the construction environment. These concerns are being addressed by industry, with the result that some advanced costing concepts are being written into negotiated contracts. The concepts are also being addressed by the CASB. CASB Standards 414 and 417 address the cost-of-money concept. CASB Standard 414 approaches this concept as an element of the cost-of-facilities capital; CASB Standard 417 regards it as an element of the cost of capital assets under construction. New cost-accounting principles may evolve from the dynamics of the industry at work in these areas.

FIG. 4-4   Equipment Division Statement of Income

| | July 19X1 | | | |
| | Combined | | Excavator #86-4 | |
	Actual	Plan	Actual	Plan
Hours used	3,468	3,714	34	67
Investment revenues	$ 74,909	$ 82,822	$ 850	$ 1,675
Investment maintenance costs:				
Repairs and maintenance (includes repair parts)	$ 6,728	$ 7,279	$ 68	$ 134
Depreciation	31,559	34,169	340	670
Other	3,304	3,617	34	67
	$ 41,591	$ 45,065	$ 442	$ 871
Contribution to investment carrying costs and profits	$ 33,318	$ 37,757	$ 408	$ 804
Investment carrying costs:				
Insurance	$ 5,426	$ 5,500	$ 17	$ 17
Taxes	4,692	4,900	17	17
Storage	9,612	8,550	64	42
Other equipment division costs	11,420	10,200	81	67
Equipment income	(2,642)	(4,200)		
	$ 28,508	$ 24,950	$ 179	$ 143
	4,810	12,807	229	661
Unabsorbed equipment division costs	$ 1,814			
Before tax return on investment	2,996	12,087	229	661
Income tax, less investment credit	1,104	5,250	(718)	(504)
After-tax return on investment	1,892	7,557	947	1,165
Cost of debt capital, less income tax benefit	4,206	4,100	200	200
Return on equity capital	$ (2,314)	$ 3,457	$ 747	$ 965
Average equity capital	$346,822	$340,000	$40,000	$40,000
Rate of return— annualized		12.2%	22.4%	28.9%

CHAPTER 5

Management Information Systems

Thomas E. Brightbill
John E. McEwen

INTRODUCTION

Having an effective management information system (MIS) is becoming increasingly important to owners and managers of construction companies in today's intensely competitive environment. The modern contractor needs timely, accurate information on which to base his operating decisions. In theory, an effective MIS could be implemented manually. In practice, however, a system that can provide information organized in a manner that will ensure timely decision-making is best implemented on a computer system. With the rapid evolution of computer technology, costs of these information systems have declined rapidly and are now within reach of contractors with annual contract volumes of $500,000 or more.

This chapter examines the applications, or functions, that comprise the contractor's MIS; the various processing alternatives that the contractor has available to implement his information system; the sources of information available for identifying computer programs; and the alternatives that present an approach to selecting an effective MIS.

COMPUTERIZED INFORMATION SYSTEMS: A STATUS REPORT

Three developments that started in the early 1970s and have continued to date have led to cost-effective, fully computerized construction MIS and have had a particularly notable affect on those contractors with contract volumes ranging between $500,000 and $250 million. These three developments are:

1 The introduction of the minicomputer in the early 1970s

2 The introduction of preprogrammed integrated software
   packages in the mid-1970s

3 The introduction of the microcomputer system in the late 1970s

In the short span of 10 years, computer systems that had previously been available only to the very largest contractors and that had needed support by staffs of operators and computer programmers became accessible to a broad range of contractors in the construction industry.

Computer technology developed at a rapid rate during the decade of the 1970s. Today, one can buy a microcomputer system for approximately $5,000 that has the power of a computer costing 20 to 40 times that amount in 1973. Not only has the cost of computer hardware dropped dramatically, but the computer of the mid-1980s can fit on a desk-top (or even in a briefcase) and requires none of the special power, air conditioning, or other environmental-conditioning apparatus that the earlier computer systems demanded. Furthermore, minicomputer and microcomputer systems no longer require specialized technical skills or experienced data-processing personnel to operate them.

In addition to the evolution of computer hardware, dramatic improvements have also occurred in the area of prepackaged software. Before the mid-1970s, companies that used data processing typically recruited a data-processing staff, which designed and programmed the company's computer programs. This process was labor intensive, and it was difficult to make accurate estimates of the effort necessary to implement the programs. The results were often less than satisfactory because the mechanisms for communicating requirements and specifications were not precise. As a result, software that today can be purchased in the local computer store for $500 or less cost $100,000 or more to develop in the early 1970s.

In the mid-1970s, software firms began to evolve by developing computer software for specific minicomputer systems. This evolution was encouraged by the computer manufacturers who sold their equipment through the software vendor, providing themselves with substantial profit, which ranged up to 35 percent on each sale. The hardware vendors effectively underwrote the development of software for their computer systems. As a result, industry experts estimate that today there are some 5,000 software firms offering some 20,000 prepackaged software products. A good estimate is that approximately 10 percent of these products—500 vendors selling 2,000 software products—are designed for the construction industry. The software products offered by these vendors can be purchased for as little as $500 or less or as much as $35,000 per application.

CONTRACTOR'S UNIQUE REQUIREMENTS

The apparent low cost, the availability of hardware and software products, and the desire to improve the quality of the contractor's information systems have caused many contractors to purchase computer systems without fully considering the implications or understanding the specific requirements of their purchases.

For example, numerous contractors have procured payroll software, which was not specifically designed for contractor use. Payroll software does not account for the variety of workers' compensation rates that one might encounter on a job. This forces the contractor to account for and report workers' compensation insurance premiums using alternative manual systems. Further, some of the available job cost systems do not provide for multiple activities or work phases on jobs, nor do they provide for automatic allocation of payroll burden. The requirements of a manufacturing job cost system or the job cost system of a professional services firm are quite different from those required by a construction contractor. The contractor must recognize this when he procures the software to satisfy his management information requirements and must understand that his requirements are different from those of other industries.

Management Style Differences

Most of the software products available today were originally designed to satisfy the requirements of a specific customer and, therefore, do not necessarily represent the result of research by the software author into a broad cross-section of the construction industry. As a result, a high percentage of the software products available to construction firms were originally designed for specific construction trades. The software purchaser must understand this and must understand the differences in management style that exist between different construction trades.

The information requirements of a heavy contractor (e.g., a highway contractor) may vary significantly from those of the specialty trade subcontractor (e.g., an electrician or plumber). For example, the highway or heavy contractor is interested in monitoring the performance of a crew operating, for example, a scraper, a loader, and three bobtail dumps. The personnel and equipment of the crew perform as a unit. Management of the company monitors the crew's performance by measuring cubic yards of material moved per day or per week as compared to the original estimate. In addition, percentage of completion may be measured by yards moved or linear feet of base prepared in relation to the estimate. Typically, the number of work items or activities accounted for in the highway or heavy job are substantially fewer than the number of activities accounted for by a general contractor putting up a high-rise building.

On the other hand, the general contractor may subcontract substantially all of the work contained in the job. A job may include as many as 1,500 or 2,000 work items or activities. In such a case, the general contractor is primarily concerned with the subcontractor's performance against the contract, back charges, retention, work schedule, and so on. A job may have as many as 100 different subcontractors. Controlling contract status from a financial and contract-performance standpoint can therefore be very difficult. The many changes to the lien laws and the liability that the general contractor assumes demand that subcontractors be monitored closely.

The service-type contractor, such as the plumbing or electrical contractor who makes residential installations and repairs, as well as subcontracts to general contractors on major jobs, has his own unique problems. The service contractor typically has a large number of service jobs in the backlog. His primary problem is the scheduling of crews. In addition, he typically marks up his labor and materials cost on the service jobs and bills them immediately upon completion of the contract. Furthermore, he may have commissioned salespeople on staff who prepare estimates and coordinate contracts with the customer. The commissions resulting from completed and billed contracts must be computed and accounted for. At the same time, the service-type contractor may have a responsibility for controlling the jobs as a subcontractor to a general contractor.

The residential developer has his own unique problems. He is concerned with accumulated costs on a per-unit basis. He is also concerned with construction loans on his projects and must attend to every aspect of the status of these loans.

Effects of Company Size

Company size influences the contractor's decisions about his MIS requirements in two ways: (1) the extent of automation and (2) the size of the computer to be employed.

A large general contractor with 1,000 subcontracts on 75 jobs in progress requires a more sophisticated subcontractor-tracking system than a general contractor with 25 subcontracts on 10 jobs in progress. The large general contractor has multiple levels of management monitoring various aspects of subcontractor performance. These might include progress on the job; subcontract financial considerations; and status of legal matters—such as lien waivers, liability insurance, and workers' compensation; as well as the reporting of summary job status and problem areas to executive management.

The smaller contractor could not justify the cost of, nor would he have the need for, the sophisticated system. He might use the automated system only to account for subcontractor invoices and retention, while the other matters relating to subcontractor relations would be handled manually.

The number and complexity of applications together with the volume of activity for each of the automated applications determine the size of the computer that must be employed. A contractor doing $500,000 to $3 million in contract volume annually can normally be accommodated by a single-user microcomputer with 10 megabytes (a measure of the number of characters that can be accommodated; usually expressed as "M") of disk storage. A contractor doing from $3 million to $8 million can normally be accommodated by a two or three terminal system and 30 megabytes of disk or mass storage. These estimates provide broad guidelines and vary according to the nature of the contractor's business. A general contractor with a small labor force and relatively few subcontracts doing $5 million may be satisfactorily accommodated by a single-user, 10 megabyte system. In summary, the fact that software was designed for the construction industry does not necessarily mean that it satisfies the requirements of every contractor in any given situation.

The purpose of this chapter is to guide the contractor through the process of defining his requirements so that he may identify the appropriate software and computer system for his needs and thus will be able to implement a suitable MIS.

THE APPLICATIONS

There are a number of applications or functions that constitute the contractor's MIS. The following discussion involves those applications unique to the construction industry and focuses on the aspects of these applications unique to contractors. Obviously, a payroll system accepts time-card data, computes an employee's pay, and produces the payroll check and the employer's quarterly and annual tax reporting requirements; virtually every payroll system does this. However, there are unique aspects of an MIS that relate specifically to the contractor.

The typical contractor's MIS consists of four basic applications that form the core of the system. These are:

1 Job cost reporting
2 Payroll
3 Accounts payable
4 General ledger—financial reporting

In addition to these four key applications, contractors often automate other related applications. The importance of these other applications depends on the specialty and size of the construction firm. These additional applications may consist of one or more of the following:

- Billing—accounts receivable
- Subcontracts payable

- Equipment accounting
- Purchasing
- Inventory
- Estimating
- Scheduling

Job Cost Reporting

An example of a job cost report is illustrated in Figure 5-1. This report was developed to illustrate various concepts in job cost reporting systems. The contractor's job cost reporting system must have the ability to account for and report on multiple jobs. Each job has a number of activities or work items; Figure 5-1 suggests three. The typical activities used by plumbing and HVAC (heating, ventilating, and air conditioning) contractors are illustrated in Figure 5-2. The numbers may vary significantly depending on the management style of the contractor.

The activities reported in the job cost report may be established as company standards or be dictated by the owner or lender for presentation on progress bills. The number of activities will vary from 5 or 10 to more than 2,000, depending on the size of the job, the type of work being performed, and the reporting requirement. Frequently, the contractor who is just starting out will want to use a limited number of activities to give his operating and office personnel time to become familiar with the activities and their use. As the employees gain experience, the number of activities accounted for can be increased. The activities on some jobs vary because of the scope of the work. For example, on one particular job, only the plumbing portion of the activities, included in Figure 5-2, would be used to account for job costs.

Cost items or cost centers must be defined for each activity. In Figure 5-1, the cost items are:

- Labor and burden
- Materials
- Equipment
- Subcontractors
- Overhead

The cost items for any activity vary. For example, if the activity is performed completely by a subcontractor, only that cost item would

FIG. 5-1 Job Cost Report

JOB NAME _____   Start Date: XX/XX/XX   End Date: XX/XX/XX

Activity	ESTIMATE				ACTUAL TO DATE							VARIANCE	
	Original	Estimate Revisions	Change Orders	Total	Percent Complete	Total to Date	Committed Costs	Costs	Total	Estimate to Complete	Total	To Date	At Completion
Activity 1:													
Labor and burden	$4,000	$1,000	$200	$5,200	60%	$3,120	$600	$4,500	$4,500	$3,000	$7,500	$(1,380)	$(2,300)
Materials	2,600		150	2,750	90	2,475		1,425	2,025	225	2,250	450	500
Equipment													
Subcontractors	1,250		50	1,300	60	780		1,125	1,125	750	1,875	(345)	(575)
Overhead													
Total	7,850	1,000	400	9,250		6,375	600	7,050	7,650	3,975	11,625	(1,275)	(2,375)
Activity 2:													
Labor and burden													
Materials													
Equipment													
Subcontractors													
Overhead													
Total	$ ══	$ ══	$ ══	$ ══	%	$ ══	$ ══	$ ══	$ ══	$ ══	$ ══	$ ══	$ ══
Activity 3:													
Labor and burden													
Materials													
Equipment													
Subcontractors													
Overhead													
Total													
Total activities	$ ══	$ ══	$ ══	$ ══	%	$ ══	$ ══	$ ══	$ ══	$ ══	$ ══	$ ══	$ ══

FIG. 5-2 Activity Descriptions

- Submittals
- Detailing
- Mobilization
- Underground cast-iron sanitary sewer and storm drain
- Above-ground cast-iron sanitary sewer and storm drains
- Inserts and sleeves
- Hangers
- Drains
- Carriers

- Fixtures
- Plumbing—supervision
- HVAC—inserts and sleeves
- HVAC—hangers
- HVAC—equipment
- HVAC—copper tubing
- HVAC—steel pipe
- HVAC—coil connection and mechanical room
- HVAC—underground piping
- HVAC—supervision

---

appear. Software products must have flexibility in definition of cost items. For example, the contractor may elect to separate labor and labor burden. He may also elect to account separately for rented and owned equipment charged to jobs. Further, he may want to report overhead as one item at the bottom of the report rather than compute and report it as part of each activity.

The contractor's original cost estimate, bidding, cost reporting, and progress billing forms should all relate to each other. Ideally, the cost estimate should be summarized in a format that could be used for cost reporting purposes. To the extent that the contractor has control of the bidding form, this same format should be used; the billing form and the bidding form should be the same. This continuity is helpful when preparing these documents and also when comparing like activities across different jobs for the purpose of evaluating the performance of the estimating and operating departments of the company.

The job cost reporting system should have the ability to report results both in terms of dollars and in units of work. Some levels of management may find it preferable to work with budgets and actual performance expressed in units rather than in dollars. This is particularly useful where the type of work performed lends itself to this kind of measurement, such as linear feet of pavement, curb, and gutter; cubic yards of earth; and so on.

One of the principal objectives of the job cost reporting system is to identify problems and to take corrective action on activities that show negative variances, either to date or at completion. The principal value of the job cost report is to present the estimate or job budget to date with the actual costs to date in order to compute the variance. In addition, most systems project the budget and the estimated cost at completion based on historical cost, in order to show the probable variance at completion.

There are two basic procedures for computing and reporting the activity variance. The first procedure involves inputting a "percent complete" at the cost-item level and having the system compute total cost to date based on the estimated percentage of completion and compare that with the total cost incurred to date. Resulting variances are then computed. The second procedure is to input the estimated costs for completion in the actual to-date costs section and have the system compute costs and the variance at completion. An effective job cost system should have the ability to accept either of these two procedures at the cost-item or activity level and to compute the other numbers, including the variance to date and the variance at completion. In all likelihood, a percentage of completion would be inputted during the early job phases because accurate costs for completion cannot be determined. Subsequently, during the latter phases of the job, the superintendent's or engineer's estimated cost for completion would be used.

The job cost system should have the ability to accommodate change orders either at the activity level, which represents modifications to the established work scope, or as additional activities representing new activities in the job. In order to provide the contractor with the most current information, the system should have the facility to introduce change orders, whether approved or not. Thus, an accurate picture of the job status can be obtained under the condition that there is a high likelihood that the owner will approve the changes.

The job cost report, as demonstrated in Figure 5-1, has a column for committed costs. Ideally, committed costs are all costs incurred to reach the reported percentage of completion including those that have not yet been billed or recorded into the system. For example, if a concrete pour occurred the day before the end of the job cost reporting period, the invoices would not be included as actual costs in the job cost system. Therefore, there should be a provision for getting these costs into the system so that the computations involving percentage of completion and cost to date will be more accurate. As a practical matter, vendor invoices are not forwarded that quickly. Therefore, most automated job cost systems use purchase orders, which have been recorded but not billed, as the basis for determining committed costs.

An extremely important issue in an effective job cost system is the reporting of frequency flexibility. Many contractors, particularly in some of the specialty trades, undertake jobs of very short duration—in some cases, a week or less; but more typically, less than a month. Thus, it is essential that a job cost system for these contractors have the ability to provide timely job cost information so that management can take necessary corrective action before the job is completed.

Payroll

The payroll system should be integrated (or interfaced) with the job cost and general ledger systems. When time-card data is reported by activity, the computer system should compute gross pay by activity and automatically transfer that data to the job cost reporting system. In addition, the gross payroll, together with the payroll tax liability, should be summarized and automatically transferred to the general ledger system. Also, the payroll system should assign different workers' compensation insurance rates to each activity and should have the ability to accept the pay data and compute and report the workers' compensation premium for each job. In addition to this calculation, the system should have the ability to apply payroll burden, consisting of employer taxes, employee benefits, and workers' compensation, as a direct charge based on the labor used by the activity. A percentage application to the activity (a percentage application of payroll burden, including workers' compensation) is probably more desirable because all jobs would then be burdened through the year at the same rate.

In addition to providing data to the general ledger and the job cost systems, the contractor's payroll system should also provide the certified payroll lists and the special Equal Employment Opportunity (EEO) reports that are required by local government agencies. The contractor also has a unique requirement involving field termination. The system should have the ability to accept manually prepared checks from the field and automatically post these costs to the job cost reporting system, the general ledger, and the data bases, which are used for employer tax reporting purposes.

The payroll system may also require flexibility as to reporting frequency (either daily or weekly) and may need multistate or multijurisdiction tax computations, depending on the type and location of the contractor's jobs.

Accounts Payable

In addition to reporting accrued accounts payable and providing disbursement checks for material purchases on a timely basis, the automated accounts payable system should transfer material equipment rental and, in some cases, subcontractor costs, to the job cost system by activity. In addition, the accounts payable system should summarize and transfer these costs to the general ledger system.

The accounts payable system should accommodate partial payments on invoices as well as record manually prepared disbursement checks, so that these costs can be transferred to the job cost and general ledger systems. The accounts payable system should also provide the federal Forms 1099 where required. In some cases, the accounts payable system also accounts for subcontractors, including the original contract amounts, any contract change orders, backcharges, retention, and such.

An accounts payable system is frequently interfaced with a purchasing system, so that purchase-order data can be transferred to accounts payable and reported in the job cost system as committed cost. Such systems must relieve the committed cost and substitute actual cost when the vendor invoices are eventually recorded.

General Ledger: Financial Reporting

The construction contractor has some unusual financial reporting requirements that cannot be accommodated by many of the automated general ledger systems. These include the following:

- The contractor capitalizes job costs into the balance sheet rather than by the more traditional method of posting them to cost of sales in the income statement. These costs carry forward from year to year while the job is in progress. The more traditional business environment, for which most automated systems are designed, assumes that costs will be expensed into cost of sales and closed out at year-end.

- The contractor does not use the traditional method of recognizing revenue when invoicing the customer. Additional computations are necessary to relate costs to date to total expected costs in order to determine job revenue. Most automated systems not designed specifically for contractors accommodate this approach.

- The contractor typically has two additional balance-sheet accounts: (1) Cost in Excess of Billings and (2) Billings in Excess of Costs. The schedules necessary to perform the computations to obtain the balance in these two accounts are not typically available in most general ledger systems.

- A contractor may have a mixture of short- and long-term jobs. This is particularly a problem with the service contractor who makes residential service calls to repair or install appliances, roofs, and so forth. For the large volume of short-term jobs, it may be desirable to use a combination of capitalizing job costs and expensing them into cost of sales. This approach requires specialized software that can accommodate both approaches.

- Many contractors are involved in multiple companies, which may take the form of joint ventures with other contractors. The firm with the automated system and good accounting and record-keeping practices usually performs the accounting for the joint venture. Developers also have a requirement for multiple general ledgers because each project may be undertaken by a different company, particularly where the developer acts as a general partner in limited partnerships.

Billing: Accounts Receivable

This application is usually not a major consideration for most contractors. Frequently, even though the rest of the MIS is automated, the billing and accounts receivable activities will be maintained manually. There are, however, a number of issues relating to the billing and accounts receivable application that a contractor must consider.

- The American Institute of Architects (AIA) "Application and Certificate for Payment," forms G702 and G703 (see Figures 5-3 and 5-4) are being used more frequently throughout the industry. These forms and the manner in which the data is presented represent a convenient billing device, including all of the information which is necessary for the owner to track the contract against contractor progress bills. In a few instances, software authors have implemented the AIA billing form as part of their billing and accounts receivable systems. More commonly, however, contractors are using one of the popular spreadsheet programs, such as VisiCalc, Lotus 1-2-3, or Symphony to prepare their contract progress bill in the AIA format.

- Accounting for accounts receivable and retention receivable is a necessary part of the specialty trade subcontractor's accounts receivable system. If the contractor has a large number of these contracts in progress, with a number of retention provisions in each contract, he may need an automated system to monitor these receivables.

FIG. 5-3 Application and Certificate for Payment

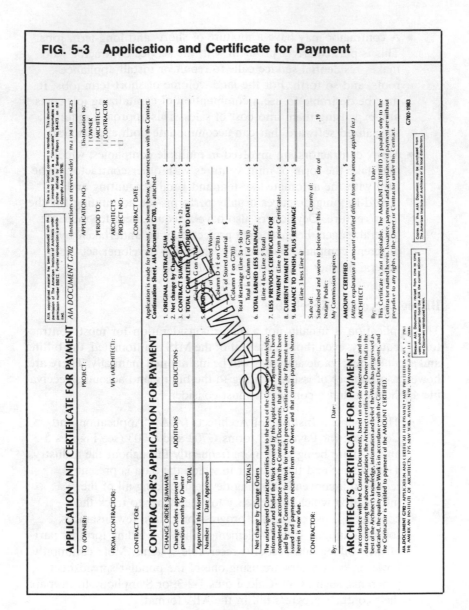

FIG. 5-4 Continuation Sheet

CONTINUATION SHEET AIA DOCUMENT G703 *Instructions on reverse side) PAGE OF PAGES

AIA Document G702, APPLICATION AND CERTIFICATE FOR PAYMENT, containing
Contractor's signed Certification is attached
In tabulations below amounts are stated to the nearest dollar.
Use Column 1 on Contracts where variable retainage for line items may apply.

APPLICATION NUMBER:
APPLICATION DATE:
PERIOD TO:
ARCHITECT'S PROJECT NO:

A	B	C	D		E	F	G		H	I
ITEM NO.	DESCRIPTION OF WORK	SCHEDULED VALUE	WORK COMPLETED			MATERIALS PRESENTLY STORED (NOT IN D OR E)	TOTAL COMPLETED AND STORED TO DATE (D + E + F)	% (G ÷ C)	BALANCE TO FINISH (C − G)	RETAINAGE
			FROM PREVIOUS APPLICATION (D + E)	THIS PERIOD						

AIA copyrighted material has been reproduced with the permission of The American Institute of Architects under permission number 85012. Further reproduction is prohibited.

There is no implied consent to reproduce. This document is intended for use as a "consumable" (consumables are further defined by Senate Report No. 94-473 on the Copyright Act of 1976).

Because AIA Documents are revised from time to time, users should ascertain from the AIA the current edition of the Document reproduced herein.

Copies of this AIA Document may be purchased from The American Institute of Architects or its local distributors.

G703-1983

AIA DOCUMENT G703 • APPLICATION AND CERTIFICATE FOR PAYMENT • MAY '83 EDITION • AIA° • © 1983
THE AMERICAN INSTITUTE OF ARCHITECTS, 735 NEW YORK AVENUE, N.W., WASHINGTON, D.C. 20006

- Partial payments on contract progress bills are common, particularly where a number of disputes arise. An automated system aids in monitoring these receivables.

- The service contractor who makes residential service calls has special billing and accounts receivable problems. These include use of marked up billing rates for labor and material. In addition, there is a requirement for timely billings for work performed and there may be a substantial number of outstanding bills that must be monitored. Finally, the system must make the computations and report the estimated sales staff commission compensation.

Subcontracts Payable

Automated subcontractor payable systems may either be incorporated into an accounts payable system or function separately. The general contractor who uses a large number of subcontractors typically has more sophisticated requirements for the subcontractor systems. These requirements extend beyond accounting for subcontractor bills retention and payments to the subcontractors. The significant attributes of the more comprehensive subcontractor payable systems include:

- A provision within the subcontractor payable system for a details ledger for each subcontractor that includes:

  a Original contract amount

  b Contract change orders (including date and amount)

  c Invoices rendered (including invoice date and amount)

  d Payments (including payment date and amount)

  e Back charges (including date and amount)

  f Balance due to the retention payable

  g Balance due on the subcontract

- Accounting for a variety of nonfinancial information relating to each subcontract, such as tracking subcontractor lien waivers and maintaining records on insurance expiration.

- Automatic preparation of two-party disbursement checks containing both the subcontractor and material suppliers as payees. Due to the changes in lien law in recent years, the general contractor has increasing responsibility for payments to material suppliers of subcontractors.

- The ability to interface with the general ledger by summarizing financial transactions relating to subcontractors and automatically posting this information to the general ledger. In addition, the details should be summarized by job activity and posted to the job cost system for reporting job status.

Equipment Accounting

There are a variety of equipment-accounting applications that perform the functions of:

- Charging jobs for equipment use
- Accumulating and reporting maintenance costs
- Scheduling preventive maintenance

These functions may be incorporated into a single system. For the more sophisticated contractors, however, the functions may be organized into separate systems that provide a variety of management-reporting capabilities to aid in the control and to assure the profitable operation of these assets.

Most of the equipment applications provide the detailed accounting necessary to establish the rental rates to be used for charging equipment to the job. These applications also capture equipment-cost data, such as depreciation, gas, oil, repairs, tires, and maintenance cost. The application also accumulates the equipment-usage data to provide a basis for management to establish adequate rental rates to ensure that the owner recovers all costs and earns a desired profit. A variable rental rate may be used in some cases. For example, the highway and heavy contractor may use multiple rates to reflect unique repair and maintenance experience from using the equipment in different job environments. The contractor may have a standard rate for the normal type of job and a higher rate for work in precarious environments.

Equipment-accounting systems are also frequently used to assist the contractor's equipment-maintenance activities. Based on usage data, which is accumulated as a result of equipment operation and maintenance schedules (established as a result of prior experience), the system can produce maintenance work-orders for performing periodic preventive maintenance.

The equipment systems generally include the operating costs detail (including depreciation) for each piece of equipment together with the associated revenue. This is used by company management to modify or establish equipment-billing rates. These systems also provide operating statistics that allow management to monitor equipment performance between major overhauls; miles operated between tire changes are typical of this type of statistic.

Purchasing

A purchasing system is frequently used by the larger contractor. When the contractor is awarded a job, purchase orders are issued to "buy out" the materials for a job in order to protect prices and quantities. In addition, contracts have to be issued to the subcontractors. The purchasing system is the source of the committed costs, which are reported in the job cost system. When the vendor's invoices are received and recorded, the committed costs are relieved and the invoice costs are posted as actual costs. The purchasing system is also used for expediting materials and for evaluating the performance of material suppliers. Suppliers who continually ship late, back order, short-ship, or ship unacceptable substitutes can be monitored by the system.

Inventory

Although contractors do not normally maintain inventories of materials, from time to time they may find it desirable to purchase a large quantity of materials to use on specific acquired jobs. Because the contractor anticipates future price increases for material shortages, there are basically two approaches to these inventory systems: (1) the materials may be shipped to the job-site or retained in the yard and charged to the job; or (2) the materials may be held in the yard, capitalized as inventory, and charged to the job as they are used. The contractor's inventory system must have the ability to account for these inventories, regardless of where they are located. In addition, the costs and the selling price (which assure adequate return on investment) must be accounted for. These inventory systems are typical of distributor-inventory systems (such as the wholesaler), except that these systems must have provisions for automatically summarizing inventory transactions and posting them to the appropriate jobs.

Estimating

There are a variety of automated estimating systems available in the marketplace today. For the most part, these systems are designed to satisfy the needs of specific trades, such as electrical, sheet metal, and concrete contracting. The function of these systems can be divided roughly into three basic classes of systems. The first is used to accumulate takeoff data, accept material and labor prices, perform extensions, and summarize estimate data. These types of estimating tools are very similar to the so-called "spreadsheet programs" commonly in use. There are a variety of templates that are used with the spreadsheet programs to assist in this estimating function.

The second type of estimating tool uses a variety of data bases, including the following:

- *The automated price book.* This is the most popular type of data base. These price books are designed to be used by some of the specialty trades—particularly the electrical, mechanical, sheet metal, and plumbing trades. The price book supplier provides the user with magnetic media containing the most current prices for materials, including the price for the installation of labor. The program-user makes local area pricing adjustments to these tools. Using the estimating programs, the estimator takes off the quantities that are priced by the automated pricing system.

- *Standard bills of material.* Another type of data base is the development of standard bills of material; these are common components or assemblies used by the contractor throughout the job and frequently referred to as work packages. They consist of groups of similar items that may be repeated. For example, for the layout, the materials required for all of the restrooms in a high-rise will be the same. The electrical layout, in similar component units, or modules, will also be the same. These can be set up and used repeatedly throughout the estimate.

- *User-developed price files and item files.* These are probably the oldest and most common data bases used in estimating systems. They are developed and maintained by the contractor's purchasing and estimating staff.

Some of the more recently developed estimating packages provide a digitizer that allows the estimator to electronically take off quantities directly from the working drawing. To use this device, which looks like a pencil or stylus, the estimator inputs the type of component and the appropriate unit of measure (linear feet, cubic feet, etc.) to be taken off, together with the type of material to be used. The drawing is placed on a sensitive table that is designed to recognize the impulses from the digitizer. The estimator positions the digitizer at the extremes of the components and depresses a key at each of the corners or other extremities. The computer system then calculates the dimensions or volumes. Some programs allow the estimator to use the digitizer to trace components, such as three-inch cast-iron pipes, across the drawing. The system automatically provides where appropriate such elements as the elbows, hangers, nipples, and couplings.

Another type of program used by estimators is the bid-day assistance program. This program summarizes the contractors bids and accepts them by component. Contractor management uses the program and inputs various overhead profit rates, testing hypothetical assumptions under various scenarios. The software also accepts variations of subcontractor bids and permits testing of the effects of these on the final bid.

Scheduling

For many years, contractors doing government work have been required to provide the contracting agency with project schedules. Out of this effort came a variety of network-type scheduling systems utilizing the critical path method (CPM) and the Program Evaluation and Review Technique (PERT). Later versions of the PERT technique included accumulating costs by activity and comparing these with the activity budgets. Except for government work, these scheduling systems were not generally accepted by the construction industry. Developing initial project schedules and subsequent maintenance of the schedules is a time-consuming task. Software vendors have developed a variety of packages, that aid in automating schedule maintenance and, in some cases, provide a graphic CPM or PERT presentation in addition to the variety of reports used by project management. More recently, microcomputer software authors have developed a variety of scheduling systems, many of which are used over the network concept and provide many of the same reports used by the earlier networking system. In addition, some have the capability of providing graphic presentation, by either using the computer-system's printer or a plotter, which can be coupled to the microcomputer system. Some systems also provide a modified chart capability and others use the precedent network to describe a project. These software packages on the microcomputer system are relatively inexpensive and can accommodate as many as 2,000 activities for a job. Some also accept the project work schedule and calendar, reflecting time-off for holidays, weekends, and so forth. Other systems provide a daily crew schedule by trade. As a result, the construction industry is initiating broader acceptance of microcomputer systems, which tend to be less complicated and easier to maintain than larger computer systems.

PROCESSING ALTERNATIVES

The contractor has a number of alternatives for automating his MIS. The following discussion describes these alternatives and presents an examination of the advantages and disadvantages of each. The alternatives are:

- Service bureau processing
- The time-sharing (remote batch or interactive) approach
- Micro- or minicomputer systems implementation

Selection of the appropriate processing alternative is largely a function of cost. In general, the contractor pays additional amounts for more timely information and control over the functionality and design of the system employed.

Service Bureau

The Service Bureau provides computer services using a computer system on its premises. The servicer may be a commercial bureau or a commercial bank. The computer programs may be provided by the service bureau or developed on behalf of the contractor. In most instances, input forms are prepared by the contractor and couriered to the service center, where they are recorded and processed by the system. The completed reports and original input data are then returned to the service bureau customer. In some cases, the servicer provides additional functions, such as the preparation, and even the filing, of the quarterly payroll tax reports and deposits.

The principal advantage of using the service bureaus is cost savings. Monthly costs are generally predictable and are based upon the service provided. Since no equipment is involved, the spending of initial capital is avoided. Therefore, the contractor can acquire an automated data-processing capability without the generally attendant risk of failure. All of the administrative responsibilities relating to the data-processing activity are performed by the service bureau.

The principal disadvantage of the service bureau is the lack of control over the timeliness and quality of the resulting product. A major problem in any data-processing environment is the errors that occur during daily recording. Although the service bureau takes responsibility for corrections, any errors result in longer lead time before the reports are available to accompany management because the errors are ordinarily not discovered until the reports are received and reviewed by company personnel.

Time Sharing

Using the time-sharing approach, the resources of a large computer system are available to the contractor. The computer system is accessed by each of the users through a terminal in each user's office. The terminal may be a multiple CRT-type (cathode-ray tube) device or a typewriter-like hard copy terminal. The terminals access the computer via telephone lines, which may be leased lines or standard dial-up lines. The output reports can either be printed at the contractor's office or printed at the computer center and couriered to the contractor's office. In some cases, large time-sharing services have national communication networks and branch offices in major cities. The time-share system user accesses the computer vendor's communication network via the local office, thereby avoiding long distance calls to the vendor's primary computer center. A printed output may be produced on the printer at the local branch office and couriered to the contractor's office.

The time-sharing approach is particularly attractive to contractors with job-sites in different geographic areas. The contractor can install terminals and printers at the job-site for entering the data required by the information system. Using the communication network provided by the time-share vendor, job-site-related reports are returned to the job-sites and consolidated with the other job-site reports for relay to the corporate-office level.

Time-share systems can be implemented by interactive or remote-batch processing. The primary distinctions between these two approaches are found in the turnaround from input data submitted on the terminal and the time when the resulting reports are available in the contractor's office. By using the remote-batch-processing technique, data is entered through the terminal and accumulated either at the terminal or the computer-site. Processing is performed at the convenience of the vendor, usually during off hours. The processed data is then transmitted back to the user's terminal or printer. The computer vendor is able to control the work flow through the computer system and can use the batch data for fill-in. Normally, when the backlog is eliminated, this can occur overnight. Most remote-batch vendors guarantee a specified turnaround.

Interactive time sharing provides the user with continuous access to the computer system. Data is entered and processed immediately. The user has virtually complete control over when jobs are processed. In most respects, interactive time sharing provides most of the benefits of an in-house computer system.

One advantage of time sharing is the relatively low start-up costs, consisting principally of the cost of the terminal and the printer—if a printer is used. In some cases, this equipment can be rented or leased from the time-share vendor. Ordinarily, the service can be terminated with 30-days notice to the vendor. In addition to the initial equipment cost, there may also be a fee for setting up the system in the contractor's office. Many of the aspects of operating a data-processing installation are avoided using the time-sharing system. The time-share vendor's communication networks typically expedite many processing activities, particularly if some of the job-sites are remote.

The major disadvantage of time sharing can be the cost. The time-share vendor takes responsibility for running the data-processing center, including the capital equipment costs, the operating costs, the cost of development and maintenance of the software, the communication network staffing, and the management of the data-processing center. The operation of an in-house computer system, including amortization of the initial investment, may appear to be a less costly alternative when compared with the time-share vendor's costs. This is frequently the case because many costs relating to operation of an in-house computer system may be hidden costs. These include the cost of utilities, management, space, and so forth.

Microcomputer/Minicomputer Systems

The bulk of the prepackaged construction industry application software has been developed for use on either the minicomputer or microcomputer system by some 5,000 different software authors. As a result, implementation of MIS on minicomputer or microcomputer systems is the most popular approach to MIS in the construction industry. For the most part, it is difficult to distinguish between the minicomputer and the microcomputer system, as they rely on the same basic technology. Both systems consist of a computer processing unit (a CPU), a CRT-type terminal input data-display device, and a mass storage unit for storing files or data bases. The mass storage units are either hard disks or diskettes (frequently referred to as floppy disks). The following material describes in some detail a suggested methodology for evaluating and selecting an appropriate computer system. Basically, the suggested approach involves identifying and selecting appropriate computer software. Once this decision is made, the number of hardware options becomes limited. In addition to selecting the appropriate software and hardware, there are several other factors that the potential purchaser must consider before making a final decision:

- The availability and adequacy of equipment maintenance
- The adequacy of system-training aids, such as classes and manuals
- The apparent financial stability of the hardware vendor

The principal advantage of owning the computer on which the company's MIS is implemented is the level of control that the company is able to exercise over processing schedules and priorities. Although the start-up costs of owning and operating a computer are higher than they are for the other alternatives, ownership need not necessarily be an expensive proposition. Once the initial investment has been amortized, the company continues to receive the benefits of the system for relatively minor additional costs.

The major disadvantages of owning a computer are the initial costs of acquiring the equipment and software; the retraining of personnel; and, in some cases, the necessary leasehold improvement needed to accommodate the computer. These potential problems are compounded if the contractor selects the wrong software or computer system. This type of error can occur as the result of selecting the wrong software, selecting an inadequate computer system from a capacity point of view, or purchasing a system from a hardware vendor who subsequently experiences financial difficulties and either significantly curtails or discontinues production of the purchased product.

SOURCES OF INFORMATION

Once the contractor has made the decision to investigate prepackaged software to be used to automate his MIS, one of the more difficult problems he faces is identifying prospective software suppliers. As indicated earlier, there are some 5,000 identified software authors marketing various software products. In addition, many of these authors distribute their products through a network of dealers that are also software companies. Approximately 10 percent of the software authors and associated dealers are targeted toward the construction industry. It is thus very fortunate that the construction industry has a rather sophisticated system of identifying potential suppliers. Several of these identifying methods are discussed herewith.

Software Directory Subscription Services

There are a number of services that publish brief abstracts or descriptions of a variety of software products. Most of these subscription services have a section containing construction industry software. These services are frequently available in the public libraries. However, if potential servicers desire to purchase any of the services, it would cost between $100 and $400 per year. The principal software services are:

- *Directory of System Houses and Computer OEM's*; Technical Publishing, Hudson, Mass. 01749

- *Data Sources*; Ziff-Davis Publishing Co., New York, N.Y. 10016

- *ICP Software Directory*; International Computer Programs, Inc., Indianapolis, Ind. 46240

- *Data Pro* (Minicomputer and Microcomputer Software Series); Data Pro Research Corporation, Delran, N.J. 08075

- *Construction Computer Applications Directory*; Construction Industry Press, Silver Spring, Md. 20910

Construction Industry Publications

Industry suppliers frequently advertise in the industry's publications. This tends to be somewhat unreliable because the costs of this type of advertising make it prohibitive for many software vendors. However, the larger software vendors do advertise in these publications.

Trade Association Surveys

Because of the interest in automation, many of the trade associations have become involved in periodically reviewing the software available for the industry and publishing the results of these surveys either in the association publications or as special reports. For example, the Association of General Contractors' special ad hoc committee on automation published abstracts on 160 software suppliers in a recent issue of the association's publication.

Trade Shows

Industry trade shows, such as the annual meetings of the major associations, often have exhibits. At the most recent Association of General Contractors meeting, a large number of software vendors were exhibiting. At the annual meeting of the Association of Builders and Contractors, 33 of the 100 exhibitors were demonstrating software products and at a recent west coast construction-industry trade show, 40 of the 120 exhibitors were demonstrating industry software.

Computer Manufacturers

Many computer manufacturers provide directories of software products that operate on their computer systems. Although in some cases the prospective purchaser may not be able to acquire them, the directories generally are available to the manufacturer's sales personnel, who provide this information to prospective purchasers.

Other Industry Computer-Users

Satisfied computer-users are a source of referral to prospective users. However, the potential purchaser must recognize that the management style and/or the size of the referring construction company may well be different from that of his own.

SYSTEM SELECTION

The following discussion is concerned with the typical process for selecting application software and the computer hardware upon which it is run.

The Selection Process

The selection process for minicomputers and mainframe computers begins with a feasibility study to determine the potential benefits and costs of installation. The lower cost of microcomputers (less than $10,000) is such that feasibility studies are rarely done, except when consideration is being given to the purchase of multiple units.

An important step in the selection process is to educate people in the organization regarding which systems are available and what their capabilities are. It is essential to attend vendor demonstrations, obtain promotional literature (particularly on various software packages), talk to people in the same segment of the industry who are already automated, and consult an accountant before starting the feasibility study.

The feasibility study should do the following:

1 Identify the specific areas to be automated and the objectives to be achieved (i.e., reduce payroll lead time by two days, provide complete financial reports two weeks after month-end, etc.)

2 Develop rough- or first-cut resource estimates for automating the various functions including:

  a Software costs:

  - Packages
  - Custom-developed software

  b Computer hardware costs

  c Testing, installation, and conversion costs

  d On-going operational costs:

  - Hardware maintenance
  - Software maintenance
  - Communications
  - Personnel

3 Determine the desirability of installing a system. While it is generally not possible to demonstrate a cost savings, it is important to develop an awareness of what the reasonable expectations for the system are. The primary reason that most systems are not strictly cost-justified is management's inability to quantify the benefits of more timely and accurate information; this is often a major result of automation.

If the contractor decides to move ahead based on knowledge of the general level of costs and benefits, the next step is the development of application requirements. These requirements, or functional specifications, identify the specific features each application must have to meet the organization's needs. In many areas, the specifications are expressed as expected capabilities, such as, "The payroll package must be capable of deducting state tax in up to three states for the same employee in a pay period," or, "The general ledger must allow at least six operating divisions and corporate consolidations at three levels." Five to ten key requirements for each application are often sufficient if the contractor is planning to purchase an existing software package. It is also necessary to identify the current and future volumes of transactions to be processed by each application.

Although the functional specifications are precise, they do not identify how the capability is to be provided. This level of specification is adequate to allow evaluation of various packages. Functional specifications should be classified as essential, desirable, and beneficial, and candidate packages should be evaluated against them. A ranking or grading scheme can be used to quantify the results.

For custom-designed applications, that is, those for which the contractor does not anticipate finding a pre-existing package, it is usually desirable to develop a general system design of how the application will function. The design would include: definition of the data to be inputted; the sources of the data; the method by which it would be automated, (i.e., keyboard entry, direct from other systems, and so forth); the key processes or calculations; the major data files; and the system outputs. Of particular interest is the question of whether the job-cost system should get a tape from a bank-payroll system or whether the total labor by job should be inputted on a terminal.

Given the necessary specifications that define what the contractor wants his system to do, there are three basic ways to proceed: (1) choose a system based on personal knowledge/feel; (2) develop a formal request for proposal (RFP) to solitict bids from a number of vendors; or (3) use specifications to select several potential vendors and negotiate with then until one is selected. The *best* approach depends on: (1) the amount of data-processing expertise available in the contractor's organization, (2) the size of the system and its probable impact on the particular operation, and (3) the time/resources available to the contractor. The table of contents for a typical RFP is shown in Figure 5-5.

It is often useful to prequalify vendors simply to control the amount of time spent on the selection process. Assuming that the contractor has identified several vendors or has issued an RFP, he is faced with deciding how to evaluate their responses. The most common evaluation approach is to develop a list of criteria, consisting of the application specifications and other factors considered essential and, then, to weigh the criteria to indicate the relative importance of each criterion to the organization. This is normally done by allocating 100 or 1,000 points among the criteria. Thus, a criterion assigned 25 or 250 points would be five times as important as one that was assigned 5 or 50 points. Each candidate vendor is then ranked on each criterion; the rankings are multiplied by weighting of the criterion and a total weighted score for each vendor is developed. Figure 5-6 illustrates this procedure.

Theoretically, the vendor with the highest total score is the best overall choice. Generally, however, the ranking process is an iterative one with much discussion and further data gathering until the decision-makers are satisfied that the relative rankings are appropriate. Even then, other judgmental factors not reflected in the criteria may appropriately influence the final decision. For example, it is often desirable to obtain additional demonstrations from the two or three highest-ranked vendors. The selected vendor's references also must be checked. In addition, the con-

---

FIG. 5-5 Typical Table of Contents for an RFP

SECTION

I. INTRODUCTION
- **A.** General Information
- **B.** Company Background
- **C.** Current Data-Processing Environment

II. OVERVIEW OF DATA PROCESSING REQUIREMENTS
- **A.** General Electronic Data Processing (EDP) Environment
- **B.** Applications Requirements
- **C.** Hardware Requirements
- **D.** Estimated Data-Processing Volumes
- **E.** Vendor Proposal Evaluation Criteria
- **F.** Schedule of Procurement Events

III. INFORMATION REQUESTED FROM VENDORS
(Schedules to Be Completed by Vendor)
- **A.** Software Fact Sheet
- **B.** Hardware Fact Sheet
- **C.** System and File Management Software Fact Sheet
- **D.** Application-Cost Fact Sheet
- **E.** Application Software Fact Sheet

IV. APPLICATION DESCRIPTIONS*
- **A.** General Overview
- **B.** General Ledger and Financial Reporting
- **C.** Accounts Payable
- **D.** Payroll
- **E.** Job Cost
- **F.** Equipment Accounting
- **G.** Billing and Accounts Receivable

  *Others as appropriate

V. VENDOR REFERENCES
- **A.** List of Present Users
- **B.** Sample Contract

FIG. 5-6 Illustrative Computer System Vendor Analysis

Criteria	Maximum Points	Vendor 1	Vendor 2	Vendor 3	Vendor 4
Application software:					
Job cost reporting	200	175	160	190	200
Payroll	100	70	65	85	90
Accounts payable	50	40	30	45	45
General ledger—financial reporting	25	15	15	25	25
Accounts receivable	10	10	10	10	5
Subcontract control	95	60	75	90	80
Estimating	20	12	16	19	19
	500	392	411	464	464
Software:					
Installation support	50	10	10	50	50
User training	30	15	15	30	30
Documentation	35	35	35	35	35
Continuing support	40	0	0	40	40
Firm size and stability	15	15	15	15	15
Industry experience	30	30	30	0	0
	200	105	105	170	170
Hardware:					
Installation support	25	10	10	25	25
Expansion potential	75	30	75	50	75
Equipment performance	50	38	50	40	50
Ease of operation	50	45	45	45	45
Equipment maintenance	50	45	45	45	45
Data communications capabilities	50	42	50	40	50
	300	210	275	245	290
Total	1,000	707	791	879	924

tractor should be specifically concerned with the vendor's ability to support him in his geographical location.

Contractual Considerations

The contractual relationship between buyers and sellers of data-processing products and services differs from that of other business contracts because of a unique mixture of technical and legal issues affecting the language employed. This necessitates continual scrutiny and advice from counsel. A contractual relationship, however, is not a guarantee that the services and functions will be provided or performed in a satisfactory manner.

In general, most data-processing contracts are too general to provide real protection for the user. This is particularly true if standard vendor contracts are used. Vendors do accept nonstandard contracts, although contract review and acceptance by the vendor may take longer.

The contractor must go through the difficult process of reducing his requirements to contract language in order to effect the most specific contract possible. This need to reduce expectations to contract language requires that the contractor be as familiar as possible with the technical and legal jargon. This process often helps clarify the contractor's thinking and can assist in the obtaining of agreements with the vendor on who is responsible for what and when the results are to be available. Because data-processing contracts contain such a mixture of technical and legal language, they should be signed only with the advice of counsel in conjunction with the guidance of adequate data processing personnel.

SYSTEM INSTALLATION

The activities required to successfully implement a single application or a complete system depend on the degree to which the system contains custom programs or custom modifications to preexisting packages. The following discussion outlines the necessary activities where custom programming is required. If no custom programming is needed, the system design and programming activities are not necessary and the nature of the testing activities changes.

System Design and Programming

The system-design process for a computer application is very similar to the design process used for physical structures. The designer first analyzes the user's needs to identify what information is required; to decide in what form, when, and where that information can be obtained; and to determine what must be done to make the information more helpful to the user. This process, when complete, is used to develop a general system design that is eventually refined to a level of detail that identifies such things as what data fields will be inputted to the system, how this process will occur, what forms and procedures are needed for manual support activities, what the output records will look like, and how frequently these records will be produced.

It is essential that user-management take an active role in understanding and approving the results of this process at every juncture. This responsibility cannot be delegated to, or assumed by, the data-processing analyst, because it is the user who has to live with the resulting system in carrying out his responsibilities.

Following approval of the system design, programming specifications are developed. These specifications, which play the same role as working drawings in the construction of physical structures, direct a programmer (who plays a role similar to the contractor) in the preparation of the actual computer programs. Programs can be developed in a number of different languages, although it is best to use generally accepted business languages such as COBOL (Common Business Oriented Language) or RPG (Report Program Generator), unless there are unusual requirements. There are a number of system design and implementation methodologies that identify, in detail, all of the necessary steps and provide a means for nontechnical individuals to manage the system-development process. Such methodologies should be used if significant customization is essential.

Testing

Testing takes on different forms, depending on the degree of customization involved. The objective, however, is always to ensure that the programs to be implemented are as error-free as possible. For custom programs, testing begins with small self-contained modules, or portions, of each program and becomes progressively more complex until high-volume tests of all interrelated programs are conducted. At each level of test complexity, it is essential to develop test data that reflects the various possible combinations of events that may occur and to test with that data. It then becomes necessary to document and evaluate the results, make essential corrections to the programs being tested, and retest the programs until there are no errors.

In identifying possible combinations of events, it is necessary to consider illogical conditions as well as logical ones. For example, a program could process all possible numerical values in a dollar field correctly but may not be able to recognize alphabetical characters as an error. As a result of the vast number of combinations of conditions in complex systems, no system, regardless of how well-tested, is ever considered 100 percent error-free.

With packaged software, testing can start at a higher level because it is generally assumed that the supplier has adequately tested the basic workings of the package. Because most generalized packages allow users to specify various options, it is necessary to thoroughly test the package to determine that the desired results of the selected options are achieved. For example, a general ledger package may provide for financial statements with different levels of summarization. Testing must ascertain that the appropriate accounts are indeed being summarized. As with custom programs, one should begin with simple tests and progress to more complex, high-volume tests to ensure as error-free a situation as is possible.

Testing provides an opportunity for training those who will run the system under actual conditions. Thus, in addition to testing the computer programs, it also trains those involved and tests the manual-support procedures associated with the new system. Tests should use actual data, in addition to that constructed for the testing, to test extreme and illogical situations.

Security and Controls

Increasing awareness of the dependence of a company on its automated systems is reflected in increased concern with physical security and controls over systems and data. While incidences of physical disaster are rare, they can literally mean the end of a company that has not taken proper precautions.

Security and control considerations include:

- Physical control over the computer, data-storage media, and programs

- Control over remote access to the computer by frequently changing passwords or using other, more sophisticated methods

- Backup arrangements to allow processing in the event of extended unavailability of the system

- Business interruption insurance to reimburse the company in the event of loss

- Segregation of duties to prevent unauthorized actions
- Management review and approval of changes to programs
- Controls over data input and data transmission

Effective controls are dependent on the execution of various actions by different people. However, there can never be 100 percent assurance that these controls are functioning as intended. Because of this dependency, it is necessary to periodically review the intended controls to determine that they are functioning as planned. If there is an independent audit by a Certified Public Accountant, he must test certain controls during the audit to determine the degree of reliance that he can expect when performing the audit procedures.

Documentation

The key to the long-term viability of any system is quality, up-to-date documentation. Without accurate documentation, the company is dependent upon the personal knowledge of a limited number of employees (usually one) to maintain and operate the system. Such dependency puts the company in a compromising position regarding disciplinary actions for such employees. It also exposes the company to an interruption of key activities, should such persons become unavailable or, even worse, become vindictive toward the company and take adverse actions—such as losing the current version of the programs.

In addition to acting as insurance against loss, good documentation decreases program-maintenance costs, reduces training time for new employees, and provides a form of quality assurance over system operations. Thus, the quality of documentation is an important item to consider in the evaluation of packaged software.

It is also important to allocate sufficient resources to develop documentation in custom software development projects. All too frequently, management agrees to let the development team focus on getting the system running, with the understanding that the documentation will be completed right after implementation. Clearly, the success of such an approach is limited.

Training

The training required for system implementation depends on such factors as the following:

- The degree of change to individual duties represented by the new system,
- The data-processing experience of the people affected, and
- The complexity of individual functions and transactions.

Training needs should be identified as part of the overall implementation-planning effort. That effort should identify specific requirements (such as new terminal procedures, use of new forms, reports, and so on) and determine who, when, and how the affected individuals will be trained.

Clerical Procedures

Revised or new clerical procedures are closely related to the training requirements associated with a new system. Often the development of these procedures is put off until after implementation and never gets completed. As a result, system implementation is more difficult than necessary because clerical errors compound other problems.

Clerical procedures should provide easy-to-read, step-by-step directions regarding specific actions. Many organizations find that a form of procedural writing, called playscript, used in conjunction with well-thought-out examples of actual transactions is affective in communicating what is required. The playscript style of procedure writing identifies who is responsible for a given activity (e.g., the accounts-payable clerk, the accounting supervisor, the foreman, and so on), and gives a brief narrative of what is required and what the results should look like for an entire procedure. Procedures should also be complete and contain the various codes required to eliminate the need to reference other documents.

System Conversion

The implementation of all new systems requires some type of conversion. The conversion may be from a completely manual system to an automated one or from one automated system to another.

Careful planning is required to ensure that the initial data in the new system is complete and accurate. Frequently, certain types of transactions span different time periods (e.g., all job-cost transactions may be through the end of a month, except for payroll, which is always through Saturday). Such timing differences may cause conversion problems if not identified and planned for. Also, it is common to have the new system treat some transactions differently than the old one did. For example, the new system may have a different chart of accounts or number of cost centers, thus requiring new procedures to be planned for.

Conversion may also require special programs to convert existing automated data files to a new format. Such programs must be carefully tested to ensure that new data files will be as intended. Temporary clerical procedures may also be needed.

To guarantee that the new system is working as intended, it is generally desirable to operate the new system simultaneously with the old one for some limited period, usually one "cycle." Where the two systems produce roughly comparable outputs, results can be directly compared. Obviously, parallel operations require extra effort, as both systems are operating on the same input data. Thus, this requires twice the amount of work in terms of data input, balancing, and so on.

As the degree of similarity between the old and new systems decreases, the value of parallel testing also declines. In the extreme case, where the new system produces outputs that are not comparable to the old, there is little value in parallel operations. Thus, alternative procedures must be designed to ensure that the conversion is accurate and complete.

OTHER AUTOMATION CONSIDERATIONS

In addition to computerized MIS, there have been significant advances in office automation and business communication systems, both of which are briefly examined in the following discussion.

Office Automation

There is no singular, precise definition of office automation. Generally, it is defined as the automation of the information input, processing, storage, and output functions of an office. Thus, it is substantially more comprehensive than just the installation of word processors, microfilm equipment, or any combination of independent systems designed to increase the efficiency of a given function or area of an office. However, such limited efforts can be an effective catalyst for the change to office automation.

Office automation includes the following:

- *Data input methods:*
  a Optical scanning
  b Keyboard entry

    c Digitizers
    d Graphic input
    e Voice input

- *Processing of data:*
  a Formatting
  b Indexing and retrieval
  c Automatic typesetting
  d Communication with other devices/systems
  e Graphic processing

- *Data storage media:*
  a Magnetic media
  b Hard copy
  c Microfilm

- *Output:*
  a Electronic copier
  b Hard copy
  c Electronic mail
  d Voice mail
  e CRTs

Office automation in its most complete form is a somewhat elusive concept; however, this should not deter companies from automating those functional areas (such as word processing) that offer cost savings in an independent mode. Depending on the size of the organization and its future plans, it may be desirable to limit the choice of equipment vendors to those who have a commitment to office automation and a history of developing open-ended systems that allow further growth and integration within a product line as it evolves.

Communications

Today, business communication systems encompass more than the traditional PBX (private branch exchange) equipment that was once only available from the local telephone-operating company. In the deregulated, competitive, communications environment of today, there are numerous communications options available, including:

- Sophisticated PBX systems that integrate voice and digital data transmission capabilities;

- Alternative suppliers of long-distance facilities, including value-added networks, resale sources, and alternative networks such as MCI, Sprint, and Allnet;

- Sophisticated handsets that offer such things as speed dialing, call forwarding, and executive interrupt;

- Video-text services, which allow terminal access to varied data bases, such as stock market reports, historical newspaper data, airline schedules, and economic data;

- Electronic mail services; and

- Voice mail services.

CHAPTER 6

Internal Controls

G. Barry Wilkinson

INTRODUCTION

This chapter explores the internal control systems and procedures that should be maintained by the construction contractor. Internal control is the system in any organization that assures that management's plans and intentions are being carried out. Internal controls, which are installed by management based upon their measurement of need and importance, act both as a means and a monitor for the continued proper functioning of an organization.

In 1958, the American Institute of Certified Public Accountants (AICPA) stated that internal control is composed of two elements: administrative controls and accounting controls. Since then, the significance of managerial-based as well as financial-based controls has been recognized. The overlap that exists between accounting and administrative controls contributes to the complexities management faces in formulating the specific controls it wishes to apply to a particular situation.

Any control system is limited by the basic cost-benefit principle, which states that the benefit derived from a system should exceed the cost of implementing that system. An efficient and effective system of internal control defines risk at a level deemed acceptable by the contractor's management team. Every organization accepts the concept that risk is necessary in order to achieve business goals. The nature of the construction contractor is such that the risks are greater than in other businesses and, as a result, the potential rewards are greater as well.

Risk, including both internal and external factors, exists because construction management is not able to control all of the factors inherent in the construction environment. Some examples of these factors are:

Internal:
- Organization restructuring
- Employee turnover
- Cash and credit policies
- Job-site location
- Reputation of the firm

- Backlog
- Productivity changes

External:
- Inflation, stagnation, recession
- Government regulation
- Labor and supplier
- Weather
- Technological changes
- Environmental impact
- Competition

Construction managers should recognize and attempt to control as many risks as possible through a properly designed system of internal controls. Control of both external and internal factors is possible to some degree. However, the cost of control over external factors may, and usually does, exceed the benefits derived. Even internal factors may be far too expensive to manage completely and, thus, management must develop a philosophy of measuring risks against costs. As a result, the final implemented system of internal control becomes a source of information that allows management to avoid, or at least reduce, unintentional and unnecessary exposure to construction risks.

ELEMENTS OF THE INTERNAL CONTROL SYSTEM

Broad Objectives

The broad objectives of internal accounting controls are:

- Safeguarding of assets, and
- Creation of reliable financial information.

The broad objectives of administrative controls are to:

- Promote operational efficiency, and
- Verify adherence to managerial policies.

Internal Control Principles

Every system of internal control that is properly planned should have as its foundation provisions for the following:

- Separation of duties
- Authorization procedures
- Documentation procedures
- Accountability controls through accounting records
- Physical controls
- Independent verification

INTERRELATIONSHIPS OF TYPES OF CONTROLS

Administrative controls

In 1973, Statement on Auditing Standards (SAS) No. 1 (AU § 320.26) defined administrative controls in the following manner:

> [I]ncludes, but is not limited to, the plan of organization and procedures and records that are concerned with the decision processes leading to management's authorization of transactions. Such authorization is a management function directly associated with the responsibility for achieving the objectives of the organization and is the starting point for establishing accounting controls of the transactions.

The administrative controls use organization structure and the goals of the entity as their base. They are constituted of those methods and procedures that are concerned mainly with efficient and effective measurement of the implementation of the organization strategy. Administrative controls are only indirectly concerned with accounting and financial records. For example, administrative controls might include:

- Statistical analysis
- Performance reports
- Quality control principles and procedures
- Labor efficiency studies
- Training programs

The development and implementation of administrative controls are passed on from central management to localized project or contract management. The primary purpose for the system of checks and balances resulting from administrative controls is to ensure that the corporate mandates of operation are being adhered to properly. The ideal result of these administrative controls where they have been properly developed, organized, executed, and reviewed through feedback, is the completion of the right kind of construction projects, in the budgeted amount of time, with the proper quality and an adequate return on investment (profit).

An illustration of the widespread use of administrative controls in the construction industry is visible in the solution to the recurring problem of the increasing amount of time required to complete construction projects. In general, construction projects take longer to complete in today's environment than they did 10 years ago. One administrative control, in the form of project management, that has been implemented is the critical path method (CPM) of scheduling. CPM not only deals with the complexities of modern projects but also allows for appropriate administrative responsibility to be set so that feedback can be received and further action can be taken. Use of CPM management systems should have the effect of offsetting some of the lack of productivity inherent in the performance of a contract.

Accounting Controls

The SAS (AU § 320.27) states that:

[A]ccounting controls compromise the plan of organization and the procedures and records that are concerned with the safeguarding of assets and the reliability of financial records and consequently are designed to provide reasonable assurance that:

a. Transactions are executed in accordance with management's general or specific authorization.

b. Transactions are recorded as necessary (1) to permit preparation of financial statements in conformity with generally accepted accounting principles or any other criteria applicable to such statements and (2) to maintain accountability for assets.

c. Access to assets is permitted only in accordance with management's authorization.

d. The recorded accountability for assets is compared with the existing assets at reasonable intervals and appropriate action is taken with respect to any differences.

The internal accounting controls are more pointed than the administrative controls. The specific purposes of the accounting controls are the safeguarding of the assets and the ensuring of accurate and reliable accounting information.

Proper internal accounting control includes systems that protect against

- Failure to evaluate contract profitability on a systematic basis.
- Breakdowns in the estimating and bidding on new contracts.
- Lack of adequate contract cost records (untimely or inaccurate).
- Poor contract maintenance (performance, billings, and collections).
- Inadequate control over the physical assets (construction equipment and job materials).
- Inefficiencies in disbursements for contract costs.
- Insufficient labor cost control—especially job-site payrolls.

These are general categories; a tailor-made operating system of internal control stands up to far more detailed coverage and yields specific controls for specific assets and liabilities; if effective, it produces an ordered manner of accounting record-keeping.

Figure 6-1, an example of a questionnaire for specific areas of construction internal control, demonstrates the types of items analyzed in reviewing an accounting control system.

INTERNAL CONTROL ENVIRONMENT

The design and operation of any properly functioning internal control system is always subject to the environment in which it operates. The environment is composed of all those factors that affect the ability of the internal control system to operate in the manner in which it was designed. These environmental factors include:

- Management leadership
- Organizational structure
- Budgets and internal reports
- Internal audit functions
- Personnel and personnel policies
- Individual company circumstances

(text continues on page 6-13)

FIG. 6-1 Internal Controls Questionnaire

SYSTEM KEYS

A reliable system of internal control should have some of the following characteristics. In theory, the more of these characteristics the system has the more reliable it will be; the fewer characteristics, the less reliable.

- Separation of functions and duties exists so that no one employee can originate, record, and complete a transaction without some form of review.
- Accounting policies and procedures that are in effect have been reduced to a written form.
- Financial and operational reports are timely and present an accurate picture of the company's operation.

SYSTEM DESCRIPTION FOR INTERNAL CONTROL—GENERAL

	Procedure Employed		
	Yes	No	N/A
A. General Ledger & Journals			
1. General ledger accounts are reviewed monthly by a responsible official.	☐	☐	☐
2. All subsidiary ledgers are balanced monthly and tied in to the general ledger by personnel independent of the preparer of records.	☐	☐	☐
3. All journal entries require approval of a responsible independent official.	☐	☐	☐
4. The accounting function is completely separate from custody and control over assets.	☐	☐	☐
B. Accounting Administration			
1. A complete, descriptive, and current chart of accounts exists.	☐	☐	☐
2. A comprehensive accounting manual, including job descriptions, exists.	☐	☐	☐
3. A current organization chart exists and is understood.	☐	☐	☐
4. Adequate physical safeguards are maintained over accounting.	☐	☐	☐
5. Access to accounting records is limited to authorized personnel only.	☐	☐	☐
6. Accounting records are proper, orderly, and well maintained.	☐	☐	☐
C. Personnel Matters			
1. Accounting personnel have adequate training and experience for their job functions.	☐	☐	☐
a. Turnover rate is reasonable.	☐	☐	☐

(continued)

FIG. 6-1 (continued)

	Procedure Employed		
	Yes	No	N/A
b. In-house training is used to prepare employees to perform to standards.	☐	☐	☐
2. All personnel in positions of trust are bonded; they are required to take adequate vacations and in their absence their job is performed by others.	☐	☐	☐
3. Transactions between the company and the accounting employees are adequately documented and are reviewed by a responsible official for compliance with company and Internal Revenue Code policies prior to reimbursement.	☐	☐	☐

D. Management Accounting Reports

1. Financial and statistical reports, especially ones covering the status of jobs in progress, are prepared on a timely and frequent basis.	☐	☐	☐
2. Budgets and forecasts are used; actual results are compared with prior results; significant cost fluctuations are investigated by an individual of appropriate authority.	☐	☐	☐

INTERNAL CONTROLS QUESTIONNAIRE
ESTIMATING AND BIDDING

SYSTEM KEYS

A reliable system of internal control should have some of the following characteristics. In theory, the more of these characteristics it has, the more reliable the system will be; the fewer characteristics, the less reliable.

- Preprinted bid summary sheets are used.
- "Rule of thumb" reviews are made for reasonableness of bids.
- Clerical accuracy checks are in place and are used.
- Coding for bidding conforms with accounting codes.
- All bids are approved before submission.

SYSTEM DESCRIPTION FOR INTERNAL CONTROL OF ESTIMATING AND BIDDING

	Procedure Employed		
	Yes	No	N/A
A. Estimating Summary			
1. Preprinted summary is used.	☐	☐	☐

	Procedure Employed		
	Yes	No	N/A
2. Space is provided for items not common to most projects to be written in.	☐	☐	☐
3. All items on the summary are either filled in or lined out.	☐	☐	☐
4. Accounting codes form the basis of the cost accounting system for comparison of actual and estimated costs.	☐	☐	☐
5. Detailed estimate sheets are produced to support the estimate summary and are then retained.	☐	☐	☐
6. Signatures are required to fix responsibility for each step in the estimating process.	☐	☐	☐
7. Form is reviewed periodically for revision, as required.	☐	☐	☐
8. Cost estimates are obtained from more than one subcontractor, as required.	☐	☐	☐
9. Performance and bid bonds are obtained from subcontractors, as appropriate.	☐	☐	☐
B. Detailed Estimate Sheets			
1. Preprinted forms are used.	☐	☐	☐
2. Accounting codes are provided.	☐	☐	☐
3. Accounting codes serve as the basis for the job cost system.	☐	☐	☐
4. Responsibility for the preparation and double-checking of the detailed estimates is fixed by initials or signature.	☐	☐	☐
5. Estimates are eventually compared to actual quantities and cost.	☐	☐	☐
6. Historical unit costs are periodically tested to determine the validity of pricing policies.	☐	☐	☐
7. Adjustments are made to historical unit costs used in the estimating process for special conditions.	☐	☐	☐
C. Job Bidding Analysis			
1. The form provides for detailed summarization of bidding activity on a periodic basis.	☐	☐	☐
2. Details show at least name of project, bid amount, low bid amount, dollar difference, and to whom contract was awarded.	☐	☐	☐
3. On successful bids, difference between next low bid and successful bid is analyzed for low bidding.	☐	☐	☐
4. Management reviews of all bidding procedures and overall results are made on all bids.	☐	☐	☐
D. Cost Breakdown and Change Orders on Contract			
1. Cost breakdown provides for front-end loading.	☐	☐	☐

(continued)

FIG. 6-1 (continued)

	Procedure Employed		
	Yes	No	N/A
2. Cost breakdown provides for high margins on items where the contractor expects change orders increasing the work and low margins on items where the contractor expects change orders decreasing the work.	☐	☐	☐
3. Formal procedures exist for processing and approving contract change orders.	☐	☐	☐
4. Change orders are adequately communicated to appropriate personnel.	☐	☐	☐

CONTROL OF JOB PROGRESS AND COSTS

SYSTEM KEYS

A reliable system of internal control should have some of the following characteristics. In theory, the more of these characteristics it has, the more reliable the system will be; the fewer characteristics, the less reliable.

The contract/job accounting system provides for:

- Planning of job from bid to completion
- Subcontractor and material supplier bids
- Quality control on work performance
- A reporting of problems for immediate management reaction
- Periodic reporting of actual versus budget costs
- Control of subcontractors' performance and payments thereon
- Change order validation
- Timely billings
- Appropriate collections

SYSTEM DESCRIPTION FOR INTERNAL CONTROL OF JOB PROGRESS AND COSTS

	Procedure Employed		
	Yes	No	N/A
A. Preplanning Contracts			
1. Plans and specifications are rechecked and all items are clarified.	☐	☐	☐
2. Recommendations are made for appropriate plan changes.	☐	☐	☐
3. A cash-flow statement is prepared.	☐	☐	☐
4. The decision is made as to whether financing will be needed.	☐	☐	☐

	Procedure Employed		
	Yes	No	N/A
5. In large projects, the use of the program evaluation and review technique (PERT) and/or the critical path method (CPM) is required.	☐	☐	☐
6. A plan indicating the necessary equipment, personnel, and materials will be available at the appropriate dates.	☐	☐	☐
7. The timing, responsibility, and coordination of the subcontractors is known and agreed to by them.	☐	☐	☐

B. Subcontracts and Materials/Equipment Purchase Orders

1. The financial ability of subcontractors is reviewed and evaluated before the awarding of subcontracts.	☐	☐	☐
2. The quality of subcontractor performance on comparable jobs is verified.	☐	☐	☐
3. Steps are taken to ensure that acceptable bidders are all qualified subcontractors.	☐	☐	☐
4. Only the lowest bid from an acceptable contractor is accepted.	☐	☐	☐
5. Variances from amounts shown in cost estimate are used to measure variance of actual costs from estimated costs.	☐	☐	☐
6. Field superintendents are required to report to management on the performance of subcontractors and suppliers.	☐	☐	☐
7. Purchase orders are issued to low bidders.	☐	☐	☐
8. Purchase orders clearly specify delivery dates and acceptance by vendor.	☐	☐	☐
9. Purchase orders form the basis for subsequent payments.	☐	☐	☐

C. Job Costs

1. All charges to jobs are coded so as to ease comparison of actual with estimated costs.	☐	☐	☐
2. Time sheets are designed to accumulate hours worked by job phase.	☐	☐	☐
3. Time sheets are approved by the job foreman.	☐	☐	☐
4. Invoices are matched with purchase orders and receiving reports prior to payment.	☐	☐	☐
5. Unsatisfactory goods are promptly reported and allowances are negotiated.	☐	☐	☐
6. Invoices are account-coded by job phase corresponding to cost codes.	☐	☐	☐
7. Usage rates are established for each fixed asset for cost recovery purposes.	☐	☐	☐

(continued)

FIG. 6-1 (continued)

	Procedure Employed		
	Yes	No	N/A
8. Equipment used on each job is charged to the job by job phase and coded in cost codes.	☐	☐	☐
9. All possible chargeable time for equipment is accounted for, including idle time.	☐	☐	☐

D. Work Completed to Date and Estimated Cost to Complete

1. All elements of job costs are scheduled by day, week, or month using PERT, CPM, or other method as appropriate.	☐	☐	☐
2. Actual job progress is compared to budgeted progress.	☐	☐	☐
3. Management verifies reported progress by making actual job-site visits.	☐	☐	☐
4. Requisitions for payments for subcontractor effort are approved by the job superintendent before payment.	☐	☐	☐
5. Management is furnished with a periodic report of work completed to date by job phase that compares actual completion with scheduled completion.	☐	☐	☐
6. Additional costs to complete are recomputed periodically using current information obtained through actual job experience.	☐	☐	☐
7. Forecasts of projected gross profit are computed using an updated estimate of costs to complete. These periodic forecasts of gross profit are compared with the original estimate of gross profits and variances are analyzed.	☐	☐	☐

E. Change Orders

1. Change order forms are designed to document contract changes.	☐	☐	☐
2. Work on contract changes is not begun until signed authorization is obtained from owner or owner's representative.	☐	☐	☐
3. Documentation, including acceptance by the subcontractor, is obtained for all subcontract change orders.	☐	☐	☐
4. Changes are merged into the job estimate to enable comparison of actual cost to estimated cost.	☐	☐	☐

F. Billings

1. Management is provided with monthly reports of aged receivables by job.	☐	☐	☐

| | Procedure Employed | | |
	Yes	No	N/A
2. Progress billings are prepared in accordance with contract terms.	☐	☐	☐
3. Billings are made as soon as allowed by contract.	☐	☐	☐
4. Percentage-of-completion method is used to the extent possible to maximize cash flow.	☐	☐	☐
5. Management obtains a periodic report of over/ under billings detailed by project.	☐	☐	☐
6. Over/under billings are appropriately recorded.	☐	☐	☐

Management Leadership

The chief executive officer (CEO) and the top management of a construction firm, in concert with the board of directors, decide on the path that an organization takes. This group has what has been referred to as the primary operational control of the business. This control has direct impact on both administrative and accounting controls and is the source of the principal direction and guidance on major policy matters of the firm. Consequently, all internal controls are based on the emphasis that management places on the running of the business.

The primary operational control is the establishment of policy and basic guidelines by which the contractor-enterprise will be directed toward achieving its business objectives. The interrelationship of the primary operational control to the administrative and accounting controls is depicted in Figure 6-2.

Organizational Structure

The formal organization of any contractor-entity is the skeleton upon which the company develops. The organization chart defines the formal chain-of-command as well as the reporting responsibilities within the organization.

The formal reporting system of the company can be structured in a myriad of ways. The most common forms of formal organization are those organized by product, by function, and by matrix.

Product Line Organization. In a construction company, the organization structure by product line might have the reporting lines organized by the type of construction project the company performs. The vice-presi-

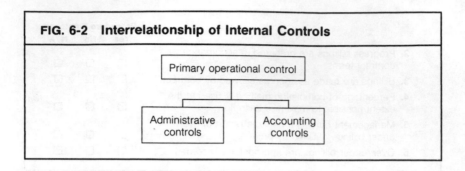

FIG. 6-2   Interrelationship of Internal Controls

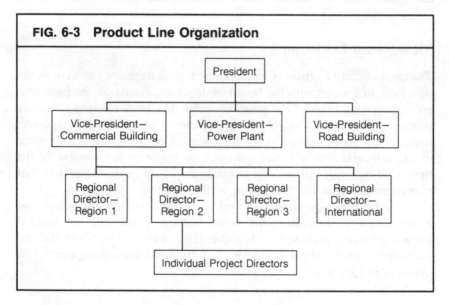

FIG. 6-3   Product Line Organization

dents might have responsibilities for different product lines, such as commercial buildings, power plants, and roads. Functions such as accounting, personnel, and production (construction) would be self-contained within the divisions and report to a vice-president. This is illustrated in Figure 6-3.

Functional Organization. This type of organization is designed to have a separation of responsibilities at the top operational level. The vice-presidents might be in charge of areas such as production, accounting, and marketing. An illustration of a functional organization chart is shown in Figure 6-4.

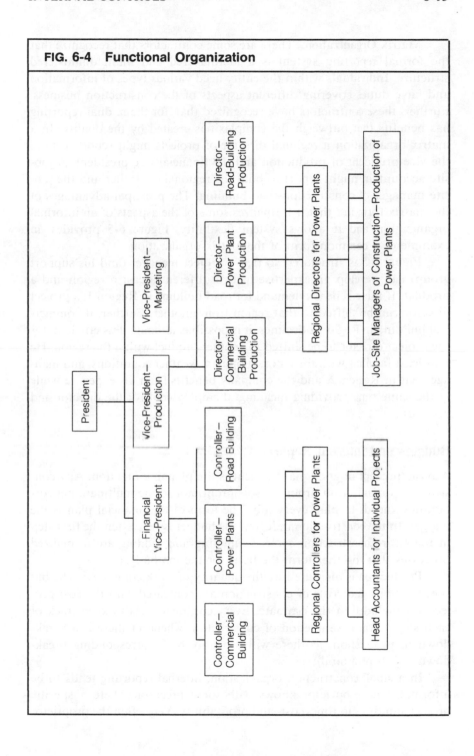

FIG. 6-4 Functional Organization

Matrix Organization. There are some contractors that recognize that the formal reporting system is a multilayered and multiadministered structure. Individuals within the entity need various types of information and have duties covering different aspects of the construction business. Further, these contractors have recognized that, for them, dual reporting has benefits that outweigh the complexities created by the duality. In a matrix organization a regional director of projects might report to both the vice-president of production and the financial vice-president. A job-site accountant might report to both the regional controller and the job-site manager of construction-road building. The principal advantages of the matrix form are that it formalizes some of the aspects of an informal organization and it allows system flexibility. Figure 6-5 provides an example of a common form of the matrix organization.

Figure 6-5 is structured so that a project manager (and his support group) can develop an expertise in two different areas: a region and a product. If there is heavy demand for road building in Region 1, a project director can be shifted to that region from another. Further, if commercial building in Region 2 declines, it is possible, although less efficient, for the project director to be shifted to another product within the region. He is at least familiar with the labor, suppliers, weather conditions, and management in Region 2, and the company benefits from his expertise while at the same time providing meaningful employment for the director and his staff.

Budgets and Internal Reports

A basic precept of good management is the planning function. Any construction organization, from the sole proprietor to the multinational corporation, needs to plan every job. The form of the financial plan is the budget. In the construction industry, the budget is very often the first step in the entire construction cycle. Estimating and bidding are formalized processes that together form the basis for the job budget.

Production is measured by the quantity of work completed. The bid specifications and costs of construction are compared with the cost projections included in the estimate. Every contractor has to keep track of each step toward completion of construction whenever there is a breakdown in production, as there will probably be a corresponding breakdown in job profitability.

In a small construction organization, internal reporting tends to be informal. The proprietor knows with some precision where a specific project stands as to time, cost, and profitability. Very often the proprietor

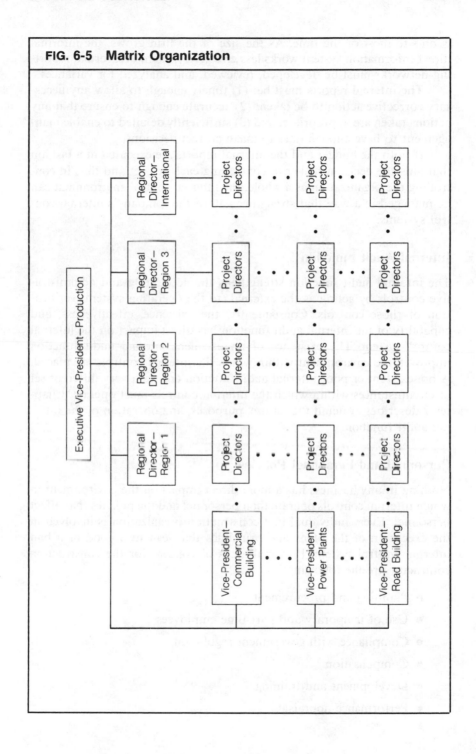

FIG. 6-5 Matrix Organization

is on-site most of the time. As the size of the firm grows, the informal direct information system works less effectively; therefore, formal reporting networks must be developed, reviewed, and analyzed for variances.

The internal reports must be: (1) timely enough to allow any necessary corrective action to be taken; (2) accurate enough to ensure that any actions taken are appropriate; and (3) sufficiently detailed to enable management to have enough data to make correct decisions.

If both the budget and the internal reports are executed in a fashion that aids the decision-maker in controlling each project and thus in controlling the organization as a whole, the internal control environment can be managed in a way that strengthens all of the company's internal control systems.

Internal Audit Function

The internal audit function strengthens the accounting and administrative controls by acting as the external (to the operating system) verification of those controls. Consequently, the existence, effectiveness, and capability of the internal audit function has direct impact on the internal control system. The existence of an excellent internal audit function improves the environment within which the internal controls are placed. A nonexistent or poor internal audit function creates a very different set of circumstances within which the internal controls must operate. Chapter 7 describes in detail the nature, purposes, and operation of the internal audit function.

Personnel and Personnel Policies

Nothing in any business has a more direct impact on the environment in which internal controls operate than personnel and the policies that affect personnel. Every individual in a construction organization is involved in the execution of the plans and principles that lead to a good or a bad internal control system. Principal areas of concern for the construction contractor are the following:

- Planning and procurement
- Use of temporary and part-time employees
- Compliance with government regulations
- Compensation
- Development and training
- Performance appraisal

- Labor relations
- Safety and security

Figure 6-6 provides some items that a contractor should consider in developing a comprehensive personnel program. This listing combined with the expertise of the contractor should allow the development of a personnel environment that by its very nature will be conducive to effective internal controls.

Individual Company Circumstances

More than by any other factor, the environment in which an internal control system operates is dictated by the company implementing the system. No two contractors are exactly alike and no two internal control systems are designed the same way; nor do they work the same once they are in place. The factors that make up the company environment include its size, area of expertise, types of contracts, and historical profitability. The capitalization of the firm, the quality and quantity of the personnel, the reliance on private or government sources for contracts, and the expertise of management also contribute to the character of the company. Each company's circumstances are individual. Therefore, every system of internal control must be designed specifically for that company.

SPECIALIZED INTERNAL ACCOUNTING CONTROLS

This section deals with those controls that are specific to the construction contractor. The areas covered herein include:

- Billings
- Construction equipment
- Contract costs
- Revenue recognition and job profitability
- Site control
- Claims, extras, and backcharges

Each of these areas must have accounting controls to deal with the systemization of accounting information through validation, completeness checks, and reperformance. In the following discussion, each of the indi-

(text continues on page 6-23)

FIG. 6-6　Personnel and Personnel Policies Planning List

A highly qualitative personnel system must consider a majority of the following areas.

A. PLANNING AND PROCUREMENT:

  1. What are our short-range/long-range needs in every job classification?

  2. What sources can be used to attract qualified candidates for these positions?
  - Internal
  - Search firms
  - Labor unions
  - Advertising
  - State and federal employment services
  - Other (e.g., referrals)
  - Any combination of the above

  3. What interviewing and selection process should be used?
  - Application forms
  - Testing
  - Interview process
  - Assessment centers

  4. What department makes decisions?
  - Department with opening
  - Personnel
  - More than one department

  5. Will any of the following investigations be made at or before hiring?
  - Credit checks
  - Reference investigations
  - Credential verification

B. TEMPORARY AND PART-TIME EMPLOYEES:

  1. Should they be used at all?

  2. If used, what is done with the following?
  - Compensation levels
  - Benefits
  - Training
  - Supervision
  - Relations with other employees

C. COMPLIANCE WITH GOVERNMENT REGULATIONS:

  1. Are pertinent documentation and records maintained to satisfy the following?
  - Fair Labor Standards Act
  - Age Discrimination in Employment Act
  - Title VII of the Civil Rights Act of 1964
  - Executive Order No. 11246 for Government Contractors

- Public Contractors Act
- Federal Insurance Contributions Act
- Other payroll tax acts
- Employees Retirement Income Security Act of 1977 (ERISA)
- Occupational Safety and Health Administration (OSHA)

2. Is there a company policy on any of the following?
   - Discrimination
   - Affirmative action
   - Equal Employment Opportunity (EEO) complaints

D. COMPENSATION:

1. Are policies set for job evaluation and pricing?

■ Description	■ Merit increases
■ Evaluation	■ Promotional increases
■ Pay grades	■ General increases

2. Are employee benefits thought through and explained, and are costs evaluated for any of the following?
   - Typical benefits:

— Social Security	— Rest periods
— Vacation	— Counseling
— Sick leave	— Transportation
— Insurance	— Parking

   - Pension benefits

E. DEVELOPMENT AND TRAINING:

1. Are programs designed to use the human resources of the contractors in the most efficient and effective manner possible? These programs should include consideration of:
   - Organization structure
   - Training
   - On-the-job/in-house programs
   - Evaluation
   - Characteristics of the evaluation

— Planned	— Continuous
— Relevant	— Specific
— Objective	— Quantitative
— Verifiable	— Cost effective

F. PERFORMANCE APPRAISALS:

1. Are performance appraisals appropriate?
   - Include in the design the steps that make the appraisal understandable to both the supervisor and employee.
   - Clarify accountability and objections.
   - Provide analysis of the rating system.
   - Avoid evaluation pitfalls such as:

(continued)

FIG. 6-6   (continued)

- Allowing one event to prejudice evaluation
- Allowing a recent event to change general performance evaluation
- Being too lenient or too harsh in evaluation
- Letting past record affect current evaluation

2. Are employees part of evaluation system?
- Right to review and discuss
- Implementation of future goals or measures

G. LABOR RELATIONS

1. Does management understand the labor union and its procedures regarding the following?
- Work rules
- Discipline
- Grivances
- Exit interviews

2. Does contractor have procedure for negotiations with union?
- Objections—short- and long-term
- Qualitative programs
- Adequate labor law representation

H. SAFETY AND SECURITY:

1. Does the contractor have a clear OSHA program to deal with the following?
- Safety records
- Filings
- Safety awareness and training
- Injury frequency
- Accident investigations

2. Has a security program been designed that includes:

Preemployment screening	Sensitive information control
Physical security	Fire prevention
Employee identification	Investigation
Asset control	Systems backup

vidual controls can be identified as a specific application of at least one of these three general control types.

Billing

Since the bulk of revenue in the construction industry comes from transactions that are represented by billings to the owner, the system of internal control should ensure that the person(s) responsible for the preparation of invoices and billing receive accurate, up-to-date information from the job-sites. The invoices must then be accurately summarized and posted to the general ledger. Where considering the cost-benefit relationship of the controls over the billing procedures, one is well advised to remember that an invoice prepared using using inaccurate information or one not in accordance with the provisions of the contract will, at best, not be paid promptly, and, at worst, not be paid at all.

Unique Aspects. Billing practices in the construction industry differ from those found in most industries. Typically, those nonconstruction companies that provide goods and services prepare invoices upon delivery of their products. Individual contractors and companies that do jobs of short duration (1–2 months) frequently prepare bills for those jobs in a similar manner. In contrast, the construction contract requires performance over a much longer time period. Very large projects can take as long as 10 years to complete. As a result, the industry has created its own unique ways of determining the amount to be billed over the life of the contract. The methods of measuring performance and billing are normally specifically provided for in the contract, and billings are not necessarily correlated with actual progress on the job. The contractor's billing system should be flexible enough to accommodate the different requirements found in the various jobs. A contractor should not agree to billing provisions in contracts unless he is satisfied that the internal systems can provide the desired information without an unreasonable amount of extra work.

Terms of the Contract. The contractor's control system should routinely prepare billings, including retentions, that are in accordance with the terms of the original contract and any approved change orders. Billing personnel must have access to, and be familiar with, the actual contract or a proper summary thereof. The billing procedure must be designed to recognize and deal with the unique terms of each contract. If any bill is not prepared in accordance with contract terms, that item should be authorized by an appropriate level of management. Although the variety of measures of performance is great, the types of measurement fall into three general categories:

- Measures of input, such as costs incurred;
- Measures of output, such as cubic yards moved, architects' estimates of completion or milestones; and
- Negotiated payment schedules, such as those on lump-sum jobs that may be related only to the passage of time.

Progress Billing. Accurate information from the construction site regarding the amount of work performed to date is a prerequisite for preparing a progress billing. Such information, where utilized in accordance with the billing specifications in the contract, eases the collection of the amounts billed.

Retentions. A contractor's receivables usually include retentions that are not due until the contract is substantially completed or until specific contract conditions or guarantees are met.

The amount retained by the owner is usually 10 percent, but it may vary. Contracts usually provide for a reduced level of retention as contract completion approaches. The contractor should have a system of preparing billings that ensures the billing of retentions in accordance with the contract. A periodic review of billed retentions receivable aids in the timely collection of retentions.

Final Billing. The final request for payment, which includes all retentions due, must be prepared in a form specified by the owner. The contract often includes very specific requirements regarding the last bill. A separate internal accounting procedure must be in place specifically for this billing. Consequently, the provisions of the contract must be strictly complied with if the final billing is to be paid on time. This billing is usually accompanied by releases, affidavits, or some other evidence of satisfaction of all lien claimants.

Liens. Contractors who routinely file liens as billings are prepared so that lien rights are protected before they expire. Lien laws are statutory provisions giving a contractor the power to obtain an interest in the owner's property if he is not paid. Work performed for the United States Government is governed by the Federal Bond Act of 1935 (Miller Act). The statutes vary from state to state, and each state's rules differ somewhat from the Miller Act. Changes in the laws and in the court decisions interpreting them are frequent, and it is therefore imperative that contractors stay current on those lien laws.

Statutes must be strictly complied with or rights are lost. The laws require formal notice within a relatively short time after the work is completed. If payment is not received when due, the filing of suit must occur within a specified time. Thus, if payment is late, the contractor must consult legal counsel immediately. (Some states have very short time-periods during which a lien can be filed.) Otherwise, any action may be brought too late and the rights may be irrevocably lost. In selecting legal counsel, care should be taken to find one with experience in construction contractors' problems.

Information Needs. Regardless of the method of measurement used to determine job progress, the billings that are prepared must be supported by valid, complete documents. Subsequent review by management, internal or external auditors, job owners, or governmental agencies is not possible without good supporting documentations. This information includes costs incurred to date, engineers' estimates of completion, architects' certifications, and any other data judged necessary by management.

Costs to Date. Each contract cost incurred should represent progress toward completion by the owner and contractor. Such costs are the information base upon which progress billings are prepared, and the contractor's control system should be designed to provide assurance that all authorized costs related to a specific contract are charged to the job. The estimated total cost to perform a job and the cost incurred to date provide the best overall estimate of the stage which physical progress on the job is at for any point of time. Contract costs are discussed in more detail later in this chapter and in Chapter 4.

Estimated Cost to Complete. The estimated cost to complete is the most difficult calculation that arises during a job. It may also be the most valuable. The estimated cost to complete must be a valid quantification or else the total estimated cost may be materially wrong. This can result in the estimated job progress being measured improperly, incorrect progress billings, and incorrect recognition of contract earnings.

Specific Controls. The preparation of appropriate invoices requires a great amount of subjective judgment as well as accurate source information. The most reliable controls are monitoring and reperformance. On pages 64–65, the AICPA Audit and Accounting Guide, known as the *Guide*, states:

> To provide continuing assurance that desirable internal accounting control is maintained over the billing function, a contractor should have assigned personnel regularly perform monitoring procedures. A contractor might assign personnel to perform the following tasks:
>
> - Relate billings, including retentions, to the terms of the original contract and of approved change orders.
> - Accumulate and retain the data necessary to prepare and support billings, including costs incurred to date, engineers' estimates of completion, architects' certifications, and other pertinent information.

Construction Equipment

Certain contractors invest a substantial amount of capital in heavy equipment. Control of equipment costs is one of the major problem areas for these contractors. Some jobs require very little equipment and, in these cases, the cost of accounting for the equipment might exceed the benefit. However, if a contractor has substantial sums invested in equipment, equipment cost control is vital to profitable operations.

Physical Control. The first step in establishing control over equipment costs is to make certain that a complete and accurate record of each piece of equipment is made at the time of purchase. One method of keeping track of this information is to list each piece of equipment on a separate page in an equipment ledger and then to verify that the total of all the individual pages agrees with the general ledger control account. Changes made to the equipment, such as major overhauls and added attachments, are also posted to the equipment ledger. Periodically, a complete physical inventory is taken and the results are compared with the ledger to verify its completeness. Income tax considerations alone justify the cost of taking this inventory.

The control problems unique to the construction industry begin when the equipment is actually delivered. The contractor should establish physical control over the equipment by requiring an inspection and receiving a report prepared and signed by an authorized employee. This report should include a description of the condition of the equipment, as well as the model, accessories, additional equipment, and serial numbers where applicable. New equipment should be given an identifying owner's number as soon as it is delivered, so that there will be no confusion as to which items are being referred to in later stages. The inspection and receiving reports are particularly important if the equipment is being leased, since the lease agreement usually specifies that the equipment be in good working order, with certain accessories attached, and must be returned in that same condition. These inspections reduce the chance of the contractor paying for repairs or for replacement of attachments when an item is, in fact, delivered in poor condition.

Cost of Equipment. Whether the equipment used on a job is owned or leased, the contractor's objective is to generate revenue from it that will exceed the cost. To achieve this goal, a contractor must have accurate cost records to evaluate his efforts. Some contractors choose to avoid the whole issue by subcontracting any portion of a job that requires the use of heavy equipment. Accounting for equipment cost is not that complicated however. Costs should be approved and accumulated by individual piece of equipment, thus simplifying management's review of costs.

Some accountants recommend charging the actual costs of equipment, including depreciation, directly to the jobs where the equipment is being used. A better alternative may be to accumulate costs for individual pieces of equipment, summarize them in one general ledger control account, and then charge the cost to jobs at standard rates based on usage. This smooths out any differences created by accelerated depreciation that are not justified by equipment performance.

Depreciation. The most significant accounting cost of equipment is depreciation. The accurate calculation of depreciation expense may be assured by periodically testing the calculation and comparing the total of the detailed equipment ledger pages to the general ledger control account. Since depreciation is the write-off of the asset over its useful life, a policy is recommended that charges the estimated costs to expense in a manner that closely approximates the actual usage of the equipment. Accelerated depreciation may increase costs that are to be used in bids, thus causing a loss of the job.

Repair and Maintenance. A contractor who has a large investment in equipment may have his own repair and maintenance facility. The members of an in-house maintenance crew should be required to account for all of their time. The cost of repairs should be charged to specific work orders and recorded in the equipment ledger. Care must be taken to assure that idle shop-time is not charged to equipment but, rather, to a repair and maintenance overhead account. Otherwise, outside repairs should be supported by authorized repair orders for all work done. When the cost of repair and maintenance exceeds the earning power of a piece of equipment, management should realize that the time for disposition has been signaled.

Rented Equipment. Substantially all equipment rentals are governed by a rental contract. These agreements must be authorized by an appropriate level of management before a contract is signed; this reduces the possibility of unnecessary or unauthorized expenditures. The contractor should review the contract in detail, noting in particular the provisions governing rates, rental period, moving, loading and unloading, condition and repairs, insurance, and idle time. The contractor is often required to keep records to substantiate information regarding usage, repairs, insurance, and idle time—if that time is not subject to full rental rates. Therefore, a system of record-keeping should be developed that provides this information.

Charges to Contracts. Whatever method is used to control equipment costs, records must be maintained on equipment usage in much the same way that time is recorded for labor. Equipment time records are the key to controlling costs and ensuring that appropriate costs are charged to contracts. Equipment rentals, as well as costs of ownership—such as depreciation, taxes, and insurance—are often based on time. Even maintenance, repairs, and provisions for major overhauls are based on usage and the passage of time. Preprinted forms should be obtained to aid in the keeping of accurate, easily controlled time records of equipment used on each job, or each phase of a job. These records can be easily tested and controlled. One example is the daily time sheet, showing each piece of equipment on the job and the hours spent on each of the various work codes. This time sheet is prepared daily, checked against the operator's time, and then sent to the person maintaining equipment records. Another example of an equipment record is a report prepared for each item of equipment that has the operator's time card, with appropriate job codes, attached. Whichever method is used, the purpose is the same: to record costs and charges to jobs (sometimes called rentals) for each piece of equipment, thus providing management with the ability to measure which pieces are productive and which are not.

Downtime (the time not chargeable either to a specific job) or idle time (such as the time it takes to move equipment from one site to another) must also be carefully accounted for. In downtime and idle time situations, costs of equipment are often charged to corporate-wide production overhead and, thus, become part of the total overhead rate. This method of accounting for indirect equipment costs is generally employed through the use of a standard rate charge to jobs and an estimate of the expected idle time. Variances are closed to total overhead for all production during the period.

Contract Costs

On page 65, the *Guide* states that:

> A contractor uses information on contract costs to control costs, to evaluate the status and profitability of contracts, and to prepare customer billings. Thus, the importance of accurate cost information cannot be overemphasized.

In general, the control system needed to properly account for contract costs should, as stated on page 65 of the *Guide*, "ensure that all appropriate costs are recorded, that only authorized costs are incurred and that errors are promptly detected and corrected." Each type of contract cost can be controlled in a cost-effective manner; only the mechanics of control differ.

The major components of direct contract costs are:

* Materials
* Labor
* Subcontractors
* Equipment

Each of these components (except for equipment, which was discussed in the preceding section) is developed in some detail in the sections that follow.

Materials. The object of any materials accounting control system is to help management recover the full value of materials. Management cannot bill for materials that are lost, stolen, or wasted; nor can management collect for materials not included in the billings to customers. Internal accounting controls should encompass necessary physical and financial controls.

The physical controls system should first be practical. While the system should be flexible enough to fit the needs of any job situation, the procedures should also be standardized enough so that it is obvious when they are not being properly followed. The internal control system should be able to satisfy the following objectives for control over the physical assets.

- When a shipment is received, the system should determine that the order is complete and in acceptable condition.

- Once the order is received, the materials should be stored where they are safe from weather, damage, or theft.

- The system should ensure that the materials can be easily located when they are needed.

- Once materials are delivered, a written record of the delivery should be retained by the field personnel.

The paperwork of the financial control over materials is equally important. Cost records should be designed to facilitate a detailed comparison of actual costs with estimated costs. The system should provide periodic detailed cost reports to project managers and top management. Management's review of the reports should help to assure early identification of problems on contracts, provide a test check on the accuracy of the cost records, and reduce the chances of unauthorized costs being charged to a contract.

One of the best overall controls on material costs is the materials budget. When the job is estimated, a list of all the materials to be used on the job is prepared. Any changes to the list of materials must be properly authorized, and the reasons for variances should be investigated. There should be periodic rechecks of the material cost records. For example, quantities and prices charged to jobs should be compared with vendor invoices, authorized purchase orders, and specified evidence of receipt. Where appropriate, inventory withdrawals and returns should be tested.

Labor. Control over labor costs is usually established by timekeeping or work records. Timekeeping serves two functions: (1) it determines the amount of time for which an employee is to be paid, and (2) it determines which cost center (or project) is to be charged. Frequent unexpected visits to jobs by timekeepers or supervisory personnel who are expected to know by sight each person working is an excellent control over actual hours worked. This is called face checking. These face checks should be supported by written documents that record the actual time worked, such as time cards and labor reports. The cost of time worked is determined by union contracts, pay authorization, or governmental agreements. Labor is the most variable of contract costs; consequently, the comparison of expected labor time to actual time is the most reliable way to identify project problems. On larger jobs, the record of time worked should be produced daily, and the amount of work units accomplished should be computed and compared to the expected work. Substantial differences must be promptly noted. On smaller jobs, this same procedure can be done weekly or less frequently.

Subcontractors. Once the subcontract is set, control of costs associated with subcontracts is relatively straightforward. Ideally, the prime contractor arranges for the subcontract bids, analyzes the bids, negotiates the terms, awards the subcontract, makes payments to the subcontractor for the agreed-upon work, and, ultimately, obtains the appropriate release from the subcontractor.

The relationship with the subcontractor is often delicate and can seriously deteriorate if either party feels that the other is taking unfair advantage. Such situations are expensive and can be avoided for the most part if the subcontract is clear, comprehensive, and understood by both parties. A prime contractor with substantial work done by subcontractors may find that using customized subcontract forms designed by competent counsel to specifically meet the prime contractor's needs is a cost-effective control effort.

Before the contract is signed, the prime contractor should determine if the selected subcontractor is physically and financially capable of performing the necessary work with appropriate quality in a timely, competent manner. Areas that should be covered in every subcontract include the following.

- *Bonds.* Contracts should usually specify that the subcontractor furnish a performance bond and often a bid bond. The contractor's control system should routinely require an inspection of these subcontractor bonds.

- *Retentions payable.* Prime contractors commonly control the retentions payable to subcontractors with contract language identical to that in the prime contract. As a result, the subcontractor does not get paid until the prime contractor does. The subcontract should provide that the retained percentage will not be paid until the subcontractor delivers to the prime contractor a valid release, with appropriate documentation. Periodically, the contractor should require comparison of the recorded retentions payable to the actual contract provisions.

- *Liens.* A contractor (or subcontractor) must be certain to protect his rights by filing a lien as soon as it is allowable by the applicable statute. Conversely, the prime contractor must insist on releases from the subcontractor before final payment is made. The contractor's system should require checking the waivers of lien for the completed work of subcontractors. Such waivers should be organized and stored in a manner that facilitates access.

- *Backcharges.* The primary control of all the problems related to backcharges is a system stating that no undocumented or improperly authorized backcharges will be honored. Fieldwork order forms, carried by the contractor's job-site superintendent, should be utilized. The superintendent should be aware that verbal authorizations are not permitted.

Equipment. The costs (discussed in detail earlier in this chapter) should include controls assuring that all of the equipment rental costs charged to contracts are authorized and supported by proper documentation, and that they are recorded accurately. The rates should be compared to the contractor's standard rates for owned equipment or for rental contracts of leased equipment. Other equipment costs charged directly to jobs should be supported by vendors' invoices, delivery reports, repair orders, and other appropriate documents. Evidence of authorization for the cost should always be part of the system.

Allocation of Indirect Costs. If indirect costs are to be allocated to jobs, management should determine the company policy, identify which costs are allocable, and decide the basis for that allocation. Allocated costs may be accumulated by job, broken down further into bid-line items, or kept by natural cost category. In any event, the control of these costs is essentially a management function and is performed by comparison of actual to expected costs.

Reimbursable Cost on Cost-Plus Contracts. The cost-plus type contract has a unique characteristic: the owner usually has the right to audit the cost incurred on the job. The contractor must make the original documents supporting the job costs available to the owner for inspections. Therefore, the system of control must provide for easy segregation and retrieval of the appropriate documents. The contractor should consider the costs involved with the accounting system when negotiating the items that qualify for reimbursement. If the government is involved, the contractor usually finds that the cost of a job-site accounting office where the records are segregated is considered a job cost, whereas the allocation of home-office costs is not as easily approved for reimbursement. Whether the records are centralized or decentralized is not important: the system should segregate job costs and generate supporting documentation.

Revenue Recognition and Job Profitability

The type of control system that best serves a contractor's needs is determined by whether the percentage-of-completion method or the completed-contract method of accounting is used. In either case, the system used should be capable of providing accurate, timely information regarding the amount of total expected revenue per job and the timing of the progress on the job.

General. Management must determine the company policy regarding which revenue recognition method is utilized. However, for financial-reporting purposes, most contractors use the percentage-of-completion method. The completed-contract method of accounting should be used only where the job will last less than one year or where some specific condition of the contract or the circumstances would render the reliability of the estimates inherently doubtful. It may appear that estimating total revenue for a construction contract is often no more difficult than inspecting the signed contract and reading the agreed-upon price; this is sometimes true. However, construction contractors sometimes begin work before the complete scope of the work is decided. This is particularly true on jobs that have cost-plus-fee provisions. Other common reasons for revenue changes are unpriced change orders, contractor claims for work outside the scope of the original contract, and contract cancellations. Through some form of contract revenue update, the contractor's system must provide for preparation of accurate current estimates of total revenue for each job in progress.

Change Orders. The control system should be able to process and approve contract change orders and properly inform personnel whose work is affected by such change orders. Supervisors should be required to approve a price for a change order, and evidence of approval should be found on the documentation supporting the change.

Claims. Estimating revenue associated with claims and unpriced change orders is a matter for top management. Since it is nearly impossible to make an estimate of these matters with certainty, many accountants prefer to take a very conservative approach where revenue equals cost. The system should prevent any recognition of estimated revenue from claims that exceeds the estimated recoverable costs. If the contract is cost-plus, no unreimbursable costs should be allowed in estimated revenue. A review by an authorized supervisor is a practical control for these claims.

Cancellation. The system should be able to deal adequately with the cancellation of a contract. It should provide timely notification to the affected personnel and provide for the required adjustments when a cancellation or postponement occurs.

Progress Measurement. The contractor's accounting system and the attendant internal controls provide the information used to estimate the progress made toward completion of the contract. Management should select an appropriate measurement method for each job. It must be understood that any one method is not appropriate in all circumstances. Whichever method of accounting estimates is used, the contractor needs periodic independent confirmation that the completed percentage is reasonable. For example, visual inspection by qualified personnel may disclose a need for reverification. Furthermore, the owner, architect, or engineer can also serve as an independent check on progress. The chosen system should routinely and accurately accumulate the information used to calculate the progress toward completion. When the units of input are labor hours expended or costs incurred to date, the contractor should periodically require that the accuracy of this information be reverified. The person who reviews this work must realize that the use of input as a measure of progress assumes a valid relationship between input and progress.

Ideally, the estimated progress on a job would be computed using units of output. Certain jobs can be easily estimated this way, particularly in cases where the contractor is paid based on units of work completed— such as cubic yards moved or square feet installed. The job supervisor and the owner's engineer or representative routinely agree on the quantity of work done each day, and the estimate of the revenue thereby earned is less difficult to ascertain. The keys to control are input measurement, verification of consistency, and accuracy of calculations.

The accuracy of revenue computations in the interim portion of a job is dependent primarily upon accurate cost data and correct application of the estimating procedures. Given this data, an individual, chosen by management, can perform the task of calculating accurate revenue amounts quickly and consistently. This person should be rechecked so that possible errors can be detected and corrections can be made.

A good internal control system provides accurate cost data. If the basic data is reliable, the estimator of revenue to be recognized must apply the cost information properly. The percentage of costs incurred can only be meaningful if the total estimated costs to be incurred is accurate. This information should be reviewed in a timely manner by a person aware of the intricacies of the job.

The mechanics of the estimating procedures and computations must be reviewed in each period that estimates are made. The best cost data and estimates of total costs can be rendered worthless by computations containing mechanical or arithmetic errors. Only double checking can control this type of error.

Management is in the position to provide the best overall control over job revenue recognized to date and job profitability estimates by analyzing each job periodically. The project manager and the top management person should each be able to identify a potential problem situation on a job during a detailed review of the estimated progress to date, comparing that information with the original plan and with the actual progress observed at the job-site. Further, such management review should detect at an early stage those jobs that are likely to generate losses. If this is done early enough, corrective action can be taken.

Site Controls

General Applications. A company with a project large enough to justify establishing a job-site office to handle some or all of the accounting tasks associated with that job is always faced with the problem of how to control the activities of that office. The problem is particularly acute because the office is temporary and is usually staffed by people who have a limited knowledge of accounting. The general accounting office should exercise control over the job-site office, particularly if the company is not large enough to have an internal audit department to help oversee the controls. The following are the most difficult areas to control and, therefore, specific control systems may be needed for payroll to deal with them.

- Cash payouts
- Transient work force
- Imprest accounts

Field Equipment. Field equipment must be protected from physical damage, theft, and misuse. The controls necessary to keep the equipment safe are specific to each job-site. At a minimum, the delivery, usage, and move orders for each piece must be documented. Responsibility for these tasks must be delegated to responsible employees or accountability may never be established. When equipment is moved to a site, someone must sign for it. If possible, the receiving location should inspect the equipment for condition and accessories. Such inspection should be a matter of record and the delivering person should concur with the result of the inspection. The control over the use of the equipment is probably best implemented by comparing the reported usage of equipment to the operator's time sheet or label record. The usage records should be approved by a supervisor.

Disbursements. Central accounting should set specific requirements for job-site personnel regarding material purchases and other disbursements, such as travel expenses. A predetermined level of authorization must be required when purchases are made. The evidence of receipt of the goods should be required before any payments are made. Periodic visits to the job-site by corporate accounting staff to test the documentation strengthens the controls. If records of purchase are not required to be stored at the job-site for audit (as in cost-plus jobs), the review of the documentation can be performed at the corporate level at the time when funds are requested to replenish the job-site disbursement account. This account should be on an imprest basis to further strengthen control.

Field Materials and Supplies. The problems with field materials and supplies are much the same as they are with equipment. The supplies and materials must be physically safeguarded. Protection from weather, pilferage, waste, and misuse must be established. If the quantity of goods is substantial or the items are costly or attractive, a restricted-access storage area is cost justified. Every contractor is aware of a story about a job where a full semi-tractor trailer was delivered, but not unloaded, at five o'clock in the afternoon. The fully loaded truck was parked inside the fenced-in compound, right beside the watchman's trailer. In the morning, the truck was completely empty; the watchman had not heard a sound. The point is clear: if an item is valuable and useful, it must be securely locked up or it may disappear. When the job is nearing completion, the contractor must make provision for an inventory of the supplies, materials, and tools. The items must be collected and moved to central storage or to another job.

Claims, Extras, and Backcharges

The contractor should routinely control claims, extras, and backcharges. The system of control must require written authorization from the owner before any extra work is done. The common denominator of these items is that in each of them, some type of work is done in a manner other than that which was planned in the original budget. If one of the parties—whether owner, prime contractor, or subcontractor—requests a change, then one of the parties will incur extra costs. Months later, when everyone involved has forgotten the circumstances, only written documentation will prevent disagreement among the principals.

The desired controls clearly help the contractor to document authorization of extra work and allow segregation and accumulation of the related costs. The amount of revenue earned in connection with the extra work is not always known at the time when the work is done. Commonly, the change in revenue is negotiated after the work has been completed. The documentation of the related costs helps the contractor realize the income needed to recover those costs.

The authorization for the work must be written. The job supervisor must be held accountable for work done outside the scope of the contract. If a subcontractor is involved, the subcontractor should know that the prime contractor will not pay for work that is not authorized in writing. The documentation need not be fancy, formal, or complicated; a forms pad is usually sufficient for making minor changes.

Claims and extras are frequently the subject of disputes and litigation between owners and contractors. It is common for a change order to be unpriced when it is authorized, since neither the owner nor the contractor knows the value of the change. Subsequent negotiations should routinely resolve this situation, and a fair price for the work can usually be arrived at.

FOREIGN CORRUPT PRACTICES ACT

The Foreign Corrupt Practices Act (FCPA) was enacted into law in December 1977 as an amendment of the Securities Act of 1934. The FCPA requires companies that are required to file periodic reports under the 1934 Act to comply with certain accounting standards, sometimes called the books and records provisions. The FCPA also makes it unlawful to engage in certain corrupt practices involving foreign officials. Any contractor who believes the FCPA may apply to his business must make reference to the Securities Act of 1934 itself. Generally, the FCPA mandates the establishment of an internal control system. These codified accounting standards can in no way be linked to overseas business activities or to corrupt practices.

Company Policy

The provisions of the FCPA that address certain practices involving foreign officials are relevant to all companies doing business in foreign countries, whether they are required to report under the 1934 Act or not. Specifically, companies are required to have internal controls that provide reasonable assurance that illegal payments to foreign officials will be prevented or detected. The cornerstone of such a control system is a company policy statement that specifically prohibits such illegal payments.

Management Statements. Once management has adopted a policy statement prohibiting such payments, it may wish to obtain in writing, once a year, a statement from all managers of the company that confirms that the employee has not violated company policy regarding the FCPA.

Local Accounting Firms. Management may wish to obtain the assistance of local accounting firms in the foreign country where operations are in progress. This step is simplified where the operations are conducted by a foreign subsidiary that is audited by a local firm and consolidated with the parent company. Management of the lead auditors in the United States need only ask the foreign auditors to represent specifically that no violations of the FCPA came to their attention during the normal course of their audit work.

Securities and Exchange Commission and Material Weakness

A material weakness in internal accounting control is a condition in which the specific control procedures and the degree of compliance with them fails to reduce to a relatively low level the risk that errors or irregularities in material amounts (1) may occur and (2) may go undetected within a particular period by the company in the normal course of its business. A material weakness in internal accounting control is likely to be a de facto violation of the FCPA for publicly held companies covered by the Act. The FCPA, however, is written in general terms that have been labeled vague by its critics. Yet the penalties for noncompliance with the Act can be quite severe, including criminal liability if guilt is determined.

Reasonable Assurance. The FCPA uses the wording reasonable assurance where describing the sufficiency of their internal accounting controls. "Reasonable" involves a cost-benefit analysis. The FCPA may not allow the reasonableness test, although it is standard practice in the auditing profession. The result may be a standard of care higher than that created by the accounting industry.

Enforcement by the SEC. The Securities and Exchange Commission (SEC) has brought more than 20 injunctive actions in which it alleged that there were violations of the FCPA books and records provisions. There have been a handful of cases where corporate bribery has been alleged as well. The most noteworthy case to date has been *SEC v. World Wide Coin Investments, Ltd.* (567 F. Supp. 724 (1983)), in which the SEC has won the first litigated decision rendered under the accounting provisions of the FCPA. The court also noted that the company and its president had violated antifraud, reporting, proxy, and other requirements of the 1934 Act as well as the books and records provision. The SEC had charged that the company's record-keeping was "sheer chaos" and that, for a period of time, the company's records were "virtually ignored."

Many additional cases have been settled out of court. Typical of the settled cases is *SEC v. Hermitite Corp.* (14 Sec. Reg. & L. Rep. (BNA) 803), in which the former bookkeeper agreed to a permanent injunction requiring him to maintain the books of Hermitite or any other issuer in conformity with the provisions of the Act.

The Senate and the House of Representatives are again considering whether to amend the provisions of the FCPA. Amendments that would exempt certain types of payments from the general bribery provisions and end criminal penalties for publicly held companies that fail to maintain safeguards against hidden "slush funds" were passed by the Senate in 1981 and will likely be passed again in the near future. Such amendments would remove the SEC from partial responsibility for enforcement by making the Justice Department wholly responsible. The United States Chamber of Commerce and other pro-business groups support the amendments to the Act, saying that the provisions should be less severe.

SUMMARY

Internal controls are the keys by which the contractor is able to know that the plans and intentions of an organization are being carried out. These controls must be designed, implemented, and enforced in a manner that allows for the achievement of corporate goals. Both administrative and internal accounting controls must be in tune with the primary operation control in order to use company resources effectively and stay within the general guidelines of the cost-benefit principle.

Internal Auditing

Norman G. Fornella

The author wishes to acknowledge the contributions of Frank Wesley to this chapter.

BASIC OBJECTIVES

In the construction industry, as well as in other industries, the purpose of establishing an internal audit function is threefold. The three objectives are to:

- Review and develop a system to safeguard and control assets;
- Ensure contractor compliance with corporate administrative and accounting controls; and
- Determine the effective and efficient conduct of business.

System Review and Development

The first objective of the internal audit function is to review and develop a system to safeguard and control assets. The physical assets employed in construction—both fixed assets, such as construction equipment, and inventories, such as materials and supplies—represent a significant risk of loss. The very nature of construction, especially in major projects, dictates that assets be employed at a project location, which is often some distance from the contractor's offices. The assets are assigned to the location for a limited period of time. Most of the work force involved at a jobsite are craftspeople hired locally for the project rather than the contractor's long-term employees. Thus, these "people" assets are in a state of flux and a contractor rarely draws on the same employee pool. These factors mandate the need for effective systems to safeguard and control assets.

The second basic objective of the internal audit function is to ensure contractor compliance with various procedures, guidelines, and regulations. This includes clearly documented compliance with corporate written procedures; customer guidelines; state, local, and federal law; union agreements; and generally accepted accounting principles (GAAP). Statistically valid compliance tests should be used to determine that the proper methods and procedures are in place and being utilized on a consistent and effective basis.

The third objective of the internal audit function is to determine that the project manager being audited is effectively conducting his business, with optimal use of allocated resources, to produce an acceptable return on the investment. The auditor should determine if the decisions being made and actions being taken on all phases of the project are the most sensible ones and that the results of these decisions are reflected in the financial results. Although this type of approach does not mandate heavy technical expertise on behalf of the audit team, it implies that the team has a working knowledge of the construction business as well as an ability to analyze the particular project involved.

Authority and Reporting Responsibility

The authority and reporting responsibility of the auditor should be clearly stated in the contractor's organization chart. As in any business enterprise, the internal auditor must be given sufficient authority and management to meet his responsibilities. The auditor must function independently of the object of the audit. To provide this independence, the internal auditor should report directly to a person at a level in the organization outside of the accounting function, such as, the president or executive vice-president. The internal audit function should also have direct access to the audit committee of the board of directors.

The most effective implementation of the audit function requires that the auditor have extreme sensitivity and the ability to interface effectively with all parties involved in the company's financial administration. Any auditor, especially an internal auditor, should attempt to:

- Enhance the view that the audit function is both necessary and supportive; this may require making a marketing effort.
- Avoid irresponsible use of his reporting relationship, so that he can maintain effectiveness.
- Discuss the results of his audit with the project manager or department manager before issuing any reports, except in cases of fraud or misfeasance.
- Be responsible in his reporting of the audit results to management and the audit committee, noting clearly those items requiring attention, commending those items properly handled, and using care so that only those items requiring significant attention appear in the audit report.

Independent Reassurance for Owners and Lenders

The necessity for a strong internal audit function in the construction industry has grown out of the need for construction firm owners to seek independent reassurance on the safety of their investments. At one time, projects were small enough that owners could be directly involved and, thus, maintain most of the required controls themselves. However, the technological advances of the last 50 years have significantly increased the size of construction firms, thereby precluding hands-on control. The owners of the construction project along with the financial institutions providing the financing for the projects require independent reassurance.

Expanding Need for Internal Audit Function

The construction industry has become international in scope and, therefore, is now complicated by time differences, language barriers, exchange requirements, and legal requirements. A major firm may have its various activities in locations extremely far from one another. Some major contractors have even grown to the proportions of other major multinational firms. In addition, the projects themselves have become so sophisticated that engineering and design have become increasingly significant, and a number of different firms are often required to handle one project, necessitating the use of joint ventures and subcontractors.

Along with the changes in growth, sophistication, individual project size, and distances between individual projects, there has been a parallel growth in society's demand for the accountability of those responsible for the construction efforts. Both the government and the private sector have influenced the increase in needed accountability. The demand for accountability is one reason that in 1977 Congress enacted the Foreign Corrupt Practices Act (FCPA). The FCPA requires among other things, that reporting entities maintain adequate safeguards over assets and keep accurate and reliable corporate records.

AUDIT APPROACH

The internal auditor can approach the audit responsibility from a variety of directions, all of which may be combined to provide a complete audit function. These audit approaches are often classified in the following groups:

- *Contract audit.* Since construction projects are often significant in size, a particular contractor may not be working on a large number of contracts at the same time, despite a significant dollar volume. Therefore, the auditor must focus on a particular contract, taking into consideration all pertinent aspects including income recognition, direct costing, overhead, cash flow, forecasts, and profitability.

- *Functional review.* Sometimes referred to as operational auditing, this entails a detailed review of the functions involved in the company, such as purchasing, accounts payable, payroll, accounts receivable, contract accounting, proposals, planning, and scheduling.

- *Procedural review.* This audit approach entails obtaining the company's written procedures, determining the adequacy of these procedures, and then testing to be sure that the procedures are being followed in practice.

The pre-audit preparation is also discussed here. It should take place prior to field work, and includes scheduling, work-paper review, and operational/management responsibilities including project files and the organization of the project in general. A discussion follows concerning the internal accounting and administrative control reviews, the steps required for the various components of the audit, and the functional reviews covering the operating cycle of the project. Each of these operations is an important part of the audit function. The last part of this chapter discusses reports issued by the internal auditors subsequent to their field effort.

Contract Audit

All major contracts should be subject to internal audit examination one or more times during their lifespan. The audit scope, although subject to customization by the auditor, should include comprehensive substantive testing of at least the following areas:

Direct Costs. The auditor should review those direct costs specifically charged to the individual contract, including labor, materials, and any direct out-of-pocket costs. The auditor should determine that a direct audit trail exists from the contract cost and billing records back to the original payroll data. An actual payroll test may be used to determine that cost charged equals cost paid.

For materials and direct out-of-pocket costs, the audit function includes tracing the cost charged to the contract back through the accounts payable and purchasing system to the initial purchase order. Project management must review and approve the expenditures for both labor and materials, and, on certain contracts, the customer must be informed and concur with any items that vary from contracted amounts.

Overhead. The auditor should also review indirect costs to determine that the correct amount of overhead has been absorbed by the contract on a reasonable basis in accordance with the work performed. Bases for overhead allocation are explained in detail later in this chapter.

Systems. Cash flow is probably the single most important measurement of the success of a particular project. The auditor must test and review cash flow, assigning the cost of money where the cash flow is negative.

Forecasts are critical to the project and should be thoroughly reviewed by the auditor in terms of their preparation and the actual results. The auditor must compare the actual results with the estimate and obtain explanations for any significant variances. Forecast audits should include a review of man-hours, costs incurred, revenue recognition, and cash flow.

Since the auditor is concerned with the contractor's profit reporting, he should review revenue recognition, keeping in mind the importance of matching revenue with costs, and limiting, to the extent possible, the degree of uncertainty inherent in percentage-of-completion reporting on unfinished projects.

A review of bottom-line profitability ensures the auditor that revenue is recognized properly, that all direct costs have been accounted for, and that sufficient overhead and any other indirect expenses are assigned to the contract on an absorption basis. This allows the creation of a bottom-line, which should be a true representation of the particular project and a valid basis for measuring company performance.

At the conclusion of the audit work, the auditor must review the results with the managers in charge of each procedure to ensure that there have been no misunderstandings and that any variations that have been discovered have been duly noted.

Functional Audits

Functional audits involve reviewing the duties of a specific department within the company. This department should be one whose function effects all projects on which the contractor is working, such as a purchasing, personnel, or central data-processing department. The functional audit involves reviewing the procedures and methods employed in the department being examined and audit testing the relationship between this department and the other departments of the contractor. This allows the auditor to evaluate the effect of the department on the contractor's operations. Later sections of this chapter explore the specifics of particular functional departments.

Procedural Audits

Every construction firm has extensive policies and procedures, which are normally set forth in writing, describing basic policies and how various functions are performed (e.g., personnel, purchasing, estimating, etc.). In addition to audits of construction projects and functional reviews, the auditor should undertake reviews to determine the adequacy of these prescribed procedures and the extent of compliance. This will establish whether the prescribed policies and procedures are adequate and meet current operating requirements in an efficient, effective manner and whether or not they are being followed in practice.

The first objective is to determine the applicability of the written procedure to actual company performance. This is the primary test of internal controls and is usually accomplished by use of one or more of the following steps:

1 Preparation by the auditor of a systems narrative emphasizing the differences between the system described in company procedures and the actual system in use.

2 Flowcharting of the actual system and comparison of that flowchart with the procedural guidelines: this identifies any additional weaknesses created by the actual operating system.

3 Use of questionnaires to identify system weaknesses and the examination of variations between actual performance and written procedures.

After reviewing the actual procedures employed within the department or function as compared to written policies and procedures, the review may indicate one or more of the following conclusions for inclusion in the report:

• The procedures and policies set forth are effective and the review indicated substantial compliance.

• There is substantial compliance with company policies and procedures; however, opportunties for improved efficiencies were noted and are incorporated in the report for consideration by management.

• Written policies and procedures have not been revised to reflect changes in the firm's operation, such as the installation of new computers, etc.

• Actual operations are not in compliance with company policies and procedures, and recommendations relating to corrective action should be incorporated in the audit report.

The performance of procedural audits is an integral part of the internal audit function and aids in assuring compliance with policies and procedures designed to operate effectively and maximize control over the company's operations.

PRE-AUDIT PREPARATION

Planning

The internal audit schedule must be prepared well in advance to allow the auditor sufficient preparation time and to ensure that the function being audited will be ready to service the auditor during the office or field visit. The internal audit manager should prepare an audit schedule each year, soliciting input from the operating departments concerned so that he can determine what functions, projects, or departments should be audited. The audit manager can then select, on either a judgmental or a random basis, the audits to be performed. Using this procedure, he can be reasonably certain that a significant amount of the total contractor's business is viewed at some time during the year.

Having chosen the particular projects to be examined, the internal audit manager should issue a draft schedule to appropriate operating management for review and comment prior to final approval. The audit manager must then coordinate the schedule based upon the availability of his staff and the contractor's operational requirements.

Controlling

Prior to the start of an audit assignment, the internal audit staff member must be provided with adequate time to familiarize himself with the object of the audit. This process should include as a minimum the following steps:

1 Introduction of the audit staff person to the department or project management to be audited by appropriate audit or supervisory personnel familiar with the area.

2 Review by the auditor of the prior work papers or documents.

3 Coordination by means of a pre-audit conference of all the staff assigned to the audit so that the nature and purpose of the specific engagement and individual assignments are understood.

4 Preparation of a time budget for the purpose of specifying the relative importance of each of the parts of the audit.

5 Set up of current year's work-paper file.

6 Contact with the site manager or function director to allow that individual adequate time to prepare necessary documentation.

Proper preplanning utilizing work-paper review and document preparation can greatly expedite the successful completion of an audit assignment.

PROJECT OPERATIONAL REVIEW

The first major step for the auditor is to perform the operational review, which consists of three basic steps:

1 Project file review

2 Systems review

3 Profitability analysis

Project File Review

The auditor should have access to project files and should allow, in the audit time-budget, an adequate amount of time for an examination of all of the available files. It is essential that the auditor examine the contract and accompanying documents, as well as all of the correspondence files.

Reviewing the Estimating System

In the review of the estimating system, the auditor should track, using audit procedures, the estimating process from inception in the initial contract proposal to completion and final acceptance by the owner. As a start to the audit, the auditor should analyze the underlying assumptions that were made in preparation of the estimate, trace these assumptions to the total overall estimate, and determine if the estimating process has been performed in accordance with company policy and procedure. This process attempts to determine that the data used to prepare the initial estimate was included in the final estimate. If the data was not used, the auditor needs to verify that the reasons for and approvals of these variances were included in the estimate documentation.

The controls maintained over the estimating process have a direct relationship to the contractor's ability to make profits and, thus, are significant items for audit review. Positive or negative variations from budgeted amounts should be documented, and the audit review should include analysis of the impact of these variances on current and future contracts. A review of levels of approval, especially in light of relative risk and profitability on the project, is part of this process. The estimating process establishes the initial forecast on the job and, therefore, becomes the auditor's basis for measuring contract performance. The auditor should review significant changes made to estimates during the contract's life and trace them to any appropriate approvals.

Profitability Analysis

The project manager should have control over and knowledge of all costs being charged to the project. The project manager's performance should be evaluated in light of the revenues generated and his degree of control over them. The first step in profitability analysis is to review both the actual and the projected gross margin compared with the original cost projections. The auditor should then analyze the absorption of overhead costs to ascertain that they are assigned to contracts on a reasonable allocation basis. As an analysis tool, the auditor can employ financial ratio analyses, such as return on investment calculations, to determine the relative profitablilty of one project compared with industry guidelines, company expectations, or original contract projections.

For the auditor, the results of performing the project operational review should be:

- Complete understanding of the contract being examined.

- Detailed analysis of the underlying assumptions of the contract.

- Accurate monitoring of contract profitability.

- Development of a base from which a financial and accounting controls review can be initiated.

FINANCIAL AND ACCOUNTING CONTROLS REVIEW

The financial and accounting controls review is covered under the following subgroupings:

- Revenue cycle
- Expenditures cycle
- Cash
- Inventory
- Fixed assets

Each of these areas is critical to the internal audit of any construction contractor. Since each has a specific application in the construction industry, each is unusually important to the contractor. The financial and accounting controls for those items that are most important to the construction contractor are dealt with here.

Revenue Cycle

The revenue cycle is composed of the following parts:

- Contract data
- Collection information
- Revenue recording
- Billings
- Unbilled receivables

Each of the above elements of audit analysis is outlined so that the auditor would be able to collect enough information to make valid audit judgments. The auditor can use the following narrations of each of the revenue cycle areas to design worthwhile audit programs for the cycle.

Contract Data. The collecting of information on each of the contracts that were in force during the audit period is the first step in the financial and accounting controls review. The contract is the unit-of-production for the construction contractor, and all financial data naturally flows from this source. Information pertaining to all relevant details of every contract that started, continued, or ended during an accounting period should be brought together in one contract file. Each contract has its own file, and a summary file should be prepared for the key information of all of the individual files.

The components of each file vary, but the minimum requirements for any file should be:

- Basic contract information, such as owner's name and address, architect's name, phone numbers, gross contract price, original estimates, and any other data of a general nature.
- Revenue and cost data listed by current period, year-to-date, and project-to-date breakdowns.
- Billing information and revenue recognition information.

Collection Information. A review of the collection process, from prebid credit analysis to collection methods employed by the project manager or the credit manager to effect the collection of receivables, is the basis of a coordinated collection process.

An aged trial balance and the monthly accounts receivable commentaries are audit tools used by the auditor to determine if any collection problems exist on a particular project. If a problem does exist, the auditor should review for reasonableness any provisions for potential loss due to a client's financial problems or his unwillingness to pay.

If the problem is the client's unwillingness to pay, the auditor should thoroughly review the efforts of the contractor to resolve the impasse, including prospective outcomes such as work stoppage or litigation.

If there is a significant problem on collection or any other question of contract performance, the auditor may want to review the effect that this particular contract's results will have on the total financial results of the contractor.

Revenue Recording. Any analysis of the revenue cycle requires detailed investigation of the procedures used to record revenues. The analysis of this portion of the revenue cycle includes gathering information on:

- Method of accounting for this project,
- Type of contract (fixed-price, cost-plus),
- Recognition of prospective losses,
- Application of matching principal, and
- Use of appropriate completion levels in interim periods.

This partial list, amplified and expanded where necessary, provides the auditor with enough information to measure the accuracy of the recording of revenues during the reporting period.

Billings. The auditor should perform a billing test, reviewing the procedure for taking unbilled costs and applying the appropriate markups of billing terms. This conversion of costs into an invoice in compliance with the contract allows the agreed-upon amount to be billed at the correct time. The auditor should review the time lag involved between incurrence of the cost and the preparation of the billing to determine if too much time is being consumed in this process. The auditor should also determine if issuance of the bill is being delayed either by the provision that excessive amounts of detail are needed to support the bill or by the requirement that project management and/or owners approve bills prior to their issuance.

Unbilled Receivables. Since cash flow can be negatively affected by a failure to bill receivables promptly, an efficient system of billing receivables should be in place. One method of keeping track of the billing dates is a form of reverse aging of accounts receivable. This aging would list the expected billing dates for costs incurred to date and the date on which the billing will take place. The accounting department should maintain records on the status of any unbilled costs and use these records to update the reverse aging of receivables. The auditor must verify that the system is in place and that it is properly maintained.

Expenditures Cycle

A significant amount of the costs of the contract arise from purchases made outside of the company for materials and other direct costs. The auditor should perform a test of the purchasing cycle, which includes verification that:

- Purchase orders have been prepared with the necessary approval,

- Purchasing has obtained the necessary bids prior to issuance of a purchase order,

- Goods and/or materials have been received and documented in a receiving report.

- The accounting department has compared the prices charged with prices authorized.

General Expenditures Tests. As a further test of the expenditures cycle, a sample of actual cash disbursements should be examined. The review of these items should include:

- Selection of the costs to be examined taken from the contract cost records.

- Matching of the vendor invoices, purchase order, receiving report, and any other necessary documents with the amount paid or to be paid.

- Tracing contract cost records back through the documentation to the original estimate.

- Analysis of any purchase order commitment system for materials and supplies.

Specialized Expenditures Tests. The expenditure cycle contains two items that are, in many cases, the most troublesome costs to account for. These are labor costs and general overhead costs. Although the significance of these costs varies from contract to contract, an adequately functioning system that carefully allocates these costs provides the contractor with accurate and usable information.

Labor Costs. The auditor should review and test the labor accumulation and distribution methods. There should be a time card for each employee and a charge code should be assigned for each hour worked. This input document, which is the basis for allocating labor costs, should be approved by an employee's immediate supervisor. Labor charges should be accumulated by charge code and then provided to the project manager for his review and adjustment, if necessary. The document flow from the time record, through the time accumulation, and eventually to the job cost summary must be verifiable. The auditor should perform a payroll and labor test on a sampling basis using samples chosen from cost records for the individual projects being tested.

The samples, which are statistically drawn, should be examined to verify the adequacy of the records maintained. The accumulated hours must be traced back to approved time cards, and the time card distribution should be matched to the labor hours credited in the payroll register. Standard controls for the payroll test of the authorizations should include a determination that: (1) the hours worked wage rates are correct and properly authorized; (2) tax and other withholdings are correct; and (3) the net payroll disbursement is correctly made. A simple test for this last control is to examine the receipt or the endorsement on the payroll check and compare it with the signature in the employee's personnel files. In addition, in some instances it may be appropriate for the auditor to make a "payoff" and observe the distribution of wages to job-site personnel.

Social benefits or fringe benefits are normally assigned to individual labor dollars at a standard rate rather than allocated directly. The auditor's review of these indirect labor costs should include:

- Recapitulation of the method used to calculate the fringe benefit rates,
- Tracing the accumulation of all fringe benefit costs into a fringe benefit pool, and
- Verification of charges from the pool to the job.

With this data the auditor can verify the standard fringe benefit cost calculated per labor dollar (or hour) and then test its actual cost in the accounting records. An analysis of the variances from actual to standard cost should be performed and the disposition of both favorable and unfavorable variances should be traced through the records.

General Overhead Costs. Some items such as phone charges, postage, and computer charges are correctly allocable to contracts. The degree of materiality of the charges indicates whether they should be made directly or indirectly to jobs. The auditor's goal is merely to verify that the method used is both appropriate and properly executed.

Cash

An adequate internal control system for cash control is vital for any business. Accordingly, internal audit personnel spend a significant amount of the total audit time on reviews of these controls. The emphasis in this section is not on standard accounting controls and internal audit procedures over cash generally but, rather, on the internal audit procedures that should be in place to handle the special problems of field funds.

Field funds are imprest accounts used for the purpose of either satisfying payroll expenditures or providing local bank accounts for the payment of incidential expenses at the job-site. The internal audit program for imprest field funds usually includes the following steps:

1 Review of the purpose, authorization, and implementation of the imprest fund

2 Verification of deposits to the fund by an analysis of the disbursements from the general operating account.

3 Scrutiny of disbursements for authorization, signature, back-up documentation, and acceptance through payee deposit.

4 Test of the cash reconciliations.

5 In the case of imprest payroll funds, comparison of the disbursements made, often in cash against net payroll.

Inventory

Materials and supplies used in the construction business often constitute a significant portion of the total cost of the project. Materials that are not properly handled and controlled can easily be wasted. A system of controls should be able to keep track of the location, quantity, and cost of all items appropriate for inventory. The receiving process is of primary concern. Ideally, materials received should come into a central location and be signed for by an authorized person. Counts of the number of items received, comparisons with the ordered quantities and types, examinations of the items' physical condition, and preparation of appropriate documentation are component parts of the receiving function and must be analyzed by the internal audit group.

Inventory records, generally perpetual in nature, should be maintained. If the volume warrants, a computerized system can be used for this process. All movement of inventory or changes in status should be recorded into these inventory records, and internal audit should verify that these changes in inventory are done as prescribed in the company's accounting procedures and properly recorded in the cost records of the project.

Other procedures that the internal auditor should make part of the standard inventory system review include:

- Proper use of Economic Order Quantity (EOQ) for acquisition of materials and supplies

- Security system check to verify that materials are not subjected to undue risk of loss

- Periodic review of physical inventories to validate the accuracy of perpetual inventory records

- Procedures to control and account for unused materials and supplies at project completion

Fixed Assets

The capital equipment required to do construction work is often a significant expenditure. The contractor should have a capital expenditure authorization process whereby the need for the additional equipment is justified on a financial basis. The auditor should review the capital expenditure approval process for selected test items. The internal auditor's key components of concern in the fixed asset cycle are:

- Asset requisition

- Asset disposition

- Accounting for depreciation

- Preparation allocation of equipment costs to contracts

- Tax depreciation, which may include differing amounts based on varying federal and state methods and rates

- Idle time costs and allocations

- Physical inventory of fixed assets and comparisons with accounting records

The accounting for these costs is detailed in Chapters 3 and 6.

FUNCTIONAL REVIEWS

A functional review is an analysis determining whether a duty within the contracting company is being executed in line with its organizational purpose. Organization charts, department goal narratives, job descriptions, and work flow charts are the ordinary tools the auditor uses in exercising this review. The primary functional areas include:

- Contract accounting
- Proposal estimating
- Planning and scheduling
- Purchasing
- Data processing

Contract Accounting

Contract accounting, also referred to as cost engineering, is achieved by a combination of financial and engineering personnel who review the performance of the project from independent standpoints. They monitor the forecast and actual results, and, along with the project manager, explain variances from budgets. They also point out possible problem areas in the cost of the contract and check the cost commitments made on the project. Contract accounting provides independent assurance to the owner. The internal audit of this function centers around a review of the procedures used by, and the results of, contract accounting.

Proposal/Estimating

Proposals must be prepared in a manner that will avoid underbidding, which can cause losses on contract, while also avoiding overbidding, which often denies the contractor the opportunity do a particular job. Contractors should have procedures for the proposal and estimating process with a superimposed matrix concerning risk possibilities. Normally, the requirements for review and the levels of authority for sign-off are based on the degree of risk involved from a financial standpoint. The auditor should review the evolution of the proposal from the initial receipt of information for request to bid, to the final submission of the proposal, and finally to the award process. Proposals to be audited should include a sample of both accepted and rejected contracts.

Planning and Scheduling

Planning and scheduling are other important functional departments in the contractor's operation. Typically, engineers determine exactly when various aspects of the project will be scheduled, when materials should be ordered, and so on. Operations research proves useful in performing the actual scheduling process. Although the auditor may not have the necessary technical expertise, reviews to determine the general efficiency of this particular functional process are useful, as well.

Purchasing

The purchasing function can be broken down into external and internal aspects. An internal purchase is a purchase of materials and services for the company; an external purchase is the purchase of goods and services for the project, whether paid by the owner directly or paid by the contractor and billed through the billing process. The functional review of purchasing should include a testing of the company procedures that relates to the acquisition and use of internal and external purchases. The review should include compliance tests of the administrative and accounting controls of all material items in the purchasing function.

Data Processing

The construction industry, like most other businesses, has become computerized. The requirement for data from distant operations mandates the requirements for computers with telecommunication hookups. Therefore, it is important that an EDP (electronic data processing) auditor participate in reviews of the computer department, examining both the hardware and the software used and determining that the system is operated on an efficient basis with sufficient controls to meet the contractor's needs. The recruiting and training of auditors to serve as EDP specialists has become a critical factor in the overall design of the internal audit organization. If the internal audit is to achieve its goals, there must be a concentrated effort made by audit management to incorporate the need for qualified EDP specialists in departmental staffing.

There may be other functional departments within a particular contract that merit review. The auditor should work with management to analyze all of the departments and their respective functions to ascertain if other departments should be included in a functional review.

FIELD-SITE REVIEW

Although the financial accounting for a project is usually performed at the corporate headquarters or home office, a number of basic financial functions are performed at the job-site and should be subject to examination by the internal auditor. The auditor should test the imprest fund reconciliations for any field-site funds. Petty cash funds should be counted and reconciled to the imprest balance. Payroll tests, including a paycheck distribution test requiring identification from each recipient, should be carried out on a frequent, but random, basis. Purchasing and receiving, where performed at the job-site, should be subjected to functional review along with some testing of the materials control system.

Subcontractor's billings should be examined where allowed by contract. This may be done at the job-site or may require a visit to the subcontractor's office. Other factors, such as adequacy of site insurance and the control of site equipment, are part of the broad range of items the internal auditor must be prepared to include in the overall project analysis.

One of the most important reasons for the field-site visit is to physically observe the particular project being constructed. The auditor should understand generally what the contract calls for and the system used to fulfill it. A guided tour of the facility is one of the easier ways an auditor can become familiar with a specific project. Since the percentage of completion of any project is an important determining factor for the gathering of all financial information, the reported percentage of completion can be physically inspected by observation.

AUDIT REPORTS

After the completion of the field audit, the auditor should share with the project manager those preliminary findings that he anticipates will be incorporated into the report. This allows the field personnel to clarify any appropriate points and to be at least fairly well informed concerning the upcoming report. The auditor should then follow up by preparing a draft of an audit report and issuing it in draft form to the chief financial officer (CFO) for his review, comment, and clarification before the final report is circulated among other company personnel.

Subsequent to review by the CFO, the final audit report can be issued. It should begin with a description of the project and the audit work being performed and follow with audit findings and recommendations. Responses from appropriate management personnel concerning proposed corrective action should be incorporated with the recommendations. The audit report, especially if it is long, should begin with a summary highlighting the main issues contained therein.

In some cases, the internal auditor may be required to issue external reports. Most federal, state, and county regulatory agencies require a report from an independent public accountant rather than from the internal auditor. Where required, however, the external report must be reviewed so that any obligatory format changes can be made.

CHECKLISTS

Checklists are important tools in performing the audit function; they assure some measure of completeness. The checklists in Figure 7-1 are examples of checklists compiled by a major contruction contractor. They cover these topics:

- Separation of duties
- General business
- Financial systems/control
- Data processing
- The revenue cycle
- The expenditure cycle
- Labor and employee costs
- Cost control and analysis inventory
- Property, plant, and equipment
- Cash
- Marketable securities and investments
- Current and long-term debt
- Stockholders' equity
- Tax

FIG. 7-1   Checklist for Performing the Audit Function: Analysis of Separation of Duties and Control Questionnaire

PART I — ANALYSIS OF SEPARATION OF DUTIES

Name of Person or Department Performing Function

A. REVENUE CYCLE

1. Sales and Shipments:	☐	☐	☐	☐	☐	☐
a. Approves customer credit	☐	☐	☐	☐	☐	☐
b. Accounts for all order numbers	☐	☐	☐	☐	☐	☐
c. Approves sales or service orders	☐	☐	☐	☐	☐	☐
d. Accounts for all shipping orders	☐	☐	☐	☐	☐	☐
e. Checks invoices for accuracy and evidence of shipment	☐	☐	☐	☐	☐	☐
f. Prepares sales register (or similar record)	☐	☐	☐	☐	☐	☐
g. Accounts for all invoices	☐	☐	☐	☐	☐	☐
h. Reconciles invoices to recorded receivables	☐	☐	☐	☐	☐	☐
2. Cash Receipts:	☐	☐	☐	☐	☐	☐
a. Opens mail	☐	☐	☐	☐	☐	☐
3. Receivables:	☐	☐	☐	☐	☐	☐
a. Maintains detailed receivable records	☐	☐	☐	☐	☐	☐
b. Reconciles detailed records to control account	☐	☐	☐	☐	☐	☐
c. Prepares or reviews customer statements	☐	☐	☐	☐	☐	☐
d. Accounts for all credit memoranda	☐	☐	☐	☐	☐	☐
e. Approves credit memoranda	☐	☐	☐	☐	☐	☐
f. Approves discounts taken in violation of regular terms	☐	☐	☐	☐	☐	☐
g. Authorizes acceptance of notes or changes in terms	☐	☐	☐	☐	☐	☐
h. Receives returned goods	☐	☐	☐	☐	☐	☐
i. Handles disputed items	☐	☐	☐	☐	☐	☐
j. Authorizes advances to employees	☐	☐	☐	☐	☐	☐

(continued)

FIG. 7-1 (continued)

	Name of Person or Department Performing Function					

k. Maintains physical custody of notes and related collateral ☐ ☐ ☐ ☐ ☐ ☐

l. Reviews delinquent accounts ☐ ☐ ☐ ☐ ☐ ☐

m. Approves write-off of bad debts ☐ ☐ ☐ ☐ ☐ ☐

B. EXPENDITURE CYCLE

1. Purchases and Receiving: ☐ ☐ ☐ ☐ ☐ ☐

a. Accounts for all purchase orders ☐ ☐ ☐ ☐ ☐ ☐

b. Approves purchase orders ☐ ☐ ☐ ☐ ☐ ☐

c. Receives and inspects goods and prepares receiving report ☐ ☐ ☐ ☐ ☐ ☐

d. Accounts for all receiving reports ☐ ☐ ☐ ☐ ☐ ☐

e. Maintains detailed record of vendors' invoices ☐ ☐ ☐ ☐ ☐ ☐

f. Compares vendors' invoices with purchase orders, receiving reports and other data ☐ ☐ ☐ ☐ ☐ ☐

2. Payables: ☐ ☐ ☐ ☐ ☐ ☐

a. Determines accounting distribution of charges ☐ ☐ ☐ ☐ ☐ ☐

b. Approves unusual charges and credits ☐ ☐ ☐ ☐ ☐ ☐

c. Reconciles vendors' statements with reported liabilities ☐ ☐ ☐ ☐ ☐ ☐

d. Maintains detailed payable records ☐ ☐ ☐ ☐ ☐ ☐

e. Reconciles detailed payable balance with control account ☐ ☐ ☐ ☐ ☐ ☐

3. Payments: ☐ ☐ ☐ ☐ ☐ ☐

a. Approves invoices ☐ ☐ ☐ ☐ ☐ ☐

b. Prepares checks ☐ ☐ ☐ ☐ ☐ ☐

c. Signs checks and inspects supporting documents ☐ ☐ ☐ ☐ ☐ ☐

d. Mails checks ☐ ☐ ☐ ☐ ☐ ☐

e. Controls blank checks ☐ ☐ ☐ ☐ ☐ ☐

f. Maintains check register ☐ ☐ ☐ ☐ ☐ ☐

	Name of Person or Department Performing Function					
g. Reconciles bank accounts	☐	☐	☐	☐	☐	☐
h. Handles petty cash	☐	☐	☐	☐	☐	☐
i. Reviews petty cash reimbursements	☐	☐	☐	☐	☐	☐
4. Labor Cost:	☐	☐	☐	☐	☐	☐
a. Maintains personnel files	☐	☐	☐	☐	☐	☐
b. Authorizes new hires and changes in classification and pay	☐	☐	☐	☐	☐	☐
c. Compares payroll records with personnel department records	☐	☐	☐	☐	☐	☐
d. Reviews compliance with labor laws	☐	☐	☐	☐	☐	☐
e. Signs original time records	☐	☐	☐	☐	☐	☐
f. Compares time records to production schedules, etc.	☐	☐	☐	☐	☐	☐
g. Prepares payroll	☐	☐	☐	☐	☐	☐
h. Authorizes unusual overtime	☐	☐	☐	☐	☐	☐
i. Maintains individual pay records	☐	☐	☐	☐	☐	☐
j. Reconciles individual pay records with control accounts	☐	☐	☐	☐	☐	☐
k. Signs checks	☐	☐	☐	☐	☐	☐
l. Distributes checks	☐	☐	☐	☐	☐	☐
m. Maintains control over back-pay and unclaimed wages	☐	☐	☐	☐	☐	☐
n. Reconciles payroll records with tax accruals and returns	☐	☐	☐	☐	☐	☐
C. PRODUCTION OR CONVERSION CYCLES						
1. Inventory Costs:	☐	☐	☐	☐	☐	☐
a. Maintains physical control of inventory	☐	☐	☐	☐	☐	☐
b. Takes physical inventory	☐	☐	☐	☐	☐	☐
c. Maintains perpetual inventory records	☐	☐	☐	☐	☐	☐
d. Compares physical counts with perpetual records	☐	☐	☐	☐	☐	☐
e. Investigates significant differences between actual and book value	☐	☐	☐	☐	☐	☐
f. Approves adjustments to perpetual records	☐	☐	☐	☐	☐	☐

(continued)

FIG. 7-1 (continued)

Name of Person or
Department Performing
Function

g. Double checks accuracy of prices, extensions, etc.	☐	☐	☐	☐	☐	☐
h. Ties cost system into general ledger	☐	☐	☐	☐	☐	☐
i. Reviews differences between standards	☐	☐	☐	☐	☐	☐
j. Reviews slow-moving, obsolete and unsalable goods	☐	☐	☐	☐	☐	☐
2. Property and Equipment:	☐	☐	☐	☐	☐	☐
a. Maintains physical control of assets	☐	☐	☐	☐	☐	☐
b. Reviews differences between authorized expenditures and actual costs	☐	☐	☐	☐	☐	☐
c. Maintains detailed property records	☐	☐	☐	☐	☐	☐
d. Balances detailed records to control account	☐	☐	☐	☐	☐	☐
e. Takes physical inventory	☐	☐	☐	☐	☐	☐
f. Compares physical inventory with records	☐	☐	☐	☐	☐	☐
g. Approves adjustments	☐	☐	☐	☐	☐	☐
h. Determines adequacy of rates and allowances for depreciation and depletion	☐	☐	☐	☐	☐	☐

D. COMPUTER

1. Initiates transactions	☐	☐	☐	☐	☐	☐
2. Initiates master file changes	☐	☐	☐	☐	☐	☐
3. Designs systems	☐	☐	☐	☐	☐	☐
4. Writes programs	☐	☐	☐	☐	☐	☐
5. Operates computer	☐	☐	☐	☐	☐	☐
6. Approves error corrections	☐	☐	☐	☐	☐	☐
7. Reviews output records	☐	☐	☐	☐	☐	☐
8. Authorizes program changes	☐	☐	☐	☐	☐	☐

**Name of Person or
Department Performing
Function**

E. GENERAL

  1. Maintains general ledger   ☐ ☐ ☐ ☐ ☐ ☐

  2. Authorizes non-routine journal entries   ☐ ☐ ☐ ☐ ☐ ☐

  3. Compares budgets to actual operation   ☐ ☐ ☐ ☐ ☐ ☐

PART II — CONTROL QUESTIONNAIRE

Generally accepted auditing standards require that there be a proper study and evaluation of the system of internal control. The evaluation will assist the auditor in compiling and documenting accounting and operational practices through procedural write-ups and transaction flow analysis that provide a basis for determining appropriate auditing procedures.

The study and evaluation of internal accounting control encompasses:

- An overall assessment of the accounting control environment (i.e., the level of control consciousness established) and the impact of that environment on the system of internal accounting control; and
- A study and evaluation of detailed controls for each of the transaction functions such as revenue, expenditures, employee and related costs, etc.

The initial step in performing this evaluation is the completion of the Internal Control Questionnaire. Questions are phrased so that "yes" answers indicate satisfactory condition, while "no" answers suggest possible weaknesses in internal accounting and operational controls.

	Yes	No	N/A	W/P Ref.	Comment
A. GENERAL BUSINESS. Does the system of control provide assurance that					
1. Investment and corporate structures are authorized by appropriate levels of management and properly recorded on the corporate books?	☐	☐	☐	....	......

(continued)

FIG. 7-1   (continued)

	Yes	No	N/A	W/P Ref.	Comment
2. Capital expenditures and long-term lease commitments are properly authorized in accordance with the Corporate Capital Expenditure Policy?	☐	☐	☐	....	......
3. Employees are hired only when authorized and according to established policies, procedures and formal job descriptions?	☐	☐	☐	....	......
4. Employee expenses (travel, living, entertainment, etc.) comply with corporate policy and are properly authorized?	☐	☐	☐	....	......
5. Employees do not enter into arrangements or relationships, either company or personal, that would restrict free and fair competition?	☐	☐	☐	....	......
6. Illegal payments (bribes, kickbacks, etc.) are not made and that unavoidable facilitating payments are limited?	☐	☐	☐	....	......
7. The use of agents or consultants is properly authorized and that they meet legal or other applicable requirements?	☐	☐	☐	....	......
8. Corporate assets (cash, property, etc.) are accurately reflected on the corporate books, records and financial statements?	☐	☐	☐	....	......
9. Legal and ethical business practices established by the board of directors and corporate management have been observed?	☐	☐	☐	....	......

B. FINANCIAL SYSTEMS/CONTROL

	Yes	No	N/A	W/P Ref.	Comment
1. Are duties and responsibilities for all accounting functions clearly established?	☐	☐	☐	....	......
2. Is the accounting system adequate for reporting requirements?	☐	☐	☐	....	......

		Yes	No	N/A	W/P Ref.	Comment
a.	Is there a chart of accounts supplemented by definitions of items to be included in the accounts?	□	□	□	....	....
b.	Are there written instructions as to the recording of accounting transactions?	□	□	□	....	....
c.	Are all accounting records kept up to date and balanced periodically?	□	□	□	....	....
d.	Are journal entries properly supported and reviewed and approved by authorized employees?	□	□	□	....	....
e.	Are forecasts used as operational controls?	□	□	□	....	....
3.	Do procedures provide assurance that adopted policies and procedures are adhered to?	□	□	□	....	....
a.	Are accounting transactions and reports reviewed by persons who are independent of the matters being reviewed?	□	□	□	....	....
b.	Are accounting records periodically compared with records existing outside the department?	□	□	□	....	....
c.	Is there continuing vision and review to determine that					
	■ Prescribed policies are being carried out?	□	□	□	....	....
	■ Procedures are not obsolete?	□	□	□	....	....
	■ Corrective actions are promptly taken?	□	□	□	....	....
C.	DATA PROCESSING					
1.	Are programmers and analysts frequently called upon to "babysit" production runs?	□	□	□	....	....
2.	Is there a data control function associated with:					
a.	User departments?	□	□	□	....	....
b.	Data-processing department?	□	□	□	....	....

(continued)

FIG. 7-1   (continued)

	Yes	No	N/A	W/P Ref.	Comment
3. Is insurance coverage of the following types in force? What are the limits and extent of coverage?					
a. Equipment damage	☐	☐	☐	....	......
b. Program or software destruction	☐	☐	☐	....	......
c. Business interruption	☐	☐	☐	....	......
d. Fidelity on EDP personnel	☐	☐	☐	....	......
4. Are computer-generated reports issued on a timely basis according to a predetermined and published schedule?	☐	☐	☐	....	......
5. Do systems under the control of the user departments require the preparation of batch totals or at least document counts?	☐	☐	☐	....	......
6. Where computer terminals are used to enter data, does the data entry program develop control totals (by operator, session, date, etc.)?	☐	☐	☐	....	......
7. Do the user departments receive written input schedules which list earliest and latest items for submission of input?	☐	☐	☐	....	......
8. Is there a data retention program?	☐	☐	☐	....	......
9. Do the user departments play a role in the determination of which files to back up and how long they are to be retained?	☐	☐	☐	....	......
10. Are the retention requirements periodically reviewed?	☐	☐	☐	....	......
11. Describe briefly the procedures for requesting changes to an existing system or the development of a new system.	☐	☐	☐	....	......

_____

_____

_____

_____

	Yes	No	N/A	W/P Ref.	Comment

D. ORDER/BILLING/REVENUE/CREDIT

1. Does the system of control provide assurance that sales are made to acceptable credit risks? □ □ □ ....

  a. Are there well-defined, written credit policies in effect? □ □ □ ....

  b. Are sales orders for customers approved by the credit department before acceptance? □ □ □ ....

  c. Are credit limits reviewed periodically? □ □ □ ....

  d. Does the credit department advise operating units of delinquent accounts? □ □ □ ....

2. Does the system of control provide assurance that sales are authorized and properly controlled? □ □ □ ....

  a. Is a list or copy of all sales orders/contracts maintained? □ □ □ ....

  b. Are such orders under numerical control? □ □ □ ....

  c. Are unfilled sales orders or uncompleted contracts reviewed periodically? □ □ □ ....

  d. Are orders approved by authorized employees? □ □ □ ....

3. Does the system of control provide assurance that invoices are prepared and properly recorded for all shipments and progress billings? □ □ □ ....

  a. Are invoices checked for accuracy of:

    ■ Quantities? □ □ □ ....

    ■ Prices? □ □ □ ....

    ■ Extensions? □ □ □ ....

    ■ Terms? □ □ □ ....

  b. Are journals or copies of invoices maintained? □ □ □ ....

  c. Are invoices under numerical control? □ □ □ ....

  d. Are invoices issued regularly reconciled to recorded receivables? □ □ □ ....

(continued)

FIG. 7-1 (continued)

	Yes	No	N/A	W/P Ref.	Comment
e. Are progress payments invoiced according to the contract?	☐	☐	☐	....	......
f. Is escalation billed according to the contract?	☐	☐	☐	....	......
g. Are logs of progress and escalation billings maintained?	☐	☐	☐	....	......
h. Are shipments made only on approved shipping orders?	☐	☐	☐	....	......
i. Are such orders under numerical control?	☐	☐	☐	....	......
j. Are shipping orders matched to:					
■ Contract or orders?	☐	☐	☐	....	......
■ Invoices?	☐	☐	☐	....	......
4. Does the system of control provide assurance that all cash receipts are adequately controlled and accurately recorded?	☐	☐	☐	....	......
a. Are lockboxes used for cash receipts?	☐	☐	☐	....	......
b. Is a list of the currency and checks deposited in the lockbox received daily?	☐	☐	☐	....	......
c. Are lockbox locations reviewed periodically?	☐	☐	☐	....	......
d. Is the daily mail opened by a person who has no accounting responsibility?	☐	☐	☐	....	......
5. Does the system of control provide assurance that all entries to accounts receivable (control and detail) are authorized and properly recorded?	☐	☐	☐	....	......
a. Are detailed accounts and notes receivable records maintained?	☐	☐	☐	....	......
b. If so, are detailed records reconciled periodically with control accounts?	☐	☐	☐	:...	......
c. Are credit memoranda under numerical control?	☐	☐	☐	....	......
d. Is approval of an authorized employee required for:					
■ Credit memoranda?	☐	☐	☐	....	......
■ Allowances for discounts at variance with regular terms?	☐	☐	☐	....	......

	Yes	No	N/A	W/P Ref.	Comment
■ Acceptance of notes or changes in terms?	☐	☐	☐	....	......
e. Are statements regularly sent to all customers?	☐	☐	☐	....	....
f. Are such statements prepared or reviewed by someone other than the person posting the detailed accounts?	☐	☐	☐	....	......
g. Are notes and related collateral under the physical custody of the treasury department?	☐	☐	☐	....	......
6. Does the system of control provide assurance that past-due accounts are followed up effectively?	☐	☐	☐	....	......
a. Are accounts aged periodically?	☐	☐	☐	....	......
b. Are delinquent accounts reviewed periodically?	☐	☐	☐	....	......
c. Are reserves for bad debts reviewed periodically?	☐	☐	☐	....	....
d. Is approval of an authorized employee required for the write-off of bad debts?	☐	☐	☐	....	......
e. Is control exercised over bad debts after they have been written off?	☐	☐	☐	....	....
7. Does the system of control provide assurance that adequate systems and procedures exist for accurate and timely recording of revenue?	☐	☐	☐	....	......
8. Does the system of control provide assurance that miscellaneous sales, sales of tools or scrap, and sales to employees are controlled adequately?	☐	☐	☐	....	......
E. PURCHASING, PAYABLES, AND PAYMENTS					
1. Does the system of control provide assurance that purchases of goods and services are authorized?	☐	☐	☐	....	......
a. Is a list or copy of all purchase orders maintained?	☐	☐	☐	....	......
b. Are such orders under numerical control?	☐	☐	☐	....	......

(continued)

FIG. 7-1   (continued)

	Yes	No	N/A	W/P Ref.	Comment
c. Are open purchase orders reviewed periodically?	☐	☐	☐	....	......
d. Are purchase orders approved by authorized employees?	☐	☐	☐	....	......
2. Does the system of control provide assurance that goods or services received and accepted are properly recorded?	☐	☐	☐	....	......
a. Are receiving reports prepared for all goods received by the persons receiving and inspecting these goods?	☐	☐	☐	....	......
b. Are such reports under numerical control?	☐	☐	☐	....	......
c. Are unmatched receiving reports reviewed periodically?	☐	☐	☐	....	......'
d. Is a register or an invoice file of all vendor's invoices maintained?	☐	☐	☐	....	......
e. Are procedures adequate to assure proper debit for returned goods or shortages?	☐	☐	☐	....	......
f. Are vendors' invoices matched to:					
■ Purchase order?	☐	☐	☐	....	......
■ Receiving report?	☐	☐	☐	....	......
g. Are invoices checked by Accounting for:					
■ Extensions and additions?	☐	☐	☐	....	......
■ Approvals?	☐	☐	☐	....	......
h. Do procedures assure proper account distribution?	☐	☐	☐	....	......
i. Do procedures assure that invoices are approved for payment (final filed) by authorized employees?	☐	☐	☐	....	......
j. Are vendors' statements regularly reconciled to recorded payables?	☐	☐	☐	....	......
3. Does the system of control provide assurance that payments are made and properly recorded only for authorized purchases of goods or services recieved and accepted?	☐	☐	☐	....	......

	Yes	No	N/A	W/P Ref.	Comment
a. Are checks prepared only for approved invoices or check requests by designated person?	☐	☐	☐	....	......
b. Are vendor invoices effectively cancelled?	☐	☐	☐	....	......
c. Are signed checks mailed out directly without being returned to persons who requested or prepared them for payment?	☐	☐	☐	....	......
d. Are checks prenumbered and under accounting control?	☐	☐	☐	....	......
e. Is a check register or other listing of checks maintained?	☐	☐	☐	....	......
f. Is the total of the listing regularly compared to charges in the accounts payable control account?	☐	☐	☐	....	....
4. Does the system of control provide assurance that the use of consultants is properly authorized and that they meet the definition of an independent contractor for legal, tax, insurance, and other requirements?	☐	☐	☐	....	......
5. Does the system of control provide assurances that subcontractors are properly authorized, and are reviewed by appropriate corporate staffs (legal, insurance, etc.)?	☐	☐	☐	....	......
6. Does the system of control provide assurances that capital expenditures and long-term commitments are properly authorized in accordance with the corporate capital expenditure policy?	☐	☐	☐	....	......

F. LABOR AND EMPLOYEE COSTS

	Yes	No	N/A	W/P Ref.	Comment
1. Does the system of control provide assurance that employees are hired only when authorized and according to authorized rates and other established policies?	☐	☐	☐	....	......
a. Are individual personnel files maintained?	☐	☐	☐	....	......
b. Is there a formal policy in effect requiring written authorization for:					
■ New hires?	☐	☐	☐	....	......

(continued)

FIG. 7-1   (continued)

	Yes	No	N/A	W/P Ref.	Comment
■ Dismissals?	☐	☐	☐	....	......
■ Changes in rates and positions?	☐	☐	☐	....	......
c. Are there procedures to assure prompt reporting of any changes in payroll (e.g., rates, classifications, deductions, dismissals)?	☐	☐	☐	....	......
d. Are payroll records regularly compared with personnel department records?	☐	☐	☐	....	......
2. Does the system of control provide assurance that time worked by employees is authorized and properly recorded?	☐	☐	☐	....	......
a. Do foremen or department heads sign original time records?	☐	☐	☐	....	......
b. Are original time records regularly compared to production schedules, piecework schedules, payroll distribution records, etc.?	☐	☐	☐	....	......
c. Are clerical operations in preparation of payrolls cross-checked?	☐	☐	☐	....	......
d. Are duties of those preparing the payroll rotated?	☐	☐	☐	....	....
e. Is overtime authorized?	☐	☐	☐	....	....
f. Are individual pay records maintained?	☐	☐	☐	....	......
g. Are the individual records periodically reconciled to control accounts?	☐	☐	☐	....	......
3. Does the system of control provide assurance that payroll payments are made and properly recorded?	☐	☐	☐	....	......
a. Is adequate control exercised in preparing, signing, and distributing the checks?	☐	☐	☐	....	......
b. Is adequate control maintained over back-pay and unclaimed wages?	☐	☐	☐	....	......
c. Are evidences of receipt of pay regularly compared to payroll?	☐	☐	☐	....	......

	Yes	No	N/A	W/P Ref.	Comment
4. Does the system of control provide assurance of accurate and timely payment of payroll withholdings and of authorized and established employee benefit plans (e.g., pensions)?	☐	☐	☐	....	......
a. Are payroll records regularly reconciled with tax accruals and tax returns by a person independent of their preparation?	☐	☐	☐	....	......
b. Are periodic reviews made to determine that payments of employee benefits are in accordance with the provisions of the plans?	☐	☐	☐	....	......
c. Are there procedures to assure timely payment of payroll taxes?	☐	☐	☐	....	......
5. Does the system of control provide assurances that employee expenses (travel, entertainment, etc.) comply with corporate policy and are properly authorized?	☐	☐	☐	....	......
6. Does the system of control provide assurance that advances to employees are authorized and properly recorded?	☐	☐	☐	....	......

G. COST CONTROL AND ANALYSIS

1. Work-in-Process

	Yes	No	N/A	W/P Ref.	Comment
a. Does the system of control provide assurance that a useful, informative cost accumulation system exists upon which management can base decisions? Is the system adequate for determining the cost of inventories and the cost of contracts in process, and for computing the (P&L) cost of construction, products, or services?	☐	☐	☐	....	......
■ Are cost reports up to date?	☐	☐	☐	....	......
■ Is the cost system reconciled to the general ledger monthly?	☐	☐	☐	....	......
■ Does the cost report breakdown conform to the estimate?	☐	☐	☐	....	......

(continued)

FIG. 7-1   (continued)

	Yes	No	N/A	W/P Ref.	Comment
■ Are absorption rates and methods for shop burdens, operating expense, equipment usage, etc., reviewed to assure effectiveness for costing and estimating procedures?	☐	☐	☐	....	......
■ If a standard cost system is used:					
— Are standards revised periodically?	☐	☐	☐	....	......
— Are reports showing differences between standard and actual costs reviewed on a timely basis?	☐	☐	☐	....	......
b. Does the system of control provide assurance for a periodic review and analysis of variances of actual costs compared to original costs?	☐	☐	☐	....	......
■ Are actual and committed costs plus estimated cost to complete compared to original estimates?	☐	☐	☐	....	......
■ Are estimates revised for contract changes?	☐	☐	☐	....	......
■ Are estimated costs to complete including contingency reviewed periodically for reasonableness?	☐	☐	☐	....	......
■ Are actual manhour or equivalent utilization reports compared to estimate?	☐	☐	☐	....	......
■ Are equipment utilization reports reviewed?	☐	☐	☐	....	......
2. Cost of Construction, Products, or Services. Does the system of control provide assurance that the profit and loss amount for costs of construction, products, or services is properly matched with related revenues in the profit and loss accounts?	☐	☐	☐	....	......

	Yes	No	N/A	W/P Ref.	Comment
3. Operating, Selling, and Administrative Expense					
a. Does the system of control provide assurance that a useful, informative cost accumulation system exists upon which management can base decisions?	☐	☐	☐	....	......
b. Does the system of control provide assurance that a periodic review is completed of actual versus planned expenses and variances analyzed?	☐	☐	☐	....	......
H. INVENTORY					
1. Does the system of control provide assurances that accountability arising from purchases, production and ultimate disposition is adequate for determining the costs of inventories and for computing cost of construction, products or services?	☐	☐	☐	....	......
a. Are there well-established procedures in effect governing the accounting for work-In-process?	☐	☐	☐	....	......
b. Are there well-defined procedures for controlling receipts and shipments?	☐	☐	☐	....	......
c. Are there well-defined procedures for controlling withdrawals from inventory?	☐	☐	☐	....	......
d. Is merchandise on hand that is not company property (customer's merchandise, consignments-in, etc.) physically segregated?	☐	☐	☐	....	......
e. Is merchandise not on hand but owned under adequate control?	☐	☐	☐	....	......
f. Are all inventories physically counted periodically?	☐	☐	☐	....	......
g. Do physical inventory procedures assure that an accurate inventory is taken?	☐	☐	☐	....	......

(continued)

FIG. 7-1   (continued)

	Yes	No	N/A	W/P Ref.	Comment
2. Does the system of control provide assurances that the material in inventory (manufactured or purchased) is within the authorized limits approved by the appropriate levels of management?	☐	☐	☐	....	......
3. Does the system of control provide assurance that inventories are safeguarded from loss (e.g., fire, theft, deterioration)?	☐	☐	☐	....	......
a. Are physical controls over inventories sufficient to reduce losses to a reasonable minimum?	☐	☐	☐	....	......
b. Is inventory properly insured?	☐	☐	☐	....	......
4. Does the system of control provide assurance that all transactions are recorded?	☐	☐	☐	....	......
a. Are detailed perpetual records maintained?	☐	☐	☐	....	......
b. Are physical inventory instructions in writing?	☐	☐	☐	....	......
c. Are the following steps checked?					
■ Quantity determination	☐	☐	☐	....	......
■ Unit conversions	☐	☐	☐	....	......
■ Prices	☐	☐	☐	....	......
■ Extensions	☐	☐	☐	....	......
■ Additions	☐	☐	☐	....	......
■ Summarization of detailed sheets	☐	☐	☐	....	......
d. Are perpetual records compared periodically to physical counts and records adjusted?	☐	☐	☐	....	......
e. Are significant differences between perpetual records and physical inventory reported to management?	☐	☐	☐	....	......
5. Does the system of control provide assurance that prompt recognition is given to slow-moving, obsolete, unsalable and excessive quantities?	☐	☐	☐	....	......

	Yes	No	N/A	W/P Ref.	Comment
a. Are there well-established procedures for the periodic reporting of slow-moving, obsolete, and unsalable goods?	☐	☐	☐	....	(......)
b. Are slow-moving, obsolete parts returned to the manufacturer promptly?	☐	☐	☐	....	......
c. Are inventories periodically compared with sales, forecasts, etc., to determine excessive quantities?	☐	☐	☐	....	......

I. PROPERTY, PLANT AND EQUIPMENT

	Yes	No	N/A	W/P Ref.	Comment
1. Does the system of control provide assurance that Improvement Authorizations are obtained prior to making commitments for all capital expenditures and long-term leases?	☐	☐	☐	....	......
2. Does the system of control provide assurance that assets are safeguarded from loss?	☐	☐	☐	....	......
a. Are physical controls over assets sufficient to reduce losses to a reasonable minimum?	☐	☐	☐	....	......
b. Are periodic appraisals made for the purpose of establishing insurable values?	☐	☐	☐	....	......
3. Does the system of control provide assurance that the disposition of property, plant and equipment follows corporate policy and is properly recorded?	☐	☐	☐	....	......
4. Does the system of control provide assurance of accurate recording of transactions involving property, plant and equipment depreciation and related accounts?	☐	☐	☐	....	......
a. Are adequate detailed property records maintained?	☐	☐	☐	....	......
b. Are detailed records periodically balanced to control account?	☐	☐	☐	....	......
c. Are recorded assets compared with existing assets at reasonable intervals?	☐	☐	☐	....	......

(continued)

FIG. 7-1   (continued)

	Yes	No	N/A	W/P Ref.	Comment
d. Are periodic reviews made to determine the adequacy of rates and allowances for depreciation, amortization, and depletion?	☐	☐	☐	....	......

J. CASH

	Yes	No	N/A	W/P Ref.	Comment
1. Does the system of control provide assurance that cash is safeguarded from loss and that accountability is established?	☐	☐	☐	....	......
a. Are bank accounts authorized by the treasury department?	☐	☐	☐	....	......
b. Are physical safeguards over cash sufficient to reduce losses to a reasonable minimum?	☐	☐	☐	....	......
c. Are employees handling cash bonded?	☐	☐	☐	....	......
d. Is responsibility for cash assigned to a minimum number of people?	☐	☐	☐	....	......
e. Are dual signatures required on all checks?	☐	☐	☐	....	......
f. Are adequate controls exercised over the facsimile signatures?	☐	☐	☐	....	......
g. Are bank reconciliations prepared monthly by an individual not preparing or signing checks?	☐	☐	☐	....	......
h. Does the person reconciling accounts obtain bank statements directly from banks?	☐	☐	☐	....	......
i. Are all reconciling items followed-up?	☐	☐	☐	....	......
j. Are petty cash funds under the imprest system?	☐	☐	☐	....	......
k. Is the primary responsibility for each petty cash fund vested in one person?	☐	☐	☐	....	......
l. Are the funds counted periodically and reconciled with the general ledger control account?	☐	☐	☐	....	......
m. Is supporting evidence obtained for all disbursements?	☐	☐	☐	....	......

	Yes	No	N/A	W/P Ref.	Comment
n. Are fund reimbursements reviewed by an authorized employee?	☐	☐	☐	....	......
2. Does the system of control provide assurance that all cash receipts are adequately controlled and recorded?	☐	☐	☐	....	......
a. Is a listing of daily lockbox deposits received?	☐	☐	☐	....	......
b. Is the total of the listing compared with credits to receivables control account and other control accounts?	☐	☐	☐	....	......
c. Are deposit slips compared with cash entries?	☐	☐	☐	....	....
d. Are bank chargebacks investigated?	☐	☐	☐	....	......
e. Are direct receipts (at the office) deposited intact daily?	☐	☐	☐	....	....
f. Are effective controls provided over receipts from:					
■ Over-the-counter sales?	☐	☐	☐	....	....
■ Outside salesmen or collectors?	☐	☐	☐	....	......
■ Branches?	☐	☐	☐	....	......
■ Miscellaneous receipts (e.g., scrap sales, interest, rents)?	☐	☐	☐	....	......
3. Does the system of control provide assurance that all corporate bank accounts are authorized?	☐	☐	☐	....	......
a. Is a listing of bank accounts (active and inactive) maintained?	☐	☐	☐	____	......
b. Are the number of bank accounts maintained at a satisfactory minimum?	☐	☐	☐	....	......
c. Are inactive bank accounts closed?	☐	☐	☐	....	......
4. Does the system of control provide assurance that bank transfers are adequately controlled and authorized?	☐	☐	☐	....	......
a. Are bank transfers investigated to determine that both sides of the transaction have been properly recorded?	☐	☐	☐	....	......

(continued)

FIG. 7-1   (continued)

	Yes	No	N/A	W/P Ref.	Comment
K. MARKETABLE SECURITIES AND INVESTMENTS					
1. Does the system of control provide assurance that marketable securities are safeguarded from loss?	☐	☐	☐	....	......
a. Are physical controls over securities and negotiable paper sufficient?	☐	☐	☐	....	......
b. Are securities, with the exception of bearer securities, in the name of the company or designated nominees?	☐	☐	☐	....	......
c. Is the custody of securities and collateral held jointly by two designated employees?	☐	☐	☐	....	......
d. Are securities held as collateral or in safekeeping subject to adequate accounting controls?	☐	☐	☐	....	......
2. Does the system of control provide assurance of accountability?	☐	☐	☐	....	......
a. Are security transactions authorized by the appropriate levels of management?	☐	☐	☐	....	......
b. Is there a well-defined policy in effect for:					
■ Approving changes in terms of investments?	☐	☐	☐	....	......
■ Endorsing securities when required to effect transfer?	☐	☐	☐	....	......
■ Approving release of collateral?	☐	☐	☐	....	......
■ Authorizing the receipt and delivery of collateral?	☐	☐	☐	....	......
3. Does the system of control provide assurance that all transactions are recorded?	☐	☐	☐	....	......
a. Are adequate, detailed records maintained?	☐	☐	☐	....	......
b. Are detailed records regularly balanced to control account?	☐	☐	☐	....	......
c. Are securities and collateral regularly compared to information in detailed records?	☐	☐	☐	....	......

	Yes	No	N/A	W/P Ref.	Comment

d. Are differences reported to management? ☐ ☐ ☐ ....

e. Are periodic reports of changes in securities and investment accounts reviewed? ☐ ☐ ☐ ....

4. Does the system of control provide assurance that payments of interest and principal are received in accurate amounts and conform to the agreements? ☐ ☐ ☐ ....

L. CURRENT AND LONG-TERM DEBT

1. Does the system of control provide assurance that notes, bonds, and mortgages payable are properly authorized and the proceeds used for authorized purposes? ☐ ☐ ☐ ....

a. Are physical controls over unissued notes, bonds, debentures, etc., sufficient to reduce losses to a reasonable minimum? ☐ ☐ ☐ ....

b. Are borrowings authorized by the board of directors? ☐ ☐ ☐ ....

c. Do the authorizations specify:
- The officer empowered to negotiate loans? ☐ ☐ ☐ ....
- The maximum commitments such officers may make? ☐ ☐ ☐ ....
- Collateral that may be pledged to secure loans? ☐ ☐ ☐ ....

d. Are loan restrictions monitored? ☐ ☐ ☐ ....

2. Does the system of control provide assurance that transactions involving notes, bonds, and mortgages payable are recorded? ☐ ☐ ☐ ....

a. Are independent registrar agents employed? ☐ ☐ ☐ ....

b. If so, are periodic reports from agents compared to the general ledger? ☐ ☐ ☐ ....

c. If not:
- Are detailed records of long-term obligations maintained? ☐ ☐ ☐ ....

(continued)

FIG. 7-1 (continued)

	Yes	No	N/A	W/P Ref.	Comment

■ Are detailed records regularly reconciled with control account? ☐ ☐ ☐ ....

d. Is an official immediately notified of differences found in the reconciliations? ☐ ☐ ☐ ....

3. Does the system of control provide assurance that payments of principal and interest are made in accurate amounts in conformity with the agreements? ☐ ☐ ☐ ....

a. Are independent agents employed for the payment of principal and interest? ☐ ☐ ☐ ....

b. If not:

■ Is adequate control maintained to assure payments by due dates? ☐ ☐ ☐ ....

■ Is adequate control exercised in preparing, signing, and mailing checks? ☐ ☐ ☐ ....

■ Are paid bonds and notes effectively controlled? ☐ ☐ ☐ ....

M. STOCKHOLDERS' EQUITY

1. Does the system of control provide assurance that transactions involving capital accounts are properly authorized and that the proceeds of any additional capital are used for authorized purposes? ☐ ☐ ☐ ....

a. Is the issuance of stock authorized by the board of directors? ☐ ☐ ☐ ....

b. Is the use of proceeds of capital stock sales reviewed by a responsible official? ☐ ☐ ☐ ....

c. Is the valuation applied to assets received in exchange for stock authorized by the board of directors? ☐ ☐ ☐ ....

	Yes	No	N/A	W/P Ref.	Comment
2. Does the system of control provide assurance that transactions involving capital accounts are accurately recorded:					
a. Are independent registrar and transfer agents employed?	☐	☐	☐	....	......
b. If so, are periodic reports from agents compared to the general ledger?	☐	☐	☐	....	......
c. If not:					
■ Are unissued certificates and treasury stock in the custody of a responsible official?	☐	☐	☐	....	......
■ Are surrendered certificates effectively cancelled?	☐	☐	☐	....	......
■ Are detailed stockholders' records maintained?	☐	☐	☐	....	......
■ Are detailed records regularly reconciled with control account?	☐	☐	☐	....	......
d. Are all unusual entries in the equity accounts approved by a responsible official?	☐	☐	☐	....	......
e. Are periodic reviews of changes in the equity accounts made by a responsible employee?	☐	☐	☐	....	......
3. Does the system of control provide assurance that payments of dividends are made in accurate amounts in conformity with resolutions by the board of directors?	☐	☐	☐	....	......
a. Are independent dividend-paying agents employed?	☐	☐	☐	....	......
b. If not:					
■ Is adequate control exercised in preparing, signing, and mailing checks?	☐	☐	☐	....	......
■ Is adequate control maintained over unclaimed dividend checks?	☐	☐	☐	....	......
N. TAX					
1. Are there reconciliations for prepaid and accrued tax accounts?	☐	☐	☐	....	......

(continued)

FIG. 7-1 (continued)

	Yes	No	N/A	W/P Ref.	Comment
2. Are the following tax returns properly, accurately and promptly paid?					
a. Federal income tax withheld — weekly	☐	☐	☐	....	......
b. FICA withheld and employer contribution?	☐	☐	☐	....	......
c. Federal unemployment contribution — quarterly	☐	☐	☐	....	......
d. State income tax — withheld usually monthly	☐	☐	☐	....	......
e. State unemployment contribution — usually quarterly	☐	☐	☐	....	......
f. State use taxes — usually monthly	☐	☐	☐	....	......
3. Have payroll base cutoffs been properly applied?	☐	☐	☐	....	......
4. Has the federal and state motor fuel tax refund been claimed for non-highway use of fuel?	☐	☐	☐	....	......
5. Is the federal tax on oil and grease segregated?	☐	☐	☐	....	......

CHAPTER 8

External Auditing

Larry D. Jaynes

INTRODUCTION

The objective of the external auditing chapter is to assist the auditor in applying generally accepted auditing standards (GAAS) to examinations of the financial statements of companies in the construction industry. This chapter focuses on the effective performance of the audit. Chapter 6 covers internal controls, Chapter 7 covers internal auditing, and Chapter 9 covers relations with public accounting firms. Auditors often encounter a variety of complex problems in audits of construction contractors because of the nature of operations in the industry and because of the methods used in accounting for contracts.

The accounting authority for auditing comes from two sources. The first source is the American Institute of Certified Public Accountants (AICPA) Audit and Accounting Guide, "Construction Contractors," (1981) known as the *Guide*. The *Guide* superseded the AICPA industry audit guide, "Audits of Construction Contractors," published in 1965 as a combination of an accounting guide and an auditing guide that the AICPA had published separately in 1959.

The second source of information is the Statement of Position (SOP) 81-1, "Accounting for Performance of Construction-Type and Certain Production-Type Contracts," which was issued concurrently in 1981 with the *Guide*. Statements of position are issued by the Accounting Standards Division of the AICPA to influence the development of accounting and reporting standards in the direction that the division believes is in the public interest.

SCOPE OF CHAPTER

The focus in this chapter is primarily on the second and third standards of field work and on auditing procedures used in compliance with those standards, since these have the most bearing on the performance of audits. The standards are the following:

- There is to be a proper study and evaluation of the existing internal control as a basis for reliance thereon and for the determination of the resultant extent of the tests to which auditing procedures are to be restricted. (SAS No. 1, "Codification of Auditing Standards and Procedures," AU § 320.05.)

- Sufficient competent evidential matter is to be obtained through inspection, observation, inquiries, and confirmations to afford a reasonable basis for an opinion regarding the financial statements under examination. (SAS No. 31, "Evidential Matter," AU § 326.)

PLANNING THE AUDIT

Familiarity With Construction Industry

The initial step in planning the audit is to gain an understanding of the construction industry and the variety of complex problems therein because of the nature of operations and the methods used in accounting for contracts. The *Guide* should be thoroughly reviewed before planning the audit. The auditing section of the *Guide* deals primarily with auditing procedures unique to audits of construction contractors.

Understanding the Contractor's Business and Accounting System

The auditor should obtain a thorough understanding of the nature of the contractor's work, the types of contracts performed, the nature of internal accounting controls maintained by management, and the contractor's accounting system. After becoming familiar with the industry and the operations of the contractor, the auditor should review a representative sample of the contractor's outstanding contract and evaluate the contractor's internal accounting control.

Review of Outstanding Contracts. In audits of construction contractors, the primary focus is on profit centers—usually individual contracts—for recognizing revenues, accumulating costs, and measuring income. The auditor must obtain a thorough understanding of the contracts upon which the financial statements are based, and the procedures an auditor follows in an audit of a contractor should be related to those contracts. Evaluation of the profitability of contracts or profit centers is central to the total audit process and to the determination of whether the information in the financial statements is presented in conformity with generally accepted accounting principles (GAAP).

The methods and the bases of measurement used in accounting for contracts require the independent auditor to review estimated contract costs, measures of extent of progress toward completion, revenues, and gross profit to form a conclusion as to the reasonableness of costs, revenue, and gross profit allocated to the period examined. Thus, much of the independent auditor's work involves evaluating subjective estimates relating to future events, a process which involves highly technical data.

Evaluation of Internal Accounting Controls. In audits of construction contractors, the areas that should receive particular attention include a company's system of internal accounting control, operating systems and procedures, project management, nature of the contracting work, history of performance and profitability, as well as other relevant accounting and operating factors. Of these, the auditor's primary concern should be with internal accounting controls and those administrative controls that bear on the reliability of the financial statements. The auditor should consider tests of compliance with internal accounting control procedures and should perform substantive tests of contract revenues, costs, gross profit or loss, and related contract receivables and payables. In determining the extent of the tests, the auditor should consider the number and significance of individual contracts that appear to pose a high potential risk or that could otherwise be troublesome. The objectives should be to obtain an overview of contract status and to document findings.

REVIEW OF CONTRACTS

Selecting a Representative Sample

To obtain a general understanding of a contractor's operations, the auditor should review the terms of a representative sample of the contractor's contracts, including significant contracts currently in force. The sample should include both contracts with customers and contracts with subcontractors. Information that the auditor should expect to find includes the following:

- Original cost estimate and related gross profit

- Billing and retention terms

- Provisions for changes in contract prices and terms, such as escalation, cancellation, and renegotiation

- Penalty or bonus features relating to completion dates and other performance criteria

- Bonding and insurance requirements

The auditor should use this information in the preliminary review of contracts as well as in subsequent stages of the audit in connection with substantive testing procedures.

Original Cost Estimate and Gross Profit. The original estimate serves as the basis for comparison of estimated costs to complete by major job categories and may help identify overruns. Any areas with cost overruns should be investigated and inquiries made to management to determine the cause and extent of the problem. The original estimated gross profit should be compared to historical gross profit percentages earned by the company. Gross profit for the job should be tracked from the original estimated profit at the time of bid to the latest date prior to issuance of the report to determine any unfavorable trends or slippage in profit. Many times, trends develop that cause a company to be either consistently optimistic or consistently pessimistic with regard to the estimated costs to complete for all jobs.

Billing and Retention Terms. On large contracts, the retention becomes a substantial amount as the job progresses. Any abnormal delays in the payment of retention to subcontractors or the receipt of retention receivable after completion of the job should be thoroughly investigated because these situations may indicate a problem with the job or a dispute with the owner. An opinion from the company's legal counsel may be necessary in problem situations to evaluate the ultimate disposition and timing.

Provisions for Change Orders. It is important that a good system of internal control exists; this will ensure that the change orders to contracts are filed in the contract files and that they are entered into the accounting system on a timely basis. Without such a system, it is often very difficult for the accounting department to be aware of such changes.

Penalty Clauses. Penalty provisions in a contract can develop into substantial liabilities should the job extend past the agreed-upon number of work days. These terms should be thoroughly reviewed for all jobs that have experienced delays for any reason beyond the original estimated completion date. Any anticipated waiver or nonenforcement of these penalties should be confirmed in writing with the owner.

Bonding Requirements. For jobs that experience problems or delays, most large contracts require that the general contractor provide a performance bond for satisfactory completion of the job. It is often at the discretion of the general contractor whether or not a bond is required from the subcontractors. Special audit consideration should be given to any subcontractor who is experiencing delays and who is not bonded. The extent of potential loss should be discussed with management, and a review should be made to determine if the subcontractor is performing work for any of the company's other jobs.

REVIEW AND EVALUATION OF INTERNAL ACCOUNTING CONTROL

After obtaining a general understanding of the contractor's operations from a review of contracts, the auditor should study and evaluate the system of internal accounting control in order to establish a basis for reliance on the selected controls in determining the nature, extent, and timing of applicable audit tests in the examination of financial statements. Internal accounting control questionnaires, narrative descriptions, flowcharts, analysis of electronic data processing (EDP) procedures, and other techniques should be used in this phase of the audit because these techniques enable the auditor to approach the review of the system of internal accounting control in a systematic manner and provide an effective means of documentation. The auditor should also evaluate, through inquiries and observations, the extent to which the contractor's personnel are performing their assigned responsibilities in accordance with the established internal accounting control procedures and to what extent, if any, incompatible responsibilities are being performed by the same individual. Some portion of the review and evaluation of the internal control, determined by the auditor and based on the need in a particular audit, should be performed in connection with the visit to the job-sites.

If the contractor has an internal audit function, the auditor, in accordance with the provisions of the Statement on Auditing Standards (SAS) No. 9, "The Effect of an Internal Audit Function on the Scope of the Independent Auditor's Examination" (AU § 322.11), should take that into consideration as part of the evaluation of internal accounting control.

Since the audit program is designed based on the study of internal control, it may later require modification to reflect the results of the auditor's compliance tests concerning the effectiveness of the contractor's internal control. Special consideration should be given to controls relating to the purchase of materials, prices paid for such materials, documentation of receipt at the job-site, and the return or disposal of unused materials from a job.

TESTS OF COMPLIANCE

The objective of compliance testing (SAS No. 1, as amended by SAS No. 43, "Omnibus Statement on Auditing Standards," AU § 320.50) is to evaluate the degree to which the internal accounting control procedures described by management are in use and are operating as planned. Compliance tests are necessary if the auditor plans to rely on the contractor's internal accounting control in determining the nature, timing, or extent of substantive tests. Substantive tests (see SAS No. 1, AU § 320.70, as amended by SAS No. 43) consist of all the analytical review procedures and tests of details of the particular classes of transactions and balances that the auditor deems necessary in the circumstances. If the results of the compliance tests show that the contractor's internal accounting controls are not being applied as prescribed (and as indicated by the preliminary review), the auditor should modify the original audit program and consider alternative audit procedures that should be applied. Section 320.55 of SAS No. 1, as amended by SAS No. 43, states the following:

> The auditor may decide not to rely on prescribed procedures because he concludes either (a) that the procedures are not satisfactory for that purpose or (b) that the audit effort required to test compliance with the procedures to justify reliance on them in making substantive tests would exceed the reduction in effort that could be achieved by such reliance.

The discussion of tests of compliance relates only to those aspects of a contractor's system of internal accounting control that the auditor intends to rely on in determining the nature, timing, or extent of substantive tests. The preceding paragraphs outline desirable elements of internal accounting control for estimating and bidding, billings, contract costs, and contract revenues. The auditor only relies on those aspects of the internal accounting control that he finds acceptable. For these aspects, the auditor should satisfy himself, through the use of compliance tests, as to the extent to which the control procedures are being applied as prescribed. For some suggested procedures, see the discussions in Chapters 6 and 7 of test procedures relating to estimating and bidding, billings, contract costs, and contract revenues.

Compliance testing consists of examination of evidence, inspection of documents, inquiries of personnel, and observation to evaluate whether control procedures are being performed as prescribed. The auditor should select, on either a statistical or a nonstatistical basis, a sample of transactions subject to the internal accounting control upon which the auditor plans to rely. Testing these transactions requires the inspection of documents to obtain evidence of the performance of the required procedures.

Material weaknesses and deficiencies in internal accounting control, audit efficiency, or other factors may cause the auditor to conclude that he cannot rely on the controls upon which he had planned to rely. If the auditor concludes that little or no reliance can be placed on the contractor's internal accounting control, he will need to determine the current status of contracts by expanding substantive tests of supporting data, such as contract costs and revenues.

JOB-SITE VISITS

Job-site visits are essential for the auditor to understand the contractor's operations and to relate the internal accounting information to events that occur at the job-sites. All or part of the accounting function relating to a given project may be performed at a temporary job-site office staffed by a limited number of trained accounting personnel, and internal control at job-sites may therefore be weak. Observations and discussions with operating personnel at the job-site can also assist the auditor in assessing physical security, the status of projects, and the representations of management (e.g., representations about the stage of completion and estimated costs to complete). The auditor may therefore visit selected job-sites to meet the following three objectives:

1 To gain an understanding of the contractor's method of operations.

2 To review the system of internal accounting control over records maintained at job-sites, if the auditor expects to rely on the system.

3 To obtain information relating to job status and problems (if any) that may be useful in other phases of the examination.

These three objectives can usually be achieved during one visit to a job-site. To do so, however, requires careful planning so that the information to be obtained or examined can be identified before the visit. To meet the third objective, it is usually desirable, before selecting the job-sites to be visited, to consider (1) the quality of the contractor's internal accounting control; (2) the size, nature, significance, and location of projects; and (3) the projects that have unusual features or that appear to be troublesome. Unusual or troublesome contracts may include those accounted for under the percentage-of-completion method on the basis of estimates in ranges (SOP 81-1, ¶ 25(b)) or on the basis of zero-profit estimates (SOP 81-1, ¶ 25(c)), those that are combined or segmented for accounting purposes (SOP 81-1, ¶¶ 35–42), those with significant unpriced change orders or unsatisfied claims (SOP 81-1, ¶¶ 60–63, ¶¶ 65–67), and those subject to unusual risks because of factors such as location, ability to complete turnkey projects satisfactorily, postponement or cancellation provisions, or disputes between the parties.

To accomplish audit objectives, job-site visits may be made at any time during the year or at the end of the year. The auditor should base his decision on an evaluation of other factors in the audit, such as the quality of internal accounting controls; the number, size, and significance of projects; the existence of projects with unusual or troublesome features; and the method of accounting for revenue. However, if the contractor's internal accounting controls are evaluated as weak and if the contractor has in progress any large projects that individually have a material effect on the contractor's results of operations, that have unusual features, or that appear to be troublesome, the auditor should consider selecting those projects for visits at or near the year-end. In most cases, the job-site observation will be at year-end because of the materiality of the job in progress.

Performance of Additional Procedures

In addition to a review of the system of internal control and tests of the accounting records, the independent auditor should consider performing the following procedures during a job-site visit:

- Observation of uninstalled materials and work performed to date;
- Inspection of contractor-owned or rented equipment; and

- Discussions with project managers, supervisors, and other appropriate individuals, including, if possible, the independent architect, regarding the status of the contract and any significant problems.

Observation of Uninstalled Materials and Work Performed to Date. The observation of uninstalled materials and work performed to date may be particularly useful in reviewing costs incurred to date where the cost-to-cost method is used to determine the measure of progress under the percentage-of-completion method. Such information may point out the need to disregard certain costs (such as advance billings by subcontractors and costs of undelivered materials) and uninstalled materials in order to measure more accurately the work performed to date. The inclusion of these items could have a significant effect in determining the gross profit earned to date for these jobs.

Inspection of Contract-Owned or Rented Equipment. The observation and inspection of equipment on the job-site can be used to support the existence of equipment in connection with fixed asset work and to support the amount of equipment costs being allocated to the job. This audit procedure is easy to perform and may yield information vital to the financial statement presentation.

Discussions With Project Managers and Supervisors. Discussions with project managers and supervisors are useful and should be initiated during the tour of the job-site, since they allow for a natural dialogue. These discussions should include questions regarding general problems; delays because of weather, materials, funding, or any other reason; delays because of weather, materials, funding, or any other reason; changes in supervisors; performance of subcontractors; length of time major pieces of equipment have been on the job-site; and the time when the job is expected to be completed. The auditors should carry on conversations in such a manner that company employees do not feel that they are being interrogated. At the completion of a discussion, information concerning the organization and management of the job should be submitted to the general accounting office; information regarding the present status of the job and unusual matters affecting the estimated costs to complete the project should be reported. All job observations should be written in enough detail to allow someone at a later date to determine what progress had been made as of the observation date and to be able to relate that description to vendors' and subcontractors' statements.

ACCOUNTS RECEIVABLE

The general approach to the audit of a construction contractor's accounts receivable is similar to that followed in the audit of accounts receivable of industrial and commercial enterprises. The auditor confirms accounts receivable, including retentions. The confirmation should request other pertinent information, such as the contract price, payments made, and status of the contract. Figure 8-1 is a sample confirmation letter requiring positive confirmation. Negative confirmations can also be used.

The characteristics requiring special consideration of a construction contractor's accounts receivable are the following:

- Unbilled receivables
- Retentions
- Unapproved change orders, extras, and claims
- Collectibility
- Contract scope changes
- Contract guarantees and cancellation or postponement provision

Other characteristics that may require special consideration include approved but unpriced change orders and government contracts, under which the contractor proceeds with all phases of the contract even though government funding is approved piecemeal.

Unbilled Receivables

Unbilled receivables arise when revenues have been recorded but the amount cannot be billed under the terms of the contract until a later date. Specifically, such balances may represent (1) unbilled amounts arising from the use of the percentage-of-completion method of accounting; (2) incurred costs to be billed under cost-reimbursement-type contracts; or (3) amounts arising from routine lags in billing (e.g., for work completed in one month but not billed until the next month). It may not be possible to confirm these amounts as receivables directly with the customer; consequently, the auditor should apply alternative audit procedures, such as the subsequent examination of the billing and collection of the receivables and evaluation of billing information on the basis of accumulated cost data.

Retentions

The contractor's accounting records should provide for separate control for retainage, since they are generally withheld until the contract is com-

---

FIG. 8-1   Confirmation Request to Owner, General Contractor, or Other Buyer

RE: [*Describe contract*]

Gentlemen:

Our independent auditors, [*name and address*], are engaged in an examination of our financial statements. For verification purposes only, would you kindly respond directly to them about the accuracy of the following information at [*date*]:

     1. Original contract price: $_____

     2. Total approved change orders: $_____

     3. Total billings: $_____

     4. Total payments: $_____

     5. Total unpaid balance: $_____ including retentions of $_____

     6. [*Insert details of any claims, back charges, or disputes concerning this contract; attach separate sheet, if necessary.*]

     7. Estimated completion date:_____

We enclose a self-addressed, stamped envelope for your convenience in replying directly to our auditors. Your prompt response will be greatly appreciated.

Very truly yours,

_____

Enc.

The above information is: ☐ Correct
                             ☐ Incorrect (Please submit details of any differences.)

Date: _____   By: _____

[*Signature*]

_____

[*Title*]

pleted and, in cases of disputed completion, for even longer periods. They may also be subject to restrictive conditions such as fulfillment guarantees. The auditor should perform tests in order to evaluate whether retentions are recorded and subject to controls and to satisfy himself that they will be collected when due. These procedures usually include attaching the retention in accounts receivable to a copy of the pay request that is sent to the owner.

Unapproved Change Orders and Claims

Unapproved change orders and claims are often significant and recurring in the construction industry, and the auditor should give special attention to receivables arising from these sources. Paragraphs 62 and 65 of SOP 81-1 set forth the circumstances and conditions under which amounts may be recorded as revenue from unapproved change orders and claims. Because of the nature of these receivables, the auditor may encounter difficulties in evaluating whether the change order or claims are receivable at all and, if they are, whether the collectibility of the related additional revenue can be recognized. The auditor may be able to confirm the amounts of unapproved change orders or claims with customers; however, if confirmation is not possible or if the amounts are disputed, the auditor should obtain evidence to evaluate the likelihood of settlement on satisfactory terms and the collectibility of the recorded amounts. The conditions that should be met under SOP 81-1 before a receivable should be recorded require adequate evidence to allow for such an evaluation. To accomplish such an evaluation, the auditor should review the terms of the contract and should document the amounts through discussions with the contractor's legal counsel and with contractor personnel who are knowledgeable about the contract.

The auditor should evaluate accumulated costs underlying unapproved change orders and claims that are the basis for significant additional contract revenues. Some of the procedures that may be used in auditing such accumulated costs are:

- Tests of the accumulation of costs to underlying invoices, time records, and other supporting documentation. One of the methods used to test these records is confirmation of relevant data and related amounts with subcontractors and others. Where this method can be used, it is an effective test.

- Consideration of whether the work performed or costs incurred were authorized in writing by the customer. If they were not, additional contract revenues may not be billable and the costs may not be recoverable.

- Evaluation of whether the costs relate to work within or outside the scope of the contract. If the costs relate to work within the scope of a lump-sum contract, no basis for additional contract revenues may exist and the costs may not be recoverable. If the additional costs are not a part of the original contract requirements and they cannot be billed, there would be cost overruns that would decrease contract profitability.

The auditor should also evaluate the nature and reasonableness of claimed damages and their causes, such as customer-caused delays or errors in specifications. In such an evaluation, the auditor should consider the quality and extent of the documentary evidence supporting the claim and the extent to which management has pursued the claim; the auditor also should consider consultation with technical personnel. In circumstances where legal questions arise, it is appropriate to obtain an option from legal counsel (1) on the contractor's legal right to file such a claim against the customer; and (2) on the contractor's likelihood of success in pursuing the claim.

Collectibility

Although a claim may be properly supported, it may nevertheless be uncollectible. Many factors influence collectibility, including the relationship between the contractor and the customer. For example, a contractor may be less likely to press for collection of a claim from a major customer. In evaluating a claim, the auditor may consider the contractor's past experience in settling similar claims. If the contractor has demonstrated a reasonable degree of success in negotiating and settling similar types of claims and if the documentation supporting the claim under review appears to be similar in scope, depth, and content, the auditor may consider such prior experience in evaluating the collectibility of the claim.

Contract Scope Changes

Scope changes on contracts, particularly cost-plus contracts, are often not well documented. Large cost-plus contracts frequently evolve through various stages of design and planning, with numerous starts and stops on the part of both the customer and the contractor. As a result, the final scope of the contract is not always clearly defined. The auditor should carefully examine costs designated to be passed through to the customer under such contracts and should determine whether the costs are reimbursable or whether they should be absorbed by the contractor as nonreimbursable contract costs. Past history regarding pass-through costs should be examined to help determine whether the claims will be valid. If receivables arise from contract scope changes that are unapproved or disputed by customers, the auditor should be guided by the recommendations on claims from legal counsel.

Contract Guarantees and Cancellation or Postponement Provisions

Many contracts provide for contract guarantees, such as a guarantee that a power plant, when completed, will generate a specified number of kilowatt hours. A contract may specify a fixed completion date, which, if not met, may result in substantial penalties. For some contracts, retentions and their ultimate realization are related to the fulfillment of contract guarantees. A careful reading of a contract is required to identify guarantees or contingencies associated with a project. Many times, these guarantees are not specifically stated but are implied in the contract. By analysis of the contracts, the auditor should determine whether the contractor has provided adequately for the cost of fulfilling contract guarantees.

In reviewing significant contracts and subcontracts, the auditor should note that many contracts contain cancellation and postponement provisions. The construction contractor's internal procedures should provide for timely notification to subcontractors of contracts cancelled or postponed in order to minimize the possibility of litigation. All correspondence in the contract file should be reviewed to disclose any possible cancellations.

Cancelled or postponed contracts may be identified in the contractor's records or may be disclosed in other ways during the audit. For example, the auditor's confirmation procedures may disclose cancelled or postponed contracts. The auditor should then satisfy himself that the open balance of accounts receivable, which may in effect be a claim, is valid and collectible.

For a contract that has been cancelled, the auditor should evaluate the contractor's right and ability to recover costs and damages under the contract. If the amounts that the contractor seeks to recover under the contracts are in dispute, they should be evaluated as claims.

For contracts that have been postponed, the auditor should evaluate whether the estimated cost to complete is documented and reflects inflationary factors that may cause costs to increase because of the delay in the performance of the contract. The auditor should consider the reason for postponement and its ultimate implications, since a postponement could ultimately lead to a cancellation with the year-end outstanding balance. If there are indications that a customer may be unable to pay the contractor, the auditor should consider the extent to which bonding arrangements and lien rights will limit possible losses by the contractor. The auditor should investigate whether lien rights have been filed to protect the contractor's rights. Some of the information obtained in the evaluation of collectibility may be useful in the audit of amounts recognized as income on the contract.

LIABILITIES RELATED TO CONTRACTS

The auditor should satisfy himself that liabilities include not only amounts currently due but also retained percentages that apply to both subcontractors and suppliers who bill the contractor in that manner. The auditor should consider requesting confirmation of balances from specific suppliers and subcontractors. Figure 8-2 is a suggested confirmation form for subcontractors.

　　The auditor should satisfy himself that the contractor has made a proper cutoff and that all costs, including charges from subcontractors, have been recorded in the correct period. Charges by subcontractors should be in accordance with the terms of their contracts and the work performed by the subcontractor; they should not simply represent advances that may be allowable under contract terms. The amounts billable by a subcontractor under the terms of a contract represent the amount that should be recorded in accounts payable; however, the actual work performed on the job represents the amount that should be recorded as allowable costs in determining the extent of progress toward completion. Payment requests by the subcontractor should be reviewed, and retentions held by the general contractor should be tied into the payable shown on the records. In addition, the payables to subcontractors should be related to the job-site observation summaries. In reviewing liabilities, the auditor should be alerted to indications of claims and extras that may be billed by the subcontractor. Inquiries should be made of management to determine if the claims and extras are valid. The review may also disclose amounts that should be accounted for as back charges to the subcontractor under the terms of the contract. All back charge claims should be documented in the contract file.

　　All invoices for services rendered should be recorded as accounts payable, even though the amount may not be used in measuring the performance to date on the contract. The auditor should satisfy himself that

---

FIG. 8-2 Confirmation Request to Subcontractor

Gentlemen:

Our independent auditors, [*name and address*], are engaged in an examination of our financial statements. For verification purposes only, would you kindly submit directly to them the following information with respect to each [*or specific*] contract(s) in force at [*date*]:

1. Original contract price
2. Total approved change orders
3. Total billings
4. Total payments
5. Total unpaid balance, including retentions
6. Total retentions included in total balance due
7. Total amount and details of pending extras and claims in process of preparation, if any (Attach separate sheet if necessary.)
8. Estimated completion date

We enclose a self-addressed, stamped envelope for your convenience in replying directly to our auditors. Your prompt response will be greatly appreciated.

Very truly yours,

_____

Enc.

---

the contractor has not included unused amounts in measuring performance in both the cost incurred to date and the estimated cost to complete.

The older invoices and retentions included in accounts payable should be reviewed for an indication of defective work, failure on performance guarantees, or other contingencies that may not have been recorded on the contractor's records or included in the estimated cost to complete.

Under the Uniform Commercial Code (UCC), financial institutions and other creditors often file a notice of a security interest in personal property on which they have advanced credit. Notices may be filed with both the state and county in which the property is located. The auditor should consider sending UCC inquiry forms to states and counties in which the contractor has significant jobs. Such inquiries may disclose unrecorded liabilities and security interest, as defined in the UCC.

CONTRACT COSTS

The auditing of contract costs involves two primary areas: the accumulated costs to date and the estimated cost to complete. The auditor should keep in mind that in the audit of a contractor, the emphasis is on the contract and the proper recording of contract revenues and costs. The determination of the accuracy of both the cost incurred to date and the estimated cost to complete is necessary for each contract in order to determine whether the gross profit on a contract is recognized in conformity with GAAP.

Income for a contractor is determined by the ultimate profit or estimated profit on each contract and is not based on the billings to date or the cost incurred to date. Under the completed-contract method, profit recognition is deferred until the contract is substantially completed; therefore, the cost incurred to date on uncompleted contracts is not reflected in the determination of current income unless a loss on the contract is anticipated. Conversely, the percentage-of-completion method requires that projected gross profit on the contract be estimated before the gross profit for the period under examination can be determined.

Costs Incurred to Date

The auditor should satisfy himself that the contractor has properly recorded costs incurred to date by contracts and that he has included accumulated contract costs, identifiable direct and indirect costs, and an acceptable and consistent allocation of overhead to specific contracts. The auditor should also make sure that the costs have been charged to the proper job. Many times, the same vendor supplies materials for several jobs. Improperly charging job costs can prevent a bad job from showing up as quickly as it can reduce the profit on good jobs. In closely held companies, job costs are a natural area to conceal owner expenses that are paid by the company. Because indirect costs, such as supervision, are accounted for in different ways by different companies, consistency of application of accounting methods becomes the key test of these costs. For cost-plus contracts, the auditor should satisfy himself that the contractor has not recognized contract revenue based on unreimbursable contract costs. By charging unreimbursable contract costs to a job, the base costs for which the owner is paid a percentage of the profit can be increased. These costs can be determined through careful review of the contract provisions in connection with analyzing job costs. The extent of substantive testing depends on the evaluation and testing of internal control. See Chapter 6 for a detailed discussion of internal control.

Estimated Cost to Complete

One of the most important phases of the audit of a construction contractor relates to estimated costs to complete contracts in process, since that information is used in determining the estimated final gross profit or loss on contracts. Estimated costs to complete involve expectations about future performance, and the auditor should (1) critically review representations of management; (2) obtain explanations of apparent disparities between estimates and past performance on contracts, experience on other contracts, and information gained in other phases of the audit; and (3) document very carefully all information used to evaluate estimated costs to complete, including any relevant conversations with employees and management. Because of the direct effect on the interim and final gross profit or loss on the contract, the auditor should evaluate whether the contractor's estimate of cost to complete is reasonable. Due to its subjective nature, the estimate of cost to complete is one of the easiest ways for a contractor to attempt to change or adjust net income.

The auditor should consider using the following information in the review of estimated costs to complete:

- Summary of the review and evaluation of internal control, with particular emphasis on estimating and bidding; project management and control evaluation; contract costs and claims, extras, and back charges, including a summary of the results of internal audits and a discussion of the contractor's historical experience. An analysis of the historical experience of the contractor is an excellent indicator in determining the accuracy of estimated costs to complete.

- Comparison of costs incurred to date and estimated cost to complete against the original bid estimate. Explanation of any unusual variances and any changes in trends (profitable job becomes a loss job).

- Summary of work performed, to determine that actual or expected contract price and estimated costs to complete include price and quantity increases, penalties for termination or late completion, warranties or contract guarantees, and related items.

- Review of project engineers' reports and interim financial data, including reports and data issued after the balance sheet date, with explanations for unusual variances or changes in projections. Particularly important would be a review of revised or updated estimates of cost to complete and a comparison of the estimates with the actual costs incurred after the balance sheet date. The latest estimates prior to issuance of the report should always be reviewed.

- Review of information received from customers or other third parties in confirmations and conversations about disputes, contract guarantees, and so forth that could affect total contract revenue and estimated cost to complete.

- Discussions with the contractor's engineering personnel and project managers who are familiar with, and responsible for, the contract in process.

- Review of the reports of independent architects and engineers.

- Review of information received from the contractor's attorney that relates to disputes and contingencies.

Not all of these types of evidence are available for all audits of construction contracts. The auditor should consider the weight to be given to each type of evidence in forming an opinion. The auditor's objective is to test the overall reasonableness of the estimated cost to complete in the light of the information obtained from these and other available sources. Since estimated costs to complete change on a daily basis, the documentation should be the latest one available prior to issuance of the audit report. The support for these costs requires a combination of many different types of procedures and inquiries, as previously noted.

INCOME RECOGNITION

The amount and timing of income recognized from contracts depend primarily on the methods and bases used to account for those contracts. The auditor should satisfy himself that contracts are accounted for in accordance with the recommendations in SOP 81-1 and that the recommendations are applied consistently to all like contracts and in all periods. To form a reasonable opinion on the amount and timing of income recognized, the auditor should obtain an overview of costs and revenues by contract and should recognize that the audit emphasis should be on the audit of contracts.

The technological complexity or the nature of the contractor's work may require the auditor to consider using independent specialists, such as engineers, architects, and attorneys, to obtain competent evidential matter in various phases of the audit. For example, on some complex contracts, the evaluation of the percentage of completion or the estimated cost to complete may require the use of a specialist. SAS No. 11, "Using the Work of a Specialist" (AU § 336) provides guidance in this area.

Anticipated losses on contracts, including contracts on which work has not started, should be recognized in full at the earliest determinable date. In addition, the contractor should consider the need to recognize other contract costs or revenue adjustments, such as guarantees or warranties, penalties for late completion, bonuses for early completion, unreimbursable costs under cost-plus contracts, and foreseeable losses arising from terminated contracts.

Evaluating the Acceptability of Income Recognition Methods

The auditor should be guided by the recommendations in SOP 81-1 in evaluating the acceptability of a contractor's basic policy for income recognition. The audit procedures described in this chapter are closely interrelated, and together they assist the auditor in satisfying himself as to the acceptability of the method of income recognition and the bases of applying that method. The procedures include all those previously discussed (i.e., review of contracts; review and evaluation of internal accounting control procedures, particularly as regards costs and contract revenues; job-site visits; and procedures applied in the audit of receivables, liabilities related to contracts, and contract costs) and the procedures discussed in this section on income recognition.

Percentage-of-Completion Method

In evaluating the acceptability of the method used by a contractor, the auditor should satisfy himself that the contractor has followed the recommendations in paragraph 23 of SOP 81-1, which describes the percentage-of-completion method as the basic accounting policy in circumstances in which reasonably dependable estimates can be made and where the contracts generally meet these three conditions:

1   Contracts executed by the parties normally include provisions that clearly specify the enforceable rights regarding goods or services to be provided and received by the parties, the consideration to be exchanged, and the manner and terms of settlement.

2   The buyer can be expected to satisfy his obligations under the contract.

3   The contractor can be expected to perform his contractual obligations.

If contracts meet these conditions, a contractor generally is deemed able to make reasonably dependable estimates of contract revenue, contract costs, and the extent of progress toward completion. Normally, estimates in single amounts should be used as the basis of accounting for contracts under the percentage-of-completion method (SOP 81-1, ¶ 25(a)). If the use of estimates in single amounts is impractical, however, estimates based on ranges of amounts or on a break-even or zero-profit basis are acceptable under certain circumstances.

For some contracts, on which some level of profit is assured, a contractor may only be able to estimate total contract revenue and total contract cost in ranges of amounts. If the contractor can determine the amount most likely to occur within the range on a specific contract, that amount should be used in accounting for the contract under the percentage-of-completion method. If a most likely amount cannot be determined, the lowest probable level of profit in the range should be used until the actual final result can be estimated more precisely (SOP 81-1, ¶ 25(b)).

In circumstances where estimating the final outcome may be impractical except to ensure that no loss will be incurred, a contractor should use a zero estimate of profit, that is, equal amounts of revenue and cost should be recognized until a measurement of profitability can be estimated more precisely. A contractor should use this basis only if single amounts or estimates in ranges of amounts are clearly not appropriate. A change from a zero estimate of profit to a more precise estimate is accounted for as a change in accounting estimate (SOP 81-1, ¶ 25(c)). The auditor should satisfy himself that the recommendations for profit recognition as stated in the *Guide* have been applied in the proper manner.

SOP 81-1, in paragraphs 44 to 51, discusses the considerations that should underlie the selection of a method of measuring the extent of progress toward completion under the percentage-of-completion method. The auditor should evaluate the methods used by the contractor in accordance with those considerations.

In evaluating the acceptability of the percentage-of-completion method as a contractor's basic accounting policy as well as the acceptability of the basis used to measure the extent of progress toward completion, the auditor should consider these procedures:

- Reviewing a selected sample of contracts to evaluate whether the contracts meet the basic conditions in paragraph 23 of SOP 81-1 for use of the percentage-of-completion method.

- Obtaining, reviewing, and evaluating documentation of estimates of contract revenues, costs, and the extent of progress toward completion for the selected sample of contracts.

- Consulting with independent engineers or independent architects, if necessary.

- Obtaining and reviewing a representative sample of completed contracts to evaluate the quality of the contractor's original and periodic estimates of profit on those contracts.

- Obtaining a representation from management on the acceptability of the method.

If the contractor applies the percentage-of-completion method on the basis of estimates in terms of ranges or in terms of zero profit for any contracts, the auditor should obtain separate schedules for contracts accounted for on each of those bases and for contracts initially reported on those bases but changed to the normal basis during the period. For contracts in each of these categories, the auditor should consider the following procedures in evaluating the acceptability of these approaches to applying the percentage-of-completion method:

- Obtaining, for a selected sample of such contracts, documentation from management of the circumstances justifying the approaches.

- Discussing with management personnel and, if necessary, independent architects and engineers the reasonableness of the approaches used for the sample of contracts selected in each category.

- Obtaining representation from management on the circumstances justifying each of the approaches.

The auditor's objective in examining contracts accounted for by the percentage-of-completion method is to determine that the income recognized during the current period is based on the total gross profit projected for the contract on completion as well as on the work performed to date. The total gross profit expected from each contract is derived from an estimate of the final contract price less the total of contract costs to date and estimated cost to complete. The auditor tests these components in connection with the auditing procedures listed above.

Completed-Contract Method

Paragraphs 31 to 33 of SOP 81-1 recommend the completed-contract method as the basic accounting policy "in circumstances in which financial position and results of operations would not vary materially from those resulting from the use of the percentage-of-completion method."

The completed-contract method should also be used in those circumstances, described in paragraph 32 of SOP 81-1, in which estimates cannot meet the criteria of dependability for use of the percentage-of-completion method or in which inherent hazards (SOP 81-1, ¶¶ 26–29) make estimates doubtful. In evaluating whether these circumstances exist, the auditor should consider the use of procedures such as the following:

- Reviewing the nature of the contractor's contracts and the period required to perform them.

- Obtaining a schedule of uncompleted contracts at the beginning and end of the period and evaluating whether the volume is significant in relation to the volume of contracts started and completed during the period.

- Estimating the effect of reporting on the percentage-of-completion basis and evaluating whether the results would produce a material difference in financial position or results of operations.

The auditor should satisfy himself that, relative to the contract, the method selected and used by the contractor to measure progress (i.e., measures based on architectural or engineering estimates, cost-to-cost, labor hours, machine hours, or units produced) produces a reasonable measurement of the work performed to date. Information obtained from job-site visits may be particularly useful in reviewing costs incurred to date where the cost-to-cost method is used. Such information may point out the need to disregard certain costs (i.e., advance billings by subcontractors, cost of undelivered materials, or cost of uninstalled materials) in order to measure more accurately the work performed to date. Contract billings to customers may signify the percentage of completion if the contract provisions require that billings be associated with various stages of work performed on the contract.

The auditor should examine unbilled contract revenues to determine the reasons that they have not been billed. If such revenues relate to change orders or claims, the auditor should evaluate the collectibility of these change orders or claims.

The objectives of the auditor in examining contracts accounted for by the completed-contract method are to determine (1) the proper amount and accounting period for recognition of the profit from completed contracts; (2) the amount of anticipated losses, if any, on uncompleted contracts that should be recognized in the current period; and (3) the consistency in application of the method of determining completion.

The auditor should review events, contract costs, and contract billings after the accounting period is over to obtain additional assurance that all contract revenues and related costs are included in the period in which the contracts are deemed to be substantially completed for income recognition purposes. Generally, a contract may be regarded as substantially completed if remaining costs and potential risks are not significant in relation to the contract. Circumstances to be considered in determining when a project is substantially completed include, delivery of the product, acceptance by the customer, departure from the site, and compliance with performance specifications (SOP 81-1, ¶ 52).

Combining and Segmenting

Income recognition in a given period may be significantly affected by the combining or segmenting of contracts. In the course of the examination, the auditor may find that contracts have been, or should have been, combined or segmented. The auditor should, therefore, evaluate the propriety of combining contracts or, conversely, of segmenting components of a contract or a group of contracts in accordance with the criteria in SOP 81-1, paragraphs 35 to 42. In evaluating the propriety of combining a group of contracts and the propriety of segmenting a contract or a group of contracts, the auditor's major concern should be to obtain assurance that the gross profit on contracts is reported appropriately and consistently in accordance with the criteria in the SOP. The auditor should consider using these procedures:

- Reviewing combined contracts to determine whether they meet the criteria in the SOP and reviewing a representative sample of other contracts to determine if any other contracts meet those criteria.

- Reviewing contracts or groups of contracts that are reported on a segmented basis to determine whether they meet the criteria of SOP 81-1, paragraphs 39 to 42, and whether the criteria are applied consistently.

Review of Earned Revenue

For significant contracts, the auditor should obtain and review workpaper schedules that summarize contract information from the contractor's books and records together with audit data arising from the audit of contract activity. These schedules are valuable because they permit an orderly analysis of the relationship of costs and revenues on a contract-by-contract basis. Illustrations of such work-paper schedules, prepared for fixed-price contracts accounted for by the percentage-of-completion method, are presented in Figures 8-3 and 8-4. These figures are based on the assumption that the contractor determines the stage of completion and adjusts his accounts accordingly. Even so, as demonstrated by the illustrations, the schedules enable the auditor to pinpoint the need for further adjustments. Similar although less detailed schedules should also be considered for significant cost-plus contracts in process and for significant contracts closed during the period.

Analysis of Gross Profit Margins

Finally, the auditor should analyze gross profit margins on contracts and investigate and obtain explanations for contracts with unusually high or low profit margins in the light of present and past experience on similar contracts. The auditor should review the original estimates of any contracts in question and compare the results on those contracts for the current period with the results of prior periods. Other procedures to be considered include comparison of profit margins recognized on open contracts with the final results on similar closed contracts and comparison of the final profit on closed contracts with the estimated profit on those contracts in the prior year.

The auditor should maintain a summary of the historical information developed in the analysis as a reference in future examinations; this is very helpful in establishing trends.

REVIEW OF BACKLOG INFORMATION ON SIGNED CONTRACTS AND LETTERS OF INTENT

The *Guide* encourages contractors to present, in the basic financial statements, backlog information for signed contracts whose cancellation is not expected. It suggests that contractors may include additional backlog information on letters of intent and a schedule showing backlog at the beginning of the year, new contracts awarded during the year, revenue recognized during the year, and backlog at the end of the year. This information would help users of the contractor's financial statements to assess the contractor's current level of activity and prospects for maintaining that level of activity for future periods. This is one of the most important indexes used in the construction industry. Information on signed contracts whose cancellation by the parties is not expected is within the scope of an examination of the contractor's financial statements. If a contractor

XYZ COMPANY, INC.

FIXED-PRICE CONTRACTS IN PROCESS

Summary of Original and Revised Contract Estimates
As of Balance Sheet Date

Contract Identification	Original Contract Price	Original Estimate of Contract Costs	Original Estimate of Gross Profit		Net Changes in Contract Price	Revised Contract Price	Revised Estimate of Contract Costs			Revised Estimate of Gross Profit		Percentage of Completion Measured By
			Amount	Percentage			Cost to Date	Estimated Cost to Complete	Revised Total Costs	Amount	Percentage	
	(a)	(b)	(b)	(b)	(c)		(d)	(e)				(f)
A	$100,000	$ 55,000	$45,000	45.0%	$ -0-	$100,000	$ 42,000	$ 18,000	$ 60,000	$ 40,000	40%	Cost to cost
B	130,000	110,000	20,000	15.4	20,000	150,000	80,000	40,000	120,000	30,000	20	Cubic yards completed
C	175,000	125,000	50,000	28.6	25,000	200,000	125,000	75,000	200,000	—	—	Labor hours
D	250,000	200,000	50,000	20.0	150,000	400,000	270,000	330,000	600,000	(200,000)	—	Cost to cost

(a) Per original contract
(b) Per original bid
(c) Supported by change orders and/or claims meeting accounting criteria for inclusion
(d) Per audit of contract costs
(e) Per audit of estimated costs to complete
(f) Reviewed for appropriateness and consistency

FIG. 8-4 Workpaper Schedule Analyzing Contract Status

XYZ COMPANY, INC.

FIXED-PRICE CONTRACTS IN PROCESS ANALYSIS OF CONTRACT STATUS
As of Balance Sheet Date

	Per Contractor's Books and Records						Auditor's Adjustments				Adjusted Gross Profit					
	Contract Billings to Date	Costs Incurred to Date	Percentage Completed	Revenue Earned to Date	Gross Profit to Date Amount	Gross Profit to Date Percentage	Revised Percentage Completed	Revised Earned Revenue to Date	Revenue Adjustments	Provision for Projected Loss Adjustments	To Date Amount	To Date Percentage	Prior Periods Amount	Prior Periods Percentage	Current Period Amount	Current Period Percentage
Contract Identification																
(a)	(b)	(c)	(d)				(e)	(f)	(g)	(g)	(h)	(h)	(i)	(i)	(h)	(h)
(A)	$ 80,000	$ 42,000	70%	$80,000	$38,000	47.5%	70.0%	$ 70,000	$ (10,000) (j)	—	$ 28,000	40%	$20,250	45.0%	$ 7,750	31.0%
(B)	82,500	80,000	65	97,500	17,500	17.9	67.0	100,500	3,000 (k)	—	20,500	20	8,500	18.9	12,000	21.6
(C)	150,000	125,000	55	110,000	(15,000)	—	62.5	125,000	15,000 (l)	—	-0-	—	28,600	28.6	(28,600)	—
(D)	300,000	270,000	45	300,000	30,000	10.0	45.0	180,000	(120,000) (m)	$110,000 (m)	(200,000)	—	—	—	(200,000)	—

(a) Per audit of contract billings
(b) Per audit of contract costs
(c) Management's estimate of completion
(d) Per contract revenue accounts on books
(e) Per auditor—based on revenue and analysis of costs, billings, management's estimate of completion, job-site visits, etc.
(f) Result of applying revised percentage-of-completion to revised contract price
(g) Adjustments to be reviewed with and accepted by management
(h) Should be compared with prior periods and with similar contracts
(i) Per audit of prior periods
(j) Adjustment necessary to reduce recorded earned revenue and recognize excess billings
(k) Adjustment necessary to increase recorded earned revenue and recognize unbilled revenue
(l) Adjustment necessary to increase recorded earned revenue and reduce recorded excess billings in order to reflect projected "break-even" on contract. Remaining revenue ($75,000) now equals estimated costs to complete.
(m) Adjustment necessary to provide for balance of the total projected loss on contract. Remaining revenue ($220,000) now equals estimated costs to complete ($330,000) less provision for projected loss ($110,000).

elects to present backlog information on signed contracts in the basic financial statements, the auditor should review the information and evaluate its completeness in light of other audit procedures for contract receivables, contract-related liabilities, contract costs, and contract revenues. In order to satisfy himself as to the accuracy of the backlog information, the auditor should consider obtaining a schedule of all uncompleted signed contracts showing, for each contract, total estimated revenue, total estimated cost, earned revenue to date, costs incurred to date, and cost of earned revenue.

If a contractor elects to present backlog information for both signed contracts and letters of intent, the auditor's responsibility for the information is less clear because letters of intent are not normally within the scope of the examination of a contractor's financial statements. The auditor may, however, be able to satisfy himself regarding the completeness and reliability of the information from letters of intent. If letters of intent are to be included in backlog information, then the auditor should obtain a schedule of the executed letters of intent, confirm them with customers, and review their terms with the contractor's legal counsel.

Information from signed contracts whose cancellation by the parties is not expected is within the scope of an examination of the contractor's financial statements. If a contractor elects to present backlog information from signed contracts in the basic financial statements, the auditor should review the information and evaluate its completeness in light of other audit procedures. The presentation of backlog information by a contractor is desirable only if a reasonably dependable determination both of total revenue and of total cost under signed contracts or letters of intent can be made. The information from signed contracts should be segregated from the information from letters of intent.

CLIENT REPRESENTATIONS

The auditor should obtain written representations from management in accordance with the requirements of SAS No. 19, "Client Representations" (AU § 333). Some of the matters on which the auditor should consider obtaining written representations in the examination of the financial statements of a contractor include:

- Method of income recognition used
- Provisions for losses on contracts
- Unapproved change orders, claims, and contract postponements or cancellations

- Backlog information if presented in the financial statements
- Joint venture participations and other related party transactions
- Estimated costs to complete
- Collectibility of accounts receivable for disputed contracts

In addition to the foregoing items, the auditor should consider obtaining client representations on all the types of matters suggested in SAS No. 19 that are relevant to the engagement.

AFFILIATED ENTITIES

In the construction industry, contractors frequently participate in joint ventures or have a direct or indirect affiliation with other entities. As a consequence, they are frequently involved in related party transactions (as the term "related parties" is defined in SAS No. 6, "Related Party Transactions," AU § 335). The prevalence of such arrangements in the industry can be attributed to factors such as legal liability, taxation, competition, ownership and operating arrangements, labor and labor union considerations, and regulatory requirements. Auditing and reporting considerations appropriate under these circumstances are discussed in this section.

Participation in Joint Ventures and Partnerships

The auditor should review a contractor's participations in joint ventures to evaluate whether investments in joint ventures are reported in accordance with the recommendations in Chapter 2 of this book. The auditor should consider some of the following factors:

- Method or methods of reporting joint venture investments
- Nature of capital contributions and the method of recording capital contributions
- Timeliness of the available financial statements of joint ventures in relation to those of the reporting investor
- Appropriateness of the accounting for those joint venture losses that exceed a contractor's loans and investments
- Adequacy of joint-venture-related disclosures in the contractor's financial statements

The auditor should review joint venture agreements and document a contractor's participation. For corporate joint ventures, the documentation should consist of the following information:

- Capital contributions and funding requirements of the venture participants.
- Ownership percentages
- Profit or loss participation ratios
- Duration of the venture
- Performance requirements of the venture participants

The audit considerations for a contractor's participation in a partnership (e.g., in a real estate tax shelter partnership) are similar to those for corporate joint ventures. They may differ only in relation to the contractor's unlimited liability as a general partner for partnership obligations.

For partnership interests, the auditor's documentation should contain information such as the following:

- Extent and nature of fees and other amounts to be paid by the partnership to the contractor and the conditions and events that would require such payments
- Contractor's obligations to the partnership for capital contributions and other funding
- Contractor's performance and other requirements of the contractor as a general partner and specified penalties for nonperformance, if any
- Profit participation ratios of the partners and events or conditions that change such ratios
- Duration of the partnership

The auditor should assess the economic and tax incentives underlying the creation of the partnership, the events requiring capital contribution installments by limited partners, and the temporary and permanent financing arrangements and related costs. The auditor should also assess the extent of actual and contingent obligations that arise from the contractor's role as a general partner. To that end, the auditor should review the financial condition of the other general partners and their ability to participate in the funding of required capital contributions, partnership obligations, and partnership losses, if any. The inability of other general partners to provide their share of such funding may require the contractor to recognize additional obligations based on the contractor's legal liability as a general partner for all partnership obligations.

For any type of venture, the auditor should consider the nature of the venture, the scope of its operations, and the extent of involvement of each participant, obtaining financial statements of the venture entity for the period under examination. If the financial statements of the venture are examined by another auditor, the principal auditor should be guided by the provisions of AU section 543 of SAS No. 1. If the venture's financial statements are unaudited, reviewed, or compiled, the principal auditor should perform such procedures as he deems necessary in the circumstances. In selecting such procedures, the auditor should be guided by the provisions of AU section 332 of SAS No. 1, which furnishes guidance in applying GAAS to examinations of the financial statements of companies with long-term investments accounted for under either the cost method or the equity method. If, for any reason, the procedures required by GAAS cannot be performed, the auditor should evaluate the effect of this scope limitation on the opinion to be expressed on the contractor's financial statements.

Auditing Affiliated Companies and Related Party Transactions

An auditor engaged to examine one of a group of affiliated companies that comprises an economic unit may find that an examination of the records of that entity may not satisfy him in regard to such aspects as substance, nature, business purpose, and transfer prices of significant transactions among the parties. In identifying and reporting on related party transactions, the auditor should be guided by SAS No. 6, "Related Party Transactions," AU section 335, which states in paragraph 2 that:

> the term 'Related Parties' means the reporting entity; its affiliates; principal owners, management and members of their immediate families; entities for which investments are accounted for by the equity method; and any other party with which the reporting entity may deal when one party has the ability to significantly influence the management or operating policies of the other, to the extent that one of the transacting parties might be prevented from fully pursuing its own separate interest.

Related parties also exist where another entity has (1) the ability to significantly influence the management or operating policies of the transacting parties or (2) has an ownership interest in one of the transacting parties and the ability to significantly influence the other, to the extent that one or more of the transacting parties might be prevented from fully pursuing its own separate interests.

The auditor should give special consideration to the following areas:

- Allocation of overhead between entities such as office space and management salaries

- Basis for determining equipment rental rates

- Criteria for determining which entity performs the work under the contract

- Elimination of intercompany profits for items such as prefabricated materials and asphalt

In all of the above situations, transactions should be examined between entities, and consistency with prior years should be determined. The auditor's duty is to be sure that the proper disclosure is made.

CAPITALIZATION AND CASH FLOW

Contractors often follow practices that accelerate the cash collections to be generated from a contract. These practices include the use of unbalanced bids and other front-end loading procedures that allocate a relatively larger portion of the total contract price to phases of work likely to be completed in early stages of the contract than to phases likely to be completed later. Also, overestimating the percentage of work completed in computing billings on contracts may have a similar effect on cash collections.

If a contractor incurs substantial losses on contracts that have been front-end loaded, he may experience a cash deficiency toward the end of those contracts. The deficiency may prevent the contractor from meeting current obligations, and the contractor may use the front-end load on new contracts to generate the funds necessary to meet obligations on loss jobs. The necessity to generate cash may cause the contractor to accept jobs that are only marginally profitable.

Therefore, the auditor should review uncompleted contracts not only to assess the adequacy of provisions for losses in the current period but also to determine the effect of projected cash receipts and payments on the contractor's cash position and ability to meet current obligations. This should be reviewed very closely in connection with the analysis of gross margins referred to in the audit of the percentage-of-completion method.

TYPES OF AUDITOR'S REPORTS

In most situations the auditor, through normal or extended audit procedures, can obtain sufficient competent evidence to issue an unqualified
opinion that a contractor's accounts are presented in accordance with
GAAP. Normally, the auditor should apply extended audit procedures to
resolve reservations that arise in any area of the audit. However, scope
limitations or major reservations that cannot be resolved to the auditor's
satisfaction by the use of extended audit procedures require the auditor to
issue a modified report or even to disclaim an opinion. The following are
examples of situations in which the auditor should consider not issuing a
standard report:

- The auditor is unable to evaluate the existence or collectibility
  of significant amounts of contract revenue related to claims.
  This is treated as an uncertainty and may require the auditor to
  issue a qualified opinion or to disclaim an opinion, depending
  on the particular circumstances.

- A contractor does not maintain detailed cost records by
  contract, and the auditor is unable to perform extended auditing
  procedures to obtain sufficient competent evidence that the data
  purporting to represent accumulated costs to date are
  reasonably correct. If the auditor considers this to be so
  material as to make the company unauditable, then the auditor
  will issue either a qualified opinion or a disclaimer of opinion.

- A company has cash problems because of undercapitalization or
  because losses have eroded its net worth and threatened its
  ability to continue to operate as a viable entity. The auditor
  should consider issuing a qualified report or disclaimer of an
  opinion, because the "going-concern" basis of presentation may
  not be appropriate.

- A contractor uses the percentage-of-completion method even
  though the measures of contract profitability do not meet with
  the standards set forth in the SOP. The audit should insist that
  the contractor change the financial statements to the completed-
  contract method, or the auditor should issue an adverse opinion.

Related Parties

When reviewed as economic units, many construction operations include
several affiliated entities that are "related parties" as defined in SAS No.
6, AU section 335, and in related auditing interpretations.

Consolidated or combined financial statements for the members of a group of affiliated entities constituting an economic unit generally present more meaningful information than the separate statements of the members of one economic unit. Separate statements of members of the group usually cannot stand on their own because they may not reflect appropriate contract revenue, costs, or overhead allocations and because transactions may be unduly influenced by controlling related parties. Under GAAP, there is a presumption that entities with parent-subsidiary relationships should present consolidated financial statements. Accounting Research Bulletin (ARB) 51, "Consolidated Financial Statements," specifies the basis of presentation. Other recommended financial statement presentations for affiliated companies are presented later in this chapter.

Combined Financial Statements

For the purpose of presenting financial condition, results of operations, and changes in financial position of a group of affiliated companies that generally conduct their construction operations as a single economic unit, combined financial statements are preferable unless consolidated financial statements are appropriate under ARB 51. In determining the need for combined financial statements, a group of affiliated companies should be viewed as a single economic unit if the members of the group are under common control and if their operations are closely interrelated and economically interdependent.

In the presentation of combined statements for members of one economic unit, general practices followed in the preparation of consolidated statements should be used in matters such as transactions between members of the economic unit, minority interest, foreign operations, income taxes, and different fiscal periods. The disclosures required in consolidated statements as well as the disclosures relating specifically to combined statements should be made. These include:

- A statement to the effect that combined statements are not those of a separate legal entity
- The names and year-ends of the major entities included in the combined group
- The nature of the relationship between the companies

The capital of each entity should be disclosed on the face of the financial statements or in a note, either in detail by entity if the number of entities is small or, if detailed disclosure is not practicable, in condensed form with an explanation as to how the information was accumulated.

Presentation of Separate Statements of Members of an Affiliated Group

Although it is preferable to use consolidated or combined statements of affiliated companies as the primary financial statements of an economic unit, the needs of specific users may sometimes necessitate the presentation of separate statements for individual members of an affiliated group. The issuer of separate financial statements for a member of an affiliated group should make at least the following disclosures on related parties:

- Transactions between a parent company and its subsidiaries;

- Transactions among subsidiaries of a common parent; and

- Transactions in which the reporting entity participates with other affiliated businesses, management, or principal stockholders (or other ownership interest).

In accordance with SAS No. 6, AU section 335, financial statements of a reporting entity that has participated in material-related party transactions should disclose, individually or in the aggregate, the following:

- The nature of common control, even where no transactions between the parties have occurred.

- A description of the transactions (summarized for numerous transactions) for the period reported on, including amounts, if any, and any other information deemed necessary for understanding the effects of the transactions on the financial statements. This includes transactions between related parties even though the transactions may not have been recorded in the accounting records. For example, if an entity provides services to a related party without charge, appropriate disclosure should be made of the nature and amount of the service rendered; also, if two or more entities in the same line of business are commonly controlled by a party with the ability to increase or decrease the volume of business done by each, the nature of common control should be disclosed.

- The dollar volume of transactions and the effects of any change in the method of establishing terms of the transactions from that used in the preceding period.

- Amounts due from or to related parties and, if not otherwise apparent, the terms and manner of settlement.

LEGAL AND REGULATORY CONSIDERATIONS

State Statutes Affecting Construction Contractors

The auditor should be aware of the existence in some states of lien laws. These laws vary from state to state but generally provide that funds received or receivable by a contractor constitute trust funds that may only be used to pay specified contract-related costs. The auditor should review with the contractor and his counsel the applicable statute in each state in which the contractor operates to evaluate whether amounts that constitute trust funds under those statutes have been properly applied. Other state statutes may also have audit or disclosure requirements. The auditor should consider such statutes in the performance of an audit and in the evaluation of the adequacy of financial statement disclosures.

Governmental Prequalification Reporting

Contractors are often required to file reports with agencies of the federal, state, or county governments in order to qualify for bidding on or performing work for such agencies. The report format required by the regulatory agencies frequently includes preprinted auditors' reports, which differ from reports issued in conformity with GAAS. Paragraphs 20 and 21 of SAS No. 14, "Special Reports," AU section 621, specify the steps an auditor should consider in these circumstances. A suggested form of auditor's report that could be substituted for the preprinted report in the regulatory filing form is shown in Figure 8-5.

Audit review involves two separate, but coordinated steps: (1) basic review (see Figure 8-6) performed by or under the direct supervision of the engagement partner, and (2) independent review.

Where possible, basic review should be made in the field. Timing and frequency of field visits depend on many factors, including the scope, size, and complexity of the engagement; whether it is an initial engagement; the timing of its completion; and the target date for submitting the report. In general, the basic review should include:

- An evaluation of audit planning and supervision
- Where reliance was placed on the internal accounting control system, and compliance tests performed, an evaluation of the conclusions reached as to the effectiveness of the system

- An evaluation of the audit program, including modifications made during the engagement
- A determination that the evidential matter obtained is sufficient to support the opinion on the financial statements
- A review of work papers to see that they are complete, clearly indicate the work performed, and support conclusions reached
- A determination that the financial statements are fairly presented in conformity with GAAP consistently applied
- A determination that the accountant's report communicates the scope and results of the examination

FIG. 8-5 Auditor's Report to Be Substituted for Regulatory Filing Form

We have examined the statement of assets, liabilities, and surplus of XYZ Company, Inc. as of December 31, 19XX, and the related statements of income and changes in surplus for the year then ended. Our examination was made in accordance with generally accepted auditing standards and, accordingly, included such tests of the accounting records and such other auditing procedures as we considered necessary under the circumstances.

As described in Note 1 below, the Company's policy is to prepare its financial statements for state construction prequalification filings on the basis of accounting practices prescribed by the state in which filed. These practices differ in some respects from generally accepted accounting principles. Accordingly, the accompanying financial statements are not intended to present financial position and results of operations in conformity with generally accepted accounting principles. This report is intended solely for filing with regulatory agencies and is not intended for any other purposes.

In our opinion, the financial statements referred to above present fairly the financial position of XYZ Company, Inc. as of December 31, 19XX, and the results of its operations for the year then ended, on the basis of accounting described in Note 1, which basis has been applied in a manner consistent with that of the preceding year.

NOTE 1. It is the policy of the Company to prepare financial statements intended solely for a state authority on the basis prescribed by that state authority. This report has been prepared for [*name of state authority*] on the basis prescribed by them. This basis differs from generally accepted accounting principles in the following respects:

1. A statement of changes in financial position is not included.

2. _____

3. _____

4. _____

FIG. 8-6   Audit Review Checklist—Construction Contractors

NAME OF CLIENT _____   PERIOD ENDED _____

BASIC REVIEW

The attached Audit Review Checklist should be completed on all audit engagements for construction contractors. It is designed as an aid in determining that firm and professional standards have been met in carrying out the audit. It does not encompass all the special audit circumstances that may be encountered in the construction industry, nor are all questions in the checklist applicable to all companies. Supplements should be added for specialized industry circumstances, and the **N/A** column marked for inapplicable questions.

All **Yes** answers should be supported by information in the work papers that include permanent files. An **N/A** should not be used if the matter is present but work was omitted because the matter is immaterial; rather the question should be answered **NM** (not material) in the **No** column. **No** answers indicate undesirable conditions and should be replaced by **Yes** answers if the undesirable conditions are removed.

	Yes	No	N/A
A. GENERAL:			
1. Was the Audit Planning Checklist completed?	☐	☐	☐
2. Did the engagement partner approve audit program modifications, if any, made subsequent to completion of the Audit Planning Checklist?	☐	☐	☐
3. Was each step of the audit program initialled?	☐	☐	☐
4. Was the information obtained in gaining an understanding of the client's business documented in the work papers and permanent files?	☐	☐	☐
Analytical Review:			
5. Was financial information (expressed in dollars, rations, and/or percentages) compared with:			
a. Comparable prior periods?	☐	☐	☐
b. Anticipated results (e.g., original cost estimates and related gross profit)?	☐	☐	☐
c. Comparable industry information (where practicable)?	☐	☐	☐
6. Were the relationships of elements of financial information that would be expected to conform to a predictable pattern studied (e.g., billings to date to percent complete on contracts)?	☐	☐	☐

	Yes	No	N/A
7. Was consideration given to studying the relationship of financial information with relevant nonfinancial information (e.g., economy, interest rates)?	☐	☐	☐
8. Were significant fluctuations and other unusual items or trends investigated through inquiry of management and, where appropriate (considering the reasonableness of replies), through additional corroborating audit procedures?	☐	☐	☐

Internal Accounting Control:

	Yes	No	N/A
9. Was an understanding of the control environment obtained?	☐	☐	☐
10. Was an understanding of the flow of transactions through the accounting system obtained?	☐	☐	☐
11. If controls are not relied on, was that decision, along with the reason for that decision, included in the planning memorandum?	☐	☐	☐
12. If controls are relied on, was the study of internal control properly completed and supported by a description of the internal accounting control system, including preparation of a Summary of Internal Accounting Controls Relied On?	☐	☐	☐
13. Where reliance was placed on internal accounting controls, were appropriate tests of compliance performed?	☐	☐	☐
a. Were the test transactions selected randomly from the period using statistical sampling techniques (or judgment when appropriate) to determine sample size?	☐	☐	☐
b. Was the sampling plan described in the work papers, including identification of the controls tested and a clear definition of errors or exceptions?	☐	☐	☐
c. Were the sample items tested by obtaining evidence that the control function was (1) performed, and (2) performed properly (as appropriate)?	☐	☐	☐
d. Were errors, exceptions, and unusual conditions or transactions noted and followed up?	☐	☐	☐
e. Was a conclusion reached as to whether the controls operated as designed?	☐	☐	☐
f. Where the results of compliance tests indicated that a control cannot be relied on, was a Summary of Control Objectives Not Met prepared and was the audit program modified appropriately?	☐	☐	☐

(continued)

FIG. 8-6 (continued)

	Yes	No	N/A
14. Have material weaknesses in internal accounting control been communicated to senior management and the board of directors or its audit committee, or have plans been made to communicate this information at the earliest practicable date?	☐	☐	☐
15. Where internal auditors' work affected procedures, was the competence and objectivity of the internal auditors considered and was their work evaluated?	☐	☐	☐

B. REVENUE CYCLE:

	Yes	No	N/A
1. Are conclusions stated; do the work papers support the conclusions; and do you agree that the audit objectives for the revenue cycle were met?	☐	☐	☐
2. Were substantive tests of revenue cycle transactions performed in areas where reliance was not placed on internal accounting controls (i.e., separately, in connection with tests of balances, or both)?	☐	☐	☐
3. Was appropriate consideration given to advance planning of job-site visits? Do work papers show:	☐	☐	☐
a. That the timing and extent of job-site visits were sufficient in light of internal controls?	☐	☐	☐
b. That sufficient consideration was given to observation of uninstalled materials, work performed to date, and contractor-owned or rented equipment?	☐	☐	☐
c. That inquiries were made with appropriate individuals regarding the status of the contract and any significant problems?	☐	☐	☐
4. Have anticipated losses on contracts been recognized in full at the earliest date at which they are determinable?	☐	☐	☐
5. Do the work papers show the basis of selection, summarize results, and present a conclusion on receivable confirmations (was pertinent information such as contract price, payment made, and status of the contract confirmed)?	☐	☐	☐
a. Were notes receivable confirmed, and were second requests sent for positive receivable confirmations?	☐	☐	☐
b. Were "no replies," "unable to confirm," and reported payments and differences followed up?	☐	☐	☐

	Yes	No	N/A
c. Was the circularization sufficient in light of internal accounting controls?	☐	☐	☐
6. In reviewing the reasonableness of the allowance for doubtful accounts, was sufficient consideration given to the following:			
a. Subsequent collections?	☐	☐	☐
b. Discussion of trends and past due accounts with management?	☐	☐	☐
c. Investigation of significant charge-offs?	☐	☐	☐
d. Activity in the account and related expense?	☐	☐	☐
7. Were cutoff tests performed as to contracts, change orders, and cash receipts?	☐	☐	☐

C. EXPENDITURE CYCLE:

	Yes	No	N/A
1. Are conclusions stated; do the workpapers support the conclusions; and do you agree that the audit objectives for the expenditures cycle were met?	☐	☐	☐
2. Were substantive tests of expenditure cycle transactions performed in areas where reliance was not placed on internal accounting controls (i.e., separately, in connection with tests of balances, or both)?	☐	☐	☐
3. In connection with the examination of job costs to date, was sufficient consideration given to costs being charged to the proper job?	☐	☐	☐
4. Were substantive tests performed on the allocation of indirect contract costs to jobs, and do the work papers show the consistency of the application of the allocation policy?	☐	☐	☐
5. Was sufficient consideration given to the reasonableness of the estimated cost to complete in light of the information obtained from available sources (i.e., analytical review, inquiries with appropriate individuals, latest estimates, etc.)?	☐	☐	☐
6. Were sufficient tests of balances of accounts payable and accrued liabilities performed, and was account activity related to expense accounts?	☐	☐	☐
7. If payables were confirmed, do the work papers show basis of selection, follow-up work performed, and conclusions reached?	☐	☐	☐
8. Were procedures sufficient to determine that prepaid expenses, deferred costs, and other assets are appropriately stated and that related amortization is reasonable?	☐	☐	☐

(continued)

FIG. 8-6 (continued)

	Yes	No	N/A
9. Were cutoff tests performed as to contract costs, payables, and disbursements, including an appropriate search for unrecorded liabilities?	☐	☐	☐

Labor and Employee Costs:

	Yes	No	N/A
10. Are conclusions stated; do the work papers support the conclusions; and do you agree that the audit objectives for labor and employee costs were met?	☐	☐	☐
11. Were substantive tests of payroll transactions performed in areas where reliance was not placed on internal accounting controls (i.e., separately, in connection with tests of balances, or both)?	☐	☐	☐
12. Were sufficient tests of balances of accrued liabilities for labor and employee costs performed, and was account activity related to expense accounts?	☐	☐	☐
Was appropriate consideration given to:			
a. Pension costs and liabilities?	☐	☐	☐
b. Salaries, fees, or other benefits paid to officers and directors?	☐	☐	☐
c. Stock option and purchase plans?	☐	☐	☐
d. Executive perquisites?	☐	☐	☐
13. Were cutoff tests performed as to labor and employee costs?	☐	☐	☐

Income Taxes:

	Yes	No	N/A
1. Are consolidations stated; do the work papers support the conclusions; and do you agree that the audit objectives for income taxes were met?	☐	☐	☐
2. Were tests of tax account balances sufficient, and was account activity related to expense accounts?	☐	☐	☐
3. Do work papers show:			
a. Reconciliations of pretax accounting income with taxable income, carry backs, carry forwards, and the latest year income tax returns were examined?	☐	☐	☐
b. Computations of deferred and current income tax expense?	☐	☐	☐
4. In calculating income tax accruals, was consideration given to adjustments proposed or which may be proposed as a result of tax examinations?	☐	☐	☐

	Yes	No	N/A
5. Was the tax accrual reviewed by a tax specialist where required?	☐	☐	☐

D. PRODUCTION OR CONVERSION CYCLE:

Inventories:

	Yes	No	N/A
1. Are conclusions stated; do the work papers support the conclusions; and do you agree that the audit objectives for inventories were met?	☐	☐	☐
2. Was appropriate consideration given to advance planning of inventory observations?	☐	☐	☐
3. Do work papers show:			
a. That the timing and extent of inventory observations were sufficient in light of internal control?	☐	☐	☐
b. Test counts and related follow-up (including subsequent adjustments)?	☐	☐	☐
c. Conclusions as to the sufficiency of inventory count procedures (e.g., memorandum on observation)?	☐	☐	☐
d. Confirmation (or where appropriate, observation) of consigned inventory or inventory in custody of others?	☐	☐	☐
e. That inventory compilation is clerically accurate?	☐	☐	☐
f. That intercompany profit was eliminated?	☐	☐	☐
g. That pricing methods are appropriate, up to date, and consistent?	☐	☐	☐
h. That provision was made for obsolete items or losses due to market decline?	☐	☐	☐
4. Were cutoff tests for inventories performed and coordinated with cutoff tests for revenues and expenditures?	☐	☐	☐

Property and Equipment:

	Yes	No	N/A
5. Are the conclusions stated; do the work papers support the conclusions; and do you agree that the audit objectives for property and equipment were met?	☐	☐	☐
6. Do the work papers show the nature and extent of tests of:			
a. Balances of property, equipment, and related accounts including the review of cutoff procedures?	☐	☐	☐
b. Additions and retirements?	☐	☐	☐
c. Related costs and expenses, including repairs, maintenance, and the reasonableness of the capitalization policy?	☐	☐	☐

(continued)

FIG. 8-6   (continued)

	Yes	No	N/A
d. Depreciation expense, including the reasonableness of rates and methods?	☐	☐	☐
e. Consistency of application of depreciation methods and capitalization policies?	☐	☐	☐
7. Were leases reviewed and properly accounted for?	☐	☐	☐
8. Was sufficient consideration given to commitments existing at the balance sheet date?	☐	☐	☐
9. Do work papers show that the sufficiency of insurance coverage was considered?	☐	☐	☐

E. FINANCING CYCLE:

Cash:

	Yes	No	N/A
1. Are conclusions stated; do the work papers support the conclusions; and do you agree that the audit objectives for cash were met?	☐	☐	☐
2. Were important cash funds confirmed if not counted?	☐	☐	☐
3. Do bank reconciliations agree with bank confirmations?	☐	☐	☐
4. Were reconciling items adequately examined?	☐	☐	☐
5. Were cutoff tests performed as to receipts, disbursements, and interbank transfers?	☐	☐	☐
6. Have important cash restrictions been identified?	☐	☐	☐
7. Was consideration given to confirming bank credit arrangements, including compensating balances?	☐	☐	☐

Investments:

	Yes	No	N/A
8. Are conclusions stated; do the work papers support the conclusions; and do you agree that the audit objectives for investments were met?	☐	☐	☐
9. Do the work papers establish the existence of investments (i.e., inspection or confirmation)?	☐	☐	☐
10. Was sufficient consideration given to the accounting basis (e.g., equity, cost, or market) and classification (e.g., long- or short-term) of investments?	☐	☐	☐
11. Do the work papers show the nature and extent of tests of changes during the period and market values at year end?	☐	☐	☐

	Yes	No	N/A
12. Were calculations of income checked and correlated with recorded income and was cutoff reviewed?	☐	☐	☐
13. Was the possible cost impairment of long-term investments considered?	☐	☐	☐
14. If the equity method has been applied, have appropriate procedures been performed to substantiate the carrying value?	☐	☐	☐
15. Was the cash value of life insurance confirmed?	☐	☐	☐

Debt:

	Yes	No	N/A
16. Are conclusions stated; do the work papers support the conclusions; and do you agree that the audit objectives for debt were met?	☐	☐	☐
17. Were analyses made of changes in accounts and do the work papers show the nature and extent of tests of additions and retirements?	☐	☐	☐
18. Were calculations of interest payments, accruals, and premium or discount checked and correlated with related expense accounts, and was cutoff reviewed?	☐	☐	☐
19. Do the work papers indicate that the requirements of loan and other debt agreements were reviewed?	☐	☐	☐
20. Were loan agreements reviewed for the existence of pledged or collateralized assets?	☐	☐	☐
21. Were confirmations received for notes payable and other debt obligations?	☐	☐	☐
22. In states that honor such requests, was consideration given to using Uniform Commercial Code information requests to confirm security agreements?	☐	☐	☐

Stockholders' Equity:

	Yes	No	N/A
23. Are conclusions stated; do the work papers support the conclusions; and do you agree that the audit objectives for stockholders' equity were met?	☐	☐	☐
24. Were analyses made of changes in stockholders' equity accounts (or capital and drawing accounts for partnerships and proprietorships), and was cutoff reviewed?	☐	☐	☐
25. Were authorizations for increases or decreases in capitalization, and purchases or sales of capital stock checked (e.g., to articles of incorporation, minutes, or bylaws)?	☐	☐	☐

(continued)

FIG. 8-6 (continued)

	Yes	No	N/A
26. Were confirmations received from transfer agents and registrars (or were capital stock records examined)?	☐	☐	☐
27. Were the requirements of preferred stock (or similar items) reviewed?	☐	☐	☐
28. Were dividend payments and other distributions reviewed?	☐	☐	☐
29. Were treasury stock certificates physically inspected or confirmed?	☐	☐	☐
30. Is information included for stock options, warrants, rights, conversion privileges, and stock reserved?	☐	☐	☐

F. EXTERNAL FINANCIAL REPORTING CYCLE:

	Yes	No	N/A
1. Was a review made of items not tested in connection with other transaction cycles:			
a. Standard and nonstandard journal entries?	☐	☐	☐
b. Adjustments and write-offs?	☐	☐	☐
c. Nonrecurring transactions?	☐	☐	☐
2. Were sufficient tests performed of the accumulation, classification, and summarization of transactions in the general ledger (to the extent not otherwise tested)?	☐	☐	☐
3. If the client compiled the financial statements (and consolidation, if applicable), were appropriate tests of the compilation performed?	☐	☐	☐
4. Were earnings per share computations (where required) reviewed?	☐	☐	☐
For audited financial statements included in a document containing other information:			
a. Was consideration given to whether the other information (or the manner of its presentation) is consistent with the audited financial statements?	☐	☐	☐
b. If any material inconsistencies or misstatements of fact were noted, were appropriate follow-ups steps taken?	☐	☐	☐

G. TRANSACTIONS SUBJECT TO SPECIAL CONSIDERATION:

	Yes	No	N/A
1. Was sufficient consideration given to the identification of transactions with affiliates, directors, employees, officers, stockholders, and other related parties?	☐	☐	☐

	Yes	No	N/A
2. Were identified related party transactions examined to determine the purpose, nature, and extent of the transactions and their effect on the financial statements?	☐	☐	☐
3. Were transactions that appear to have unusual or questionable purposes investigated?	☐	☐	☐

If the examination caused us to believe that an error, irregularity, or illegal act may have occurred:

	Yes	No	N/A
a. Did we make appropriate investigation to determine whether, in fact, it did occur and what effect it had on the financial statements and our report?	☐	☐	☐
b. If necessary, was the matter reported to appropriate personnel in the client's organization?	☐	☐	☐
4. If an illegal act was involved, was consideration given to the need to make appropriate disclosure other than in the financial statements?	☐	☐	☐

H. LITIGATION, CLAIMS, AND ASSESSMENTS:

	Yes	No	N/A
1. Were procedures relating to litigation, claims, and assessments sufficient (including inquiries of management's policies and procedures for identifying, evaluating, and accounting for litigation, claims, and assessments)?	☐	☐	☐
2. Was notification given to the client's lawyer(s) of the client's representation regarding unasserted claims?	☐	☐	☐
3. Was a written reply(s) to the audit inquiry letter(s) received directly from the client's attorney(s) and reviewed?	☐	☐	☐
4. Was a supplemental "updating" letter obtained from the client's attorney(s), if closing was delayed or counsel responded too early?	☐	☐	☐

I. SUBSEQUENT EVENTS AND REPRESENTATIONS:

	Yes	No	N/A
1. Do work papers show the nature and extent of the review of subsequent events and the conclusions reached?	☐	☐	☐
2. As to areas of uncertainty (including accounting estimates for matters accounted for on the basis of preliminary or incomplete data, such as losses on sales or purchase commitments, litigation, claims, etc.) do work papers indicate that consideration was given to additional information, if any, available in the subsequent period?	☐	☐	☐

(continued)

FIG. 8-6 (continued)

	Yes	No	N/A
3. Do work papers show that minutes of stockholders, directors, and appropriate committees were reviewed both for the period under examination and the subsequent events period?	☐	☐	☐
4. Do minutes include authorization of matters included in the accounts (e.g., dividend declarations, salary authorizations, stock option plans, contracts)?	☐	☐	☐
5. Was a written representation obtained covering each applicable matter included in the firm's Illustrative Letter of Representation From Management, and any other appropriate matters?	☐	☐	☐
6. Do work papers contain memoranda describing client representations made in conferences and other discussions?	☐	☐	☐

J. WORK PAPERS:

	Yes	No	N/A
1. Do all amounts in financial statements and notes agree with adjusted trial balances (or lead schedules or assembly sheets) and do such amounts agree with underlying work papers?	☐	☐	☐
2. Are adjusting entries clearly explained, supported by, and cross-referenced to the work papers and recorded or accepted by the client?	☐	☐	☐
3. Was a Summary of Proposed Adjustments Not Made prepared, and have we determined that the aggregate effect of waived adjustments is not material?	☐	☐	☐
4. Were the contents of the permanent file reviewed and updated, and was superseded material removed and filed in a separate inactive file?	☐	☐	☐
5. Does each group of related work papers include the source of information, work performed, initials of the preparer, and the date prepared?	☐	☐	☐
6. Did the in-charge accountant initial conclusions reached?	☐	☐	☐
7. Are work papers indexed and cross-referenced?	☐	☐	☐
8. Is the meaning of symbols clearly indicated?	☐	☐	☐
9. Were all questions and exceptions arising from audit tests explained?	☐	☐	☐
10. Were the circumstances and conclusions of any problem areas discussed with the National Office, industry, or tax consultants sufficiently documented?	☐	☐	☐

	Yes	No	N/A

11. Were review and "To Do" notes followed up and appropriate documentation included in related sections of the work papers? □ □ □

[*Review and "To Do" notes should be destroyed when (1) the related work or follow-up has been completed and appropriately documented in the work papers and (2) the engagement partner is satisfied as to follow-up including documentation in the papers.*]

K. OTHER:

1. If interim work was performed, did year-end procedures consider transactions in the intervening period? □ □ □

2. Are you satisfied that there has been proper supervision (including on-the-job training) of staff assigned to the engagement, giving due consideration to their technical training and proficiency to carry out the work to which they were assigned? □ □ □

3. Are you satisfied that no factors exist that would cause you to question client continuance? □ □ □

4. Do the work papers include either (1) a memorandum setting forth points to be considered during the ensuing engagement (including suggestions for improving the effectiveness or efficiency of audit procedures) or (2) a post-audit critique memorandum? □ □ □

5. If we are using the work of other auditors as to subsidiaries, investees, or components, were the procedures followed appropriate? □ □ □
If a work specialist (e.g., engineer) was used, were steps taken to:

  a. Determine the specialist's qualifications, determine his relationship to the client, if any, and reach an understanding as to the work to be performed by him? □ □ □

  b. Review the methods and assumptions used and the test data provided to the specialist and evaluate whether the specialist's findings support the related representations in the financial statements? □ □ □

6. *For Securities and Exchange Commission (SEC) clients:*
Do the work papers indicate that appropriate procedures were applied to segment information unaudited interim information, or required supplemental information, included in or accompanying the audited financial statements? □ □ □

(continued)

FIG. 8-6 (continued)

	Yes	No	N/A
7. Have arrangements been made to report total fees for management advisory services and a description of services rendered to the audit committee or board of directors?	☐	☐	☐
8. Has the National Accounting and Auditing Office been advised of any disagreements with management that should be considered for disclosure?	☐	☐	☐

L. AUDIT ISSUES MEMORANDUM:

	Yes	No	N/A
1. Do the work papers include an Audit Issues Memorandum identifying significant accounting and auditing issues encountered during the engagement and indicating their disposition?	☐	☐	☐
2. If no significant accounting and auditing issues were identified, do the work papers include a notation to that effect?	☐	☐	☐

M. REPORT:

	Yes	No	N/A
1. Has the Report Preparation Checklist been prepared?	☐	☐	☐
Based on a reading of the report on an overall basis:			
a. Do the statements (including notes) make sense?	☐	☐	☐
b. Is the language concise and accurate?	☐	☐	☐
c. Are the notes written as representations by management?	☐	☐	☐
d. Is the opinion appropriately worded?	☐	☐	☐

	Initials	Date
PREPARED BY _____ *[engagement partner or, under his direct supervision, the manager or, in some cases, the in-charge accountant]*		_____
PREPARATION SUPERVISED BY _____ *[engagement partner should initial if the checklist was prepared by another.]*		_____

CHAPTER 9

Relationship With a CPA Firm

Kenneth R. Alderman

INTRODUCTION

The responsibility for budgeting, cash forecasting, implementing management information systems (MIS), establishing adequate internal control systems, and preparing timely and accurate interim and year-end financial statements is the primary responsibility of the contractor's management. Although this responsibility is normally delegated to financial executives (i.e., chief financial officers, controllers, and treasurers), the responsibility rests ultimately with the board of directors and the chief executive officer (CEO).

SERVICES OFFERED BY THE CPA FIRM

Every contracting entity prepares reports to its owners. These financial statements, issued annually, quarterly, or more frequently, are often attested to in some fashion by independent certified public accountants (CPAs). The purpose of this chapter is to explore the relationship between the contractor and the CPA.

The services provided by the CPA are most often classified into three categories:

1   Audit and accounting services

2   Taxation services

3   Management consulting services

The audit and accounting services are the primary focus of the relationship between construction contractors and CPAs. These services include:

- Attestation to the company's financial statements
- Reviews and compilations
- Accounting, including actual client record-keeping

The taxation services provided by CPA firms include:

- Advice and guidance regarding federal and state income taxes
- Compliance services relating to preparation of federal and state income taxes and other tax returns
- Tax consulting services for the principals of the organization
- Advice and direction when seeking international projects

The management consulting services that CPA firms provide include:

- Financial information systems (FIS), management information systems (MIS), data-processing systems (DPS), and executive compensation studies
- Business advisory services

IMPORTANCE OF THE ATTEST FUNCTION

Among the comprehensive, interrelated financial services offered by the CPA firm to its clientele, the attest, or audit, function is most critical for the contractor since financial statements, together with the report of the independent CPA, are essential for reliance by third parties. Annual examinations of contractors' financial statements by independent accountants are required for those contractors who are publicly owned and subject to the requirements of various stock exchanges. Investors and the other users of financial statements, including credit grantors such as banks and surety companies, frequently require an annual independent audit.

Although the surety industry, for competitive reasons, does not always require audited financial statements, there is no doubt that these statements are preferred and probably enhance bonding capacity. Furthermore, audited statements are often required by law to obtain licenses to do business and to prequalify for state and other governmental contracts. Moreover, many owners frequently demand audited financial statements before a contractor can be considered as a prospective bidder of private work. For all these internal and external reasons, the contractor usually engages an independent certified public accountant to make an annual audit and to provide tax and related services.

SELECTING THE CPA FIRM

In selecting the CPA firm, the contractor should use the same criteria as he would use in selecting other professional or outside services. In view of the requirements of the construction industry and its unique accounting principles, auditing requirements, and tax regulations, the selection of a firm bears considerable significance.

Some of the principal factors for a contractor to consider in evaluating and selecting a firm of public accountants are professional reputation, experience, industry expertise, scope and range of services, staffing, and fee arrangements.

Professional Reputation

Although CPAs are licensed to practice by the various state boards of accountancy throughout the United States, it is the professional standing of the CPA firm in the business community that is indicative of its reputation for competence, independence, objectivity, professionalism, and service.

The professional reputation of an accounting firm may be ascertained by considering the following factors:

- Relative size of the firm

- Specific clients served in the construction industry

- Reputation with banks and surety companies

- Membership in professional organizations such as the American Institute of Certified Public Accountants (AICPA) or the local state society of CPAs

- Membership in the Division of Firms of the AICPA, a voluntary self-regulatory body with two sections: Securities and Exchange Commission (SEC) practice and private company practice. The Division has membership requirements, such as peer review and mandatory continuing professional education standards

- Litigation record

The professional reputation of a firm is not solely a function of its size, although size may be a factor for other reasons (e.g., geographic area). The contractor should recognize that although the prestige of a multinational public accounting firm may impress a credit grantor, there are smaller national, regional, and local firms that offer excellent qualifications, reputations, and services as well. CPA firms also provide prospective clients with references from the business community and from their existing clients.

Experience and Industry Expertise

In many industries, the background and experience that an accounting firm has in a particular industry is a key factor in the selection process; this is particularly true in the construction industry. In reviewing the qualifications of CPA firms, the contractor should consider the following questions:

- Does the CPA firm have experience in the industry and with other contractor clients?

- Are personnel and partners of the firm knowledgeable about contracting?

- Is the firm known by and acceptable to surety companies and other credit grantors?

- Is the firm experienced in areas such as prequalification reporting?

Since public accountants are often the primary outside advisors to their clients—particularly for smaller organizations—the industry experience factor should be given considerable weight in the selection process.

Scope and Range of Services

In reviewing the qualifications of a prospective CPA firm, the contractor may wish to consider all of the services provided by a CPA firm. In addition to attest, or audit, services, the contractor should also consider the following:

- Tax services

- Management services capability

- Business advisory services

- Geographic (domestic and world-wide) capability

If the contractor has widely dispersed operations in the United States or includes operations in various international locations, or if such operations are even contemplated, the contractor should consider the capabilities of an accounting firm to provide services in these other locations, either through maintaining offices or through available representatives of the CPA firm.

Staffing

As in any professional relationship, it is important to understand how services are provided and by whom. In the area of staffing, the contractor should consider the following:

- Which individual within the accounting firm has ultimate responsibility for client service? What is that individual's industry expertise?

- For firms with rotation policies for engagement partners, how is the transition effected?

- Are other experienced individuals available to provide expert tax and management consulting services?

- Who are the staff to be assigned to the audit? What is their industry background?

- Does the CPA firm have adequate human resources to ensure proper staffing and service on the engagement?

- Can the CPA firm provide adequate backup support in the event of staff turnover?

Fees

Many companies have little or no understanding of the fee structures charged by CPA firms. Information pertaining to fee structures is available, however, that can clear up any mystery relating to this topic. Fees are generally based either on hourly rates for the various categories of personnel assigned to an engagement or on a flat amount for services rendered on a regular basis.

It is fairly common for the contractor client and the CPA firm to have the fee arrangements clearly set forth in an engagement letter; this helps to avoid any possible misunderstanding. Accounting firms often absorb the high initial expenses related to new clients and offer lower fees for start-up situations for small businesses and companies with fiscal years ending during nonpeak season (usually April to November).

Proposals

The contractor should be able, from the information provided in this chapter, to narrow his choice down to a manageable number of CPAs who would be qualified to provide the services needed. At this point, the contractor should invite proposals from each of these CPAs.

In the proposal, the contractor should have the CPA specify, as a minimum, the following items:

- Nature of the services to be provided (audit, review, compilation)

- Background of the partner(s) to be assigned to the engagement

- Expertise of staff personnel who will be running the field work

- Timing of the assignment (beginning of field work, completion of field work, and projected date of report issuance)

- Fee to be charged and expected timing of payments

Engagement Letter

After the contractor has selected a CPA firm, all of the relevant terms of engagement should be clearly and concisely set forth in an engagement letter, which is normally reviewed, revised, and updated annually. This letter forms the basis of the arrangement between the CPA and the contractor and generally covers such matters as the following:

- Responsibility of management relating to the financial statements
- Scope of the examination by the CPA (e.g., a full audit with the objective of issuing an unqualified opinion on the financial statements, limited examination for special purposes, and so on)
- Ancillary services, if any, to be included in the basic service and for the basic fee
- Names of the engagement partner in charge of the work and any supporting or consulting partners
- Timing of the examination
- Degree of assistance to be provided by the contractor's financial staff
- All arrangements relating to fees to be charged by the CPA firm, including the provision, if any, for progress billings.

A sample letter incorporating some of these points is provided in Figure 9-1.

PLANNING THE YEAR-END AUDIT

The annual examination calls for considerable planning by and coordination between the CPA firm and the contractor's financial management team. Planning offers the opportunity to maximize audit efficiency, reduce overall audit costs, assist in training the contractor's financial personnel, and minimize the disruption that may occur during the normal course of an examination. With good internal preplanning by CPAs and effective coordination with financial management of the contractor, both audit time and costs can be reduced and the independent accountants can schedule their work effectively in order to meet required deadlines. A checklist for effective year-end audits is given in Figure 9-2.

FIG. 9-1   Audit Engagement Letter

[Date]

Mr. John Jones, President
ABC Corporation
488 Madison Avenue
New York, New York 10022

Dear Mr. Jones:

UNDERSTANDING OF ENGAGEMENT

We are confirming our understanding of the services to be provided in connection with our planned examination of the consolidated balance sheet of ABC Corporation (the Company) at December 31, 19XX and the related consolidated statements of income, retained earnings, and changes in financial position for the year then ended.

The purpose and scope of our examination will be to express an opinion on the consolidated financial statements rather than on the separate financial statements of the parent company and each of its subsidiaries. Insofar as the consolidated financial statements include the earnings of NOP Corporation (40 percent owned by the Company), our opinion will be based on the reports of other auditors that will be furnished to us.

Our examination will be made in accordance with generally accepted auditing standards, a listing of which is enclosed. Accordingly, we will selectively test the accounting records of the Company and related data, and will perform other auditing p. cedures by the methods and to the extent that we deem appropriate.

As part of our examination, a letter of representation will be requested from management confirming (a) certain representations made orally during our examination that are not reflected in the Company's books of account or other records, and (b) certain other representations implicit in the books and records maintained by employees of the Company. That letter will deal with matters such as management's acknowledgment of its responsibilities for the fair presentation of the financial statements; information concerning related party transactions, plans, or intentions that may affect the carrying value or classification of assets and liabilities; and similar matters.

We will advise you promptly if we discover conditions that cause us to believe that material errors, defalcations, or other irregularities may exist or if we encounter other circumstances that require us to extend our work significantly. However, an examination based on selective testing is subject to the inherent risk that material errors or irregularities, if they exist, will not be detected.

At the conclusion of our examination, we will submit our report with respect to the consolidated financial statements. We will separately advise you of any material weaknesses in internal accounting control that come to our attention and make any recommendations that we may have for strengthening those controls. We will also make recommendations for improving operational procedures to the extent that such matters come to our attention.

We will also review the federal and state income tax returns prepared by your staff prior to filing for 19XX. We will not review any other tax returns prior to filing unless specifically requested to do so.

Our charges will be made at regular per diem rates, plus out-of-pocket expenses. We will obtain the assistance of Company personnel to the extent possible and otherwise endeavor to keep these charges to a minimum. We estimate that the aggregate charge will be approximately $XX,000. Accordingly, we will make monthly billings of $X,000 during the period from October 19XX to January 19XX, with a final billing in February 19XX on completion of the work.

If the above arrangements are acceptable to you and if the services outlined are in accordance with your requirements, please sign the copy of this letter in the space provided and return it to us.

Very truly yours,

CPA FIRM

_____

Enclosure

The services described in the foregoing letter are in accordance with our requirements and are acceptable to us.

Dated: _____, 19____   _____
President, ABC Corporation

GENERALLY ACCEPTED AUDITING STANDARDS

The generally accepted auditing standards (GAAS) as approved and adopted by the membership of the AICPA are as follows:

GENERAL STANDARDS

1. The examination is to be performed by a person or persons having adequate technical training and proficiency as an auditor.
2. In all matters relating to the assignment, an independence in mental attitude is to be maintained by the auditor or auditors.
3. Due professional care is to be exercised in the performance of the examination and the preparation of the report.

(continued)

FIG. 9-1 (continued)

STANDARDS OF FIELD WORK

1. The work is to be adequately planned and assistants, if any, are to be properly supervised.

2. There is to be a proper study and evaluation of the existing internal control as a basis for reliance thereon and for the determination of the resultant extent of the tests to which auditing procedures are to be restricted.

3. Sufficient competent evidential matter is to be obtained through inspection, observation, inquiries and confirmations to afford a reasonable basis for an opinion regarding the financial statements under examination.

STANDARDS OF REPORTING

1. The report shall state whether the financial statements are presented in accordance with generally accepted accounting principles.

2. The report shall state whether such principles have been consistently observed in the current period in relation to the preceding period.

3. Informative disclosures in the financial statements are to be regarded as reasonably adequate unless otherwise stated in the report.

4. The report shall either contain an expression of opinion regarding the financial statements, taken as a whole, or an assertion to the effect that an opinion cannot be expressed. When an overall opinion cannot be expressed, the reasons therefor should be stated. In all cases where an auditor's name is associated with financial statements the report should contain a clear-cut indication of the character of the auditor's examination, if any, and the degree of responsibility he is taking.

The preceding listing has been taken, verbatim, from SAS No. 1, "Codification of Auditing Standards and Procedures," AU § 150.02.

Early Audit Planning

Planning of the audit should begin immediately after the completion of the current examination. The CPA firm and the contractor's financial management should review all problems encountered during the audit and discuss how to improve audit effectiveness in future examinations. If this conference is deferred until the following year-end, its maximum potential benefit will not be realized.

Interim Audits

Depending on the size of the contractor's operation, the CPA firm may wish to perform interim audits during the course of the year to review

---

FIG. 9-2   Checklist for Effective Year-End Audits

PLANNING SPECIFICS

- ☐ Start audit planning when the current audit is complete.
- ☐ Set deadlines and assign responsibilities to internal accounting staff two to three months prior to year-end.
- ☐ Conform reconciliations in the monthly and quarterly reports in the same manner as done in the year-end audits.
- ☐ Conform the format of periodic and year-end reports.
- ☐ Have the contractor prepare all schedules, trial balances, consolidations, and the annual report.
- ☐ Make use of and coordinate efforts with the contractor's internal audit department.
- ☐ Monitor progress during the audit through periodic (e.g., weekly) meetings between the CPA firm and the contractor's financial management staff.
- ☐ Monitor deadlines and review materials as presented.

ALL-YEAR GUIDELINES

- ☐ Contact should be maintained during the year between the CPA firm and the contractor.
- ☐ Make sure the CPA firm is on the contractor's mailing list for quarterly reports, SEC filings, and other publications.
- ☐ Discuss unusual items or new authoritative requirements that arise during the year to avoid misunderstandings later.

---

internal controls and perform certain audit procedures. Interim auditing allows the CPA firm to:

- Expedite completion of the year-end examination and issuance of final reports.
- Identify accounting and auditing problems and resolve them without the year-end deadline pressures.
- Finalize planning several months before year-end.

In the event interim auditing is not practical or feasible, a year-end planning meeting should be held with the key engagement personnel of the CPA firm two or three months prior to the beginning of the year-end examination. The agenda for this meeting should include the following points:

- Current accounting issues related to the contractor, including relevant new accounting pronouncements and reporting requirements

- Timing of the examination

- Staffing by the CPA firm

- Deadlines for completion and scheduled issuance of reports

Assignment of responsibility between the contractor's personnel and the CPA firm should also be established at this meeting to minimize any duplication of effort. In this connection, all required schedules and analyses should, to the extent possible, be prepared by the contractor's financial personnel. This should result in a reduction in audit time and cost and will help to train internal financial personnel. After all, it is the responsibility of the CPA firm to audit, not prepare, original work.

Preparation of Periodic Statements

During the year, the contractor should prepare monthly and quarterly financial statements for internal management purposes and, in some cases, for outside third parties such as bonding companies. In the case of publicly owned contractors, quarterly reports to shareholders and to the SEC should be reviewed by the CPA firm. These reports should conform to the year-end reporting format to the extent possible. Furthermore, the contractor's financial management personnel should prepare the necessary analyses and reconciliations in the same format as that used in the year-end report. Such a procedure enables analyses and reconciliations to be prepared on a timely basis, avoids duplication of effort, and minimizes the possibility of year-end adjustments and surprises.

Coordination With Contractor's Internal Audit Staff

Many larger contractors have internal audit departments. Where this is the case, maximum coordination should be established between the internal and external auditing functions. The annual plan of the internal auditor should be carefully reviewed and coordinated with the CPA firm to maximize the effectiveness of the two functions and to eliminate possible duplication and overlap of the two auditing functions; this system requires cooperation on both sides. Since both groups have similar objectives, a cooperative coordinated program maximizes overall audit effectiveness. The resources and capabilities of the internal audit department should be matched with the CPA firm in planning and conducting the year-end examination.

Regular Meetings During the Year-End Audit

Throughout the year-end audit, the contractor's financial personnel should meet regularly (e.g., on a weekly basis) with the staff of the CPA firm to discuss accounting issues as they arise in order to keep the examination on track. Such a procedure can help eliminate any surprises regarding accounting issues, audit problems, or difficulties encountered that relate to the timely completion of the examination.

Regularly scheduled meetings also serve to assure that predetermined deadlines are met. Delays cost money and diminish the essential rapport between the company and its external auditors.

With effective planning and open communication during the course of the examination between the contractor and the CPA firm, problems, difficulties, and total audit costs are minimized; service is maximized; and a relationship of mutual respect and trust is engendered.

MANAGEMENT LETTERS OF COMMENT

Many independent CPA firms have established policies with regard to the submission of comment letters to management following the completion of an examination. These comments, which are based on the knowledge and expertise of the CPA firm, consist of the auditor's observations during the course of the examination and frequently make significant constructive suggestions to improve the client's operation. From the CPA's point of view, the benefits of a comment letter are:

- Strengthened client relations
- Improved accounting and auditing effectiveness
- Strengthened internal controls
- Practice development for the CPA firm by identifying those areas in which additional services may be provided

In a typical management comment letter, the CPA firm would discuss the following subjects:

- Deficiencies in internal accounting and administrative controls
- Suggestions relating to accounting matters that arose during the examination
- Recommendations to improve the efficiency and effectiveness of examinations in future years

- Modifications of systems, procedures, and organization to improve the contractor's operating efficiency
- Counseling on tax matters

Clients deserve and therefore should receive a constructive comment letter that reflects the CPA firm's accumulated background and knowledge, both about the industry generally and the contractor specifically. Contractors should carefully review and evaluate these suggestions; they are designed to be constructive tools to assist management and merit careful consideration.

PUBLICLY OWNED CONTRACTORS

The discussion in this chapter thus far has related to both privately owned and publicly owned contractors. Since the reporting requirements of the publicly held contractor create special considerations for its CPA firm, further discussions on this topic are in order.

Interim Reporting

Shareholders and Stock Exchanges. Public companies are subject to the quarterly reporting requirements of the SEC and to the stock exchanges on which their securities are listed. Quarterly earnings and earnings-per-share figures must be disclosed to a company's shareholders and to the stock exchanges as soon as they become available. Publication of quarterly earnings normally takes the form of a news release to the public press and the national newswire sources. Annual and quarterly statements are frequently sent directly to the shareholders.

Form 10-Q. In addition to such disclosures, companies subject to the periodic reporting requirements of the Securities Exchange Act of 1934 are required to file a quarterly report with the SEC on Form 10-Q. A Form 10-Q is filed within 45 days after the end of each of the first three quarters of each fiscal year. No report is required for the fourth quarter. The information provided in a Form 10-Q is much more limited than that required in annual filings with the SEC. Quarterly financial information may consist of condensed financial statements that disclose only major account classifications. Footnote disclosures do not have to be as detailed as those contained in a company's annual financial statements, since it is assumed that the annual report is available to readers of the interim statements. Disclosures should be sufficient so as not to make the interim information misleading; however, the inclusion of an inventory breakdown by major classifications and a discussion of significant commitments and contingencies are required where applicable. The financial statements should include: (1) a condensed balance sheet as of the end of the most recent fiscal quarter and the end of the most recent fiscal year; (2) an income statement for the current fiscal quarter, current fiscal year to date, and the correspendong periods for the preceding fiscal year; and (3) a condensed statement of changes in financial position for the year-to-date period in the current and preceding years.

Form 10-Q financial statements are not required to be audited or reviewed by an independent public accountant. However, it is advantageous for a company to have its financial information reviewed on a quarterly basis, instead of waiting until the required annual audit. Subjecting a company's books and records to a quarterly review allows identification and resolution of accounting issues and problems on a more timely basis. This reduces the likelihood of year-end adjustments to previously reported operating results. Material year-end adjustments require a company to provide a reconciliation in its annual report and to amend the Form 10-Q for the first three fiscal quarters, either at the end of those quarters or when reporting quarterly results for the subsequent fiscal year.

AUDIT COMMITTEES

In recent years, the trend toward the formation of an audit committee of the board of directors is noteworthy. For publicly owned companies, the AICPA and the SEC strongly encourage corporate audit committees of outside directors to provide added assurance to stockholders of the corporate financial statements and to ensure that all possible steps have been taken to provide an independent objective review of the company's financial policies and operations. The New York Stock Exchange requires that all listed companies have an audit committee. For these reasons, the vast majority of publicly owned companies and many large privately owned companies have established audit committees.

The audit committee is comprised solely of outside directors of the board and normally meets three to four times a year. Although the duties and responsibilities of audit committees vary, committee functions often include reviews of the following:

- Recommendation of appointment of the independent CPAs to the board of directors and stockholders

- Scope of the annual audit

- Reports of the independent CPAs' limited reviews of interim financial statements

- The consolidated financial statements of the company and the accountants' report thereon at the completion of the annual examination.

- Independent CPAs' management comment letters

- Operations of the internal audit staff

- Approval of audit fees

- Officers' expense reports

Audit committees are a positive constructive force in corporate governance and improve the efficiency of independent CPAs in the performance of their responsibilities. Audit committees also strengthen the role of financial management within organizations, which helps to improve the effectiveness and independence of the internal audit function.

NONAUDIT SERVICES

The CPA firm usually offers tax, management consulting, and other services in the financial area to its clientele. Service in the tax area is most commonly associated with CPAs and most widely used by the contractor's client. These services can range from actual preparation of state and federal income tax returns to consultation on complex domestic and international tax matters.

Tax Compliance

For the smaller contractor and for those with limited human resources in the financial area, the preparation of state and federal income tax returns, or tax compliance services, are greatly needed. CPA firms usually have considerable experience and expertise in the area of tax compliance and are organized to provide this service effectively. They also have the advantage of familiarity with the contractor's financial data. CPAs can also be of assistance in connection with tax examinations by the IRS and state taxing authorities.

For the larger contractor, particularly those with international operations, an international accounting firm can offer invaluable services in tax planning, including operations in foreign locations and information on dealing with foreign tax authorities.

Management Consulting Services

The planning, administration, and control of contractors and their business organizations has become increasingly complex in today's environment of rapid economic and technological growth and change. To help improve the effectiveness of their organizations, many contractors turn to CPA firms for advice and assistance as professional management consultants.

CPA firms are often uniquely qualified to provide consulting services for contractor's clients for a variety of reasons, including the following:

- CPA firms have a background knowledge of client operations that increases efficiency; and

- CPA firms value and do not wish to jeopardize their existing client audit relationships. Therefore, they strive to perform consulting engagements effectively.

Today, the range of consulting services available within CPA firms is almost unlimited and varies considerably from firm to firm. Emphasis is generally in the following key areas:

- General management
- Information management
- Financial management
- Human resource management
- Operations management
- Marketing management

Other Services

Another area where a CPA firm can be of great assistance to the contractor is in relations with bankers, other credit grantors, and sureties. Frequently, CPAs can assist in preparing presentations to these parties where outside financing is required or additional bonding capacity or coverage is needed. Since the CPA is often able to explain accounting issues that may be unclear to the surety, his involvement can make the difference between success and failure in negotiations of this nature.

CONCLUSION

CPAs are valued professional advisers to contractors; generally a mutually beneficial relationship exists between them. Relations between a CPA firm and a contractor's clients can be maximized with sound planning and communication, resulting in the provision of services in the most productive manner possible. This is effected at the least possible cost to the contractor, who receives the maximum benefit from the accounting firm. The CPA firm adds credibility to the financial statements and presentations in which it participates. Its independence, among other things, establishes its reputation and provides a degree of comfort and reliance to third parties dealing with the contractor. Thus, the unique relationship between the contractor and the CPA firm should be granted the fine professional reputation to which it is entitled.

CHAPTER **10**

External Reporting

John L. Callan

BASIC ACCOUNTING OBJECTIVE

One basic objective of accounting is to provide useful information that aids in the making of financial decisions. In financial accounting, this objective is met through the use of general-purpose financial statements that communicate relevant economic information. Accounting information is used by a wide variety of groups whose concern is the economic activity of business organizations. Those users who benefit directly from having accounting information about business organizations because their information needs are directly related to the commitment of personal or financial resources include the following:

- Management
- Owners and investors (including potential investors)
- Other third-party users:
  - a Credit grantors (including banks and other financial institutions)
  - b Surety companies
  - c Suppliers (vendors) of materials and supplies and subcontractors
  - d Customers (including owners and local, county, state, and federal agencies requiring prequalification financial information)
  - e Employees and/or labor unions
  - f The general public
  - g Public company reporting:
    - Stock exchanges
    - Securities and Exchange Commission (SEC)
    - Security analysts
  - h Government agencies:
    - Internal Revenue Service (IRS)
    - Census Bureau
    - State and local tax agencies

Internal and External Users

Users of financial statements with regard to a business organization can be conveniently separated into two groups: internal and external users. The management of a firm is regarded as an internal user; all others, such as investors and creditors, are external users of accounting information. Management uses accounting information inside the business organization to aid in making decisions that directly affect its firms. Management is an internal user of accounting information for purposes of internal decision making—primarily with regard to planning, directing, and controlling the organization.

External users are those outside the business firm who use accounting information to aid in decision making about their relationship with the business organization. Their decisions, however, do not directly affect the operations of the firm. External users are concerned with business organizations as a source of decision alternatives, such as investment decisions and credit judgments, available to them.

There are few, if any, businesses that are as dependent on external parties as contractors are. External parties in the contracting business often rely on an analysis of the contractor's financial statements as a means of establishing business dealings with the contractor. In many instances, the contractor's ability to continue as a going concern depends on an interpretation of the contractor's financial statements by external parties. For example, a surety's willingness to write a bond for a contractor (a bond that may be required to enable the firm to do work) will probably depend on the surety's appraisal of the contractor's financial capacity based on financial statements.

To serve both the contractor's external users and the contractor, financial statements prepared by the contractor and, in most cases, examined by independent certified public accountants (CPAs) should be prepared in accordance with generally accepted accounting principles (GAAP). A Statement of Position (SOP), referred to as *Accounting for Performance of Construction-Type and Certain Production-Type Contracts*, along with the American Institute of Certified Public Accountants' (AICPA) *Audit and Accounting Guide for Construction Contractors*, known as the *Guide*, provide guidance on the application of GAAP for the construction industry. This chapter provides the contractor and the CPA with guidance and assistance in the preparation of financial statements that conform with GAAP. Emphasis is placed on the unique financial reporting, the correct reporting of supplementary information, and the suitable disclosures that contractors must make.

This chapter does not purport to review and discuss all the accounting reporting and disclosure requirements that may have an impact upon individual contractors under all circumstances. Rather, the discussion is limited to the unique aspects of those requirements that generally apply to contractors alone, by virtue of the very nature of the construction industry.

Thus, the main focus of this chapter is on the following subjects:

- Balance sheet
- Working capital
- Receivables
  - a Billed receivables
  - b Unbilled contract receivables
  - c Allowance for uncollectible amounts
  - d Claims

- Contracts-in-process
  - a Percentage-of-completion method
  - b Completed-contract method
- Joint ventures
- Deferred income taxes
- Sample financial statements
- Financial reporting by affiliated entities
- SEC reporting requirements
- Special reports and supplementary reporting

BALANCE SHEET

The predominant practice in financial reporting for business, including the contracting business, is to present a classified balance sheet generally set forth in the order of liquidity. Although there are many variations in balance sheet captions, the typical classifications are:

ASSETS	LIABILITIES
Cash	Notes payable
Accounts receivable	Accounts payable
Inventories	Accrued expenses
Prepaid expenses	Federal income taxes ____
Total current assets ____	Total current liabilities ____
Property, plant, and equipment (net of accumulated depreciation)	Long-term debt
	Stockholders' equity
	Capital stock
Other assets ____	Retained earnings ____
═══	═══

The determination of the classification of current assets and current liabilities is based on paragraph 5 of Chapter 3A of Accounting Research Bulletin (ARB) No. 30 and the concept of the operating cycle set forth therein, which states that:

The ordinary operations of a business involve a circulation of capital within the current asset group. Cash is expended for materials, ...

labor, and ... services, and such expenditures are accumulated as ... costs. These costs ... are converted into ... receivables and ultimately into cash again. *The average time intervening between the acquisition of materials or services entering this process and the final cash realization constitutes an operating cycle.* A one-year time period is to be used as a basis for the segregation of current assets in cases where there are several operating cycles occurring within a year. *However, where the period of the operating cycle is more than twelve months, as in, for instance, the tobacco, distillery, and lumber businesses, the longer period should be used.* Where a particular business has no clearly defined operating cycle, the one-year rule should govern. The term current liabilities is used principally to designate obligations whose liquidation is reasonably expected to require the use of existing resources properly classifiable as current assets, or the creation of other current liabilities. [Emphasis added.]

In addition, ARB 43, paragraphs 4 and 7, defines current assets and current liabilities in terms of operating cycle:

Current assets:
   For accounting purposes, the term current assets is used to designate cash and other resources commonly identified as those which are reasonably expected to be realized in cash or sold or consumed during the operating cycle of the business. Thus the term comprehends in general such resources as (a) cash available for current operations and items which are the equivalent of cash; (b) inventories of merchandise, raw materials, goods in process, finished goods, operating supplies, and ordinary maintenance material and parts; (c) trade accounts, notes, and acceptances receivable; (d) receivables from officers, employees, affiliates, and others, if collectible in the ordinary course of business within a year; (e) installment or deferred accounts and notes receivable if they conform generally to normal trade practices; (f) marketable securities representing the investment of cash available for current operations; and (g) prepaid expenses such as insurance, interest, rents, taxes, unused royalties, current paid advertising services not yet received, and operating supplies.

Current liabilities:
   The term current liabilities is used principally to designate obligations whose liquidation is reasonably expected to require the use of existing resources properly classifiable as current assets, or the creation of other current liabilities. The classification is intended to include obligations for items which have entered into the operating cycle, such as payables incurred in the acquisition of materials and supplies to be used in the production of goods or in providing services to be offered for sale; collections received in advance of the delivery of goods or performance of services; and debts which arise

from operations directly related to the operating cycle, such as accruals for wages, salaries, commissions, rentals, royalties, and income and other taxes. Other liabilities whose regular and ordinary liquidation is expected to occur within a relatively short period of time, usually twelve months, are also intended for inclusion, such as short-term debts arising from the acquisition of capital assets, serial maturities of long-term obligations, amounts required to be expended within one year under sinking fund provisions, and agency obligations arising from the collection or acceptance of cash or other assets for the account of third persons.

These definitions are relatively straightforward and result in a high degree of uniformity and consistency evidenced by the vast majority of business entities. However, guidance is frequently useful for contractors in view of: (1) the difficulty in determining the operating cycle, and (2) the unique nature of assets and liabilities in the construction industry. Furthermore, working capital—the excess of current assets over current liabilities—is considered particularly significant by sureties and other credit grantors in their analysis and appraisal of a contractor's operations and financial capability.

Quite often, it is difficult to determine with any degree of accuracy the operating cycle for contractors; this is due to the varying periods of contracts from award to substantial completion of the project. For those specialty contractors with a fairly consistent time frame on projects from inception to completion, the operating cycle may be relatively clearly defined. However, where the contractor's backlog is diverse and the length of individual projects changes significantly from period to period, the operating cycle (generally the average duration of contracts in progress) will also vary accordingly. In addition, the operating cycle, which is defined in the *Guide* as "the average time intervening between the inception of contracts and the substantial completion of contracts" will vary from year to year, based on both the mix of contracts in process and their stage of completion.

For these reasons, it is the predominant practice in the construction industry to classify all contract-related assets and contract-related liabilities as current when preparing a classified balance sheet.

The following contract-related assets and liabilities are set forth in the *Guide* and would be shown as current in a classified balance sheet.

Contract-related assets include:

- Accounts receivable on contracts (including retentions)
- Unbilled contract receivables
- Costs in excess of billings and estimated earnings

- Other deferred contract costs
- Equipment and small tools specifically purchased for, or expected to be used solely on, an individual contract

Contract-related liabilities include:

- Accounts payable on contracts (including retentions)
- Accrued contract costs
- Billings in excess of cost and estimated earnings
- Deferred taxes resulting from the use of a method of income recognition for tax purposes different from the method used for financial-reporting purposes
- Advanced payments on contracts for mobilization or other purposes
- Obligations for equipment specifically purchased for, or expected to be used solely on, an individual contract regardless of the payment terms of the obligations
- Provision for losses on contracts

Guidance in relation to various aspects of this listing of contract-related assets and liabilities is provided later in this chapter.

Classified vs. Unclassified Balance Sheets

As noted earlier, the customary practice for most business entities is to include a classified balance sheet in their financial statements. This is also the predominant practice in the construction industry, although the *Guide* indicates that an unclassified balance sheet may be preferable for those contractors whose operating cycle exceeds one year.

The recommendations in the *Guide* follow:

- A classified balance sheet is preferable for entities whose operating cycle is one year or less. An entity whose operating cycle for most of its contracts is one year or less but that periodically obtains some contracts that are significantly longer than normal may use a classified balance sheet with a separate classification and disclosure for items relating to contracts that deviate from its normal operating cycle. For example, if a company with a normal cycle of one year obtains a substantial contract that greatly exceeds one year, it may still use a classified balance sheet if it excludes from current assets and liabilities the assets and liabilities related to the contract that are expected to be realized or liquidated after one year and discloses in the financial statements information on the realization and maturity of these items. Where appropriate, the one-year basis of classification presents information in a form preferred by many sureties and credit grantors as one of the many tools that they use to make analyses of a contractor's operations and financial statements.

- An unclassified balance sheet is preferable for entities whose operating cycle exceeds one year. An entity whose operating cycle exceeds one year may also use a classified balance sheet with assets and liabilities classified as current on the basis of the operating cycle if, in management's opinion, an unclassified balance sheet would not result in a meaningful presentation.

While these recommendations indicate a preference for a classified balance sheet for contractors for contracts of relatively short duration and an unclassified balance sheet for contractors with an operating cycle of longer duration, the guidance provided by the *Guide* is merely that— guidance. While the unclassified balance sheet is described by the *Guide* as preferable for contractors with contracts of long duration, it is not a required statement format.

Since publication of the *Guide* in 1981, there has been no discernible trend toward unclassified balance sheets in the published financial statements of publicly owned construction companies. It is not possible to state authoritatively the reasons for the lack of any trend in this direction. However, it may be attributable to the following:

- Management's opinion that an unclassified balance sheet would not result in a more meaningful presentation, or a more meaningful statement, than a classified balance sheet.

- A belief that investors, potential investors, and financial analysts would not readily accept a statement format significantly different from that used by the vast majority of other publicly owned companies.

- An unclassified balance sheet would not be generally accepted by sureties, credit grantors, and other third-party users of contractors' financial statements.

- Where an unclassified balance sheet is used by a construction company, additional footnote disclosure concerning the liquidity characteristics of various assets and liabilities is recommended, a fact that further compounds the already complex disclosure requirements.

RECEIVABLES

Receivables in the construction industry are unique as compared to receivables in other business entities. Certain of the unique characteristics of these receivables require special consideration for financial reporting purposes. A contractor's receivables generally include:

Contract receivables
  Billed:
      Completed contracts $xxx
      Contracts in process   xxx
      Retentions   xxx
  Unbilled   xxx
  Claims and change orders $xxx

The term accounts receivable is common to the financial statements of most entities; however, it is too broad for informative disclosure of a contractor's receivables. Thus, it is recommended that the appropriate presentation on a contractor's balance sheet or in the notes to his financial statement include information as to the composition of the total amounts receivable on contracts:

Contract receivables
  Billed:
      Completed contracts $xxx
      Contracts in process   xxx
      Retentions on contracts in process xxx
  Unbilled   xxx
                                   xxx
  Less: Allowance for uncollectible amounts (xxx)
                                   $xxx

This detailed presentation of the composition of contract receivables is often not included in the annual reports of those relatively few contractors who are publicly owned and who thus prepare annual reports for their stockholders. Since receivables are of particular interest to third-party users of a contractor's financial statements (particularly to sureties and other credit grantors), detailed information of this nature is appropriate and, therefore, recommended.

Billed Receivables

Completed Contracts. The amount receivable for completed contracts should be set forth separately, as these amounts are normally collectible on a relatively short-term basis. Retentions on completed contracts should be included with this balance if they become due from the owner or if they will become due in the near future when the project is accepted by the owner. In the event that retentions on completed contracts will not be realized at the completion of the project because of special contractual arrangements with the owner, consideration should be given to: (1) classifying such amounts as noncurrent if appropriate, or (2) disclosure of the payment terms if the amounts are material.

Retentions. In the construction industry, progress billing made in accordance with the terms of the contract is normally subject to a retention by the owner in amounts that frequently range from 5 to 15 percent. Retentions are normally due and collectible by the contractor from the owner upon completion and acceptance. Separate disclosure of retentions on contracts in process is required as these contract-related assets are not currently receivable by the contractor.

Retentions may be withheld by the owner in the event that certain contract guarantees regarding performance need to be fulfilled prior to his final acceptance. Under these circumstances, the contractor should make adequate provision for the costs related to the fulfillment of these contract guarantees. Thus, it is appropriate to deduct the amounts accrued for such expenditures (that do not represent definite amounts payable to specific subcontractors or suppliers) from retentions receivable, rather than to classify the amount accrued as a current liability.

The *Guide* states that retentions that will not be realized within a contractor's normal operating cycle, that is, upon completion and acceptance (perhaps because of special payment arrangements with the owner), should be classified as noncurrent assets.

Unbilled Contract Receivables

Unbilled contract receivables arise where it is appropriate to recognize revenues on a contract but the amount cannot be billed to the owner under the terms of the contract until a later point in time. These amounts typically arise from a variety of circumstances:

- Routine delays in billing (such as billing in the subsequent month for work performed in the previous month).

- Billing delays caused by the payment schedule in the contract with the owner, whereby amounts billable are less than amounts earned as revenue based on the extent of progress on the project.

- Unbilled amounts arising from the use of either of the two alternative methods of computing income on the percentage-of-completion method, as described in Chapter 3.

While unbilled receivables may exist for contractors using the completed-contract method of income recognition as well as those using the percentage-of-completion method, contractors following the former method often do not record unbilled receivables caused by billing delays. Thus, these amounts would be included in costs on uncompleted contracts until such time as the billing is rendered, rather than as a contract receivable. Since receivables are generally considered to be more rapidly converted to cash, contractors using the completed-contract method of income recognition should be concerned with recording unbilled receivables.

As described in Chapter 3 of this book and in paragraphs 79–81 of SOP 81-1, there are two widely used and equally accepted approaches to determining earned revenue and cost of earned revenue. Method *B* is more commonly used by contractors. Although both methods result in the same gross profit for financial-reporting purposes, the gross profit ratio is generally more consistent with the measurement of contract performance under Method *A*. Furthermore, if there are no significant changes in estimated contract gross profit during the performance of the contract, there will be a more consistent gross profit percentage. The two methods also result in differing balances for unbilled contract receivables, as shown in the following examples.

Example 1. Earned revenues in excess of contract billings:

Total estimated contract	$1,000,000
Percentage complete	× 75%
Total earned revenues	$ 750,000
Contract billings to date	$ 600,000
Costs incurred to date	750,000
Estimated total contract profit	$ 100,000

Computation of unbilled receivables:

METHOD *A*

Total earned revenue (75% of estimated total contract)	$ 750,000
Less: Amount billed to date	(600,000)
Unbilled receivables	$ 150,000

METHOD *B*

Costs incurred to date	$ 750,000
Estimated profit earned to date (75% of $100,000)	75,000
Total earned revenue	$ 825,000
Less: Amount billed to date	(600,000)
Unbilled receivables	$ 225,000

Example 2. Earned revenues equal to contract billings:

Total estimated contract	$1,000,000
Percentage complete	×60%
Total earned revenues	$ 600,000
Contract billings to date	$ 600,000
Costs incurred to date	600,000
Estimated total contract profit	$ 100,000

Computation of unbilled receivables:

METHOD *A*

Total earned revenue (60% of estimated total contract)	$ 600,000
Less: Amount billed to date	(600,000)
Unbilled receivables	$ -0-

METHOD *B*

Costs incurred to date	$ 600,000
Estimated profit earned to date (65% of $100,000)	60,000
Total earned revenue	$ 660,000
Less: Amount billed to date	(600,000)
Unbilled receivables	$ 60,000

Example 3. Earned revenues less than contract billings:

Total estimated contract	$1,000,000
Percentage complete	× 60%
Total earned revenues	$ 600,000
Contract billings to date	$ 700,000
Costs incurred to date	600,000
Estimated total contract profit	$ 100,000

Computation of unbilled receivables:
  METHOD *A*

Total earned revenue (60% of estimated total contract)	$1,000,000
Less: Amount billed to date	(700,000)
Unbilled receivables	$ -0-

  METHOD *B*

Costs incurred to date	$ 600,000
Estimated profit earned to date (65% of $100,000)	60,000
Total earned revenue	$ 660,000
Less: Amount billed to date	(700,000)
Unbilled receivables	$ -0-

STATEMENT EFFECT

Example 1.

	Method A	Method B
Current asset		
Unbilled contract receivable	$150,000	$225,000
Current asset (liability)		
Contract costs and estimated earnings in excess of billings:		
Contract costs	$750,000	$750,000
Estimated earnings	75,000	75,000
	$825,000	$825,000
Revenues (billed and unbilled)	750,000	825,000
Net	$ 75,000	$ -0-
Net contract balance	$225,000	$225,000

Example 2.

	Method A	Method B
Current asset		
Unbilled contract receivable	$ -0-	$ 60,000
Current asset (liability)		
Contract costs and estimated		
earnings in excess of billings:		
Contract costs	$600,000	$600,000
Estimated earnings	60,000	60,000
	$660,000	$660,000
Revenues (billed and unbilled)	600,000	600,000
Net	$ 60,000	$ -0-
Net contract balance	$ 60,000	$ 60,000

Example 3.

	Method A	Method B
Current asset		
Unbilled contract receivable	$ -0-	$ -0-
Current asset (liability)		
Contract costs and estimated		
earnings in excess of billings:		
Contract costs	$600,000	$600,000
Estimated earnings	60,000	60,000
	$660,000	$660,000
Revenues (billed and unbilled)	700,000	700,000
Net	($ 40,000)	($ 40,000)
Net contract balance	($ 40,000)*	($ 40,000)*

*Balance would be included in current liabilities under caption: "Billings in excess of costs and estimated earnings on contracts in process."

While both of these methods are equally acceptable, there is generally a balance arising from unbilled contract receivables under Method B. Where the cost-to-cost method of determining percentage of completion is used, there is usually no balance on individual contracts for costs and estimated earnings in excess of billing on uncompleted contracts, and a liability balance exists only when a contract is overbilled at statement date.

Allowance for Uncollectible Amounts

As in the financial statement of any entity, a contractor should make adequate allowance for uncollectible receivable balances. Contractors experience collection problems relating to receivables that are different from the collection problems of other commercial and industrial companies. In addition to any normal allowances that may be required due to the inability of the creditor to meet his financial obligations, additional allowances may be needed for possible billing disputes arising from such items as change orders, extra claims, or back charges. The extent of the allowance required for these items varies according to the contractor's individual accounting policies.

Claims

Paragraphs 65–67 of SOP 81-1 set forth the following guidance with respect to the accounting for claims.

Claims

65. Claims are amounts in excess of the agreed contract price (or amounts not included in the original price) that a contractor seeks to collect from customers or others for customer-caused delays, errors in specifications and designs, contract terminations, change orders in dispute or unapproved as to both scope and price, or other causes of unanticipated additional costs. Recognition of amounts of additional contract revenue relating to claims is appropriate only if it is probable that the claim will result in additional contract revenue and if the amount can be reliably estimated. Those two requirements are satisfied by the existence of all the following conditions:

(a) The contract or other evidence provides a legal basis for the claim; or a legal opinion has been obtained, stating that under the circumstances there is a reasonable basis to support the claim.

(b) Additional costs are caused by circumstances that were unforeseen at the contract date and are not the result of deficiencies in the contractor's performance.

(c) Costs associated with the claim are identifiable or otherwise determinable and are reasonable in view of the work performed.

(d) The evidence supporting the claim is objective and verifiable, not based on management's "feel" for the situation or on unsupported representations.

If the foregoing requirements are met, revenue from a claim should be recorded only to the extent that contract costs relating to the claim have been incurred. The amounts recorded, if material, should be disclosed in the notes to the financial statements. Costs attributable to claims should be treated as costs of contract performance as incurred.

66. However, a practice such as recording revenues from claims only when the amounts have been received or awarded may be used. If that practice is followed, the amounts should be disclosed in the notes to the financial statements.

67. If the requirements in paragraph 65 are not met or if those requirements are met but the claim exceeds the recorded contract costs, a contingent asset should be disclosed in accordance with FASB Statement No. 5, paragraph 17.

Contractors follow a wide range of practices in the accounting for claims, a fact that was recognized when the SOP was prepared. These guidelines set forth the appropriate accounting and reporting disclosures. When these criteria for recognizing a claim are met, claims should be recorded as a receivable—either billed or unbilled, as appropriate. In the event that the total amount of claims receivable is material, it should be set forth as a separate receivable item on the balance sheet or disclosed in the notes to the financial statements.

When claims receivable are recorded, they are generally considered to be contract-related assets and set forth as a current asset on a classified balance sheet. Nevertheless, consideration should be given to classifying claims—including both costs and revenues attributable thereto as appropriate—as noncurrent under those circumstances where the resolution of the claim and its ultimate collection may be the subject of considerable delay resulting from negotiation or litigation with the owner.

In those instances where the criteria set forth in the SOP for recognition of a claim are not met or if the claim exceeds the recorded contract costs (and the amounts are material), a contingent asset should be disclosed in the notes to the financial statements in accordance with Statement of Financial Accounting Standards (SFAS) 5, paragraph 17:

Gain Contingencies
17. The Board has not reconsidered ARB No. 50 with respect to gain contingencies. Accordingly, the following provisions of paragraphs 3 and 5 of that Bulletin shall continue in effect:
   (a) Contingencies that might result in gains usually are not reflected in the accounts since to do so might be to recognize revenue prior to its realization.
   (b) Adequate disclosure shall be made of contingencies that might result in gains, but care shall be exercised to avoid misleading implications as to the likelihood of realization.

When disclosure of the existence of a contingent asset is made in the notes to the financial statements, care must be exercised to avoid misleading inferences as to the ultimate resolution and collectibility of the claim.

CONTRACTS IN PROCESS

Costs applicable to contracts in process are recorded and accumulated by individual contracts in a manner similar to inventories and are charged to the cost of earned revenues as the related contract revenue is recognized. Contract costs normally include direct costs of materials, labor, and subcontractor's billing, as well as certain indirect costs applicable to or allocated to a particular contract. Additional information related to contract costs can be found in Chapter 4.

Percentage-of-Completion Method

Under the percentage-of-completion method of accounting, there is normally a difference between the costs accumulated and the estimated earnings recognized to date on the contract compared with the amount of billings on that contract. For those contracts on which contract costs and estimated earnings are in excess of billing, the balance represents an underbilling and would be reflected in a classified balance sheet as a current asset, with a caption similar to the following:

> *Costs and estimated earnings in excess of billings on uncompleted contracts*
>> (Note: For contractors employing Method *B*, this amount would normally be included in unbilled receivables.)

For those contracts on which billings are in excess of costs and estimated earnings, the balance represents an overbilling; on a classified balance sheet, it would be reflected as a current liability.

Billing in Excess of Costs and Estimated Earnings

Although accounting literature has for many years been abundantly clear in relation to offsetting overbillings against underbillings when reporting a single net amount for contracts in process, nevertheless, questions occasionally arise relating to this subject. A restatement of the guidance in the AICPA *Guide* is appropriate here:

Offsetting or Netting Amounts

A basic principle of accounting is that assets and liabilities should not be offset unless a right of offset exists. Thus the net debit balances for certain contracts should not ordinarily be offset against net credit balances relating to others, unless the balances relate to contracts that meet the criteria for combining in the SOP.

In paragraph 5, ARB 45 recognized the principle of offsetting in discussing the two accepted methods of accounting for long-term construction-type contracts. For the percentage-of-completion method, ARB 45 states that:

> current assets may include costs and recognized income not yet billed, with respect to certain contracts; and liabilities, in most cases current liabilities, may include billings in excess of costs and recognized income with respect to other contracts.

In commenting on the completed-contract method, ARB 45 states in paragraph 12 that:

> an excess of accumulated costs over related billings should be shown in the balance sheet as a current asset, and an excess of accumulated billings over related costs should be shown among the liabilities, in most cases as a current liability. If costs exceed billings on some contracts, and billings exceed costs on others, the contracts should ordinarily be segregated so that the figures on the asset side include only those contracts on which costs exceed billings, and those on the liability side include only those on which billings exceed costs.

According to page 50 of the *Guide*, offsetting should be applied in the same way under the percentage-of-completion method.

> Although the suggested mechanics of segregating contracts between those on which costs exceed billings and those on which billings exceed costs do not indicate whether billings and related costs should be presented separately or combined (netted), separate disclosure in comparative statements is preferable because it shows the dollar volume of billings and costs (but not an indication of future profit or loss). In addition, grantors of credit, such as banks and insurance companies, have expressed a preference for separate disclosure. Disclosure may be made by short extension of the amounts on the balance sheet or in the notes to the financial statements. Thus, under the percentage-of-completion method, the current assets may disclose separately total costs and total recognized income not yet billed for certain contracts, and current liabilities may disclose separately total billings and total costs and recognized income for other

contracts. The separate disclosure of revenue and costs in statements of income is the generally accepted practice. Only through comparable presentation of such data in the balance sheet can the reader adequately evaluate the contractor's comparative position.

When the disclosure recommended above is made in the notes to the financial statements, the following can serve as an illustration of the suggested comparative format:

Costs incurred on uncompleted contracts	$xxx	$xxx
Estimated earnings thereon	xxx	xxx
	xxx	xxx
Less: Billings applicable thereto	(xxx)	(xxx)
	$xxx	$xxx
Included in accompanying balance sheet under the following captions:		
Cost and estimated earnings in excess of billings on uncompleted contracts	$xxx	$xxx
Billings in excess of costs and estimated earnings	xxx	xxx

The presentation of contract costs, earnings, and billings where contracts are combined or segmented for purposes of determining an appropriate profit center for purposes of income recognition should be treated in a similar manner for balance sheet presentation.

Excess of Overbillings. Billings in excess of costs and estimated earnings are reflected as a current liability on a contractor's classified balance sheet. Under the infrequent circumstances where billings exceed total estimated cost at completion of the contract plus estimated earnings to date, this amount should be classified as deferred income.

The argument often advanced is that all or some portion of the billings in excess of costs and estimated earnings should not be included as a current liability on a classified balance sheet because these amounts represent deferred income. Except for those instances where billings exceeded total estimated cost at completion plus estimated earnings to date, this concept was rejected in the *Guide*, which states on page 49 that "excess billings should be regarded as obligations for work to be performed and classified as a current liability."

Although all of the authoritative accounting literature supports this position and the vast majority of contractors' financial statements follow the guidance in the *Guide*, research has indicated that a very limited number of publicly owned construction companies have classified a portion of excess billings as noncurrent, that is, as deferred credit. In those instances where footnote disclosure sets forth the basis for the classification as a deferred credit, it is generally in the following form:

> Billings in excess of costs and recognized earnings on uncompleted contracts are classified as current liabilities only to the extent that they exceed estimated unearned profit on the contracts.

Since working capital is vitally important to the contractor, particularly in relation to the determination of financial capacity by sureties and other credit grantors, caution should be exercised prior to adopting this method of financial reporting of overbilling, despite the apparent benefits.

Accordingly, in the event that (1) excess billings are classified as a deferred credit or (2) a construction company proposes to adopt this method of presenting statements of excess billings, it is important to note that it is not in conformity with current authoritative literature; therefore, the burden of justification for the method's adoption rests with the reporting company.

In the event that some portion of excess billings is not classified as current liabilities by a contractor, the following factors and criteria must be considered:

- The deferred credit should be reduced by deferred income taxes, which are considered to be a contract-related liability and, accordingly, a current liability.

- The contractor should have a strong estimating process. In fact, historical experience over a long-term period should indicate minimal variance (e.g., 10 percent) between estimated and actual project profits.

- It may be inappropriate to classify excess billings as deferred credits where excess billings are attributable to front-end loading and the contract has not progressed sufficiently to provide reasonable assurance that estimated total profit on the project is reliable.

It is probable that in financial statements issued in the next few years, an increased number of contractors will resort to classifying some portion of excess billings as deferred credit. Thus, it is important that reasonable guidelines, similar to those just suggested, be observed in order to prevent the improper or inappropriate classification of credit balances arising from excess billings.

Completed-Contract Method

For those construction companies using the completed-contract method of revenue recognition, uncompleted contracts on which contract costs exceed billings are reported as a current asset. Where billings exceed contract costs, the amount is reported as a current liability:

- *Current asset* – cost in excess of billings on uncompleted contracts.

- *Current liability* – billings in excess of costs on uncompleted contracts.

The guidelines set forth for contractors using the percentage-of-completion method apply equally to those using the completed-contract method in terms of the following:

- Offsetting over- and underbillings; and

- Disclosures on the balance sheet (or in the notes to financial statements) of the amounts of costs and billings.

For those contractors using the completed-contract method, there is no basis for the classification of any portion of excess billings as a deferred credit rather than as a current liability. The criteria for selecting the completed-contract method of revenue recognition set forth in paragraphs 31–32 of SOP 81-1 are incompatible with the suggested guidelines for classifying a portion of the excess billings as deferred credit for contractors employing the percentage-of-completion method.

Losses on Contracts. Provision should be made for the entire loss on a contract in the period when current estimates of total contract costs indicate a loss (see Chapter 3 for a discussion of income recognition). This applies to both the percentage-of-completion method and the completed-contract method.

The accrual for such contract losses should be used as a reduction of the related contract costs that are included in current assets or added to those that are included in current liabilities. An allowance for loss may change a contract that would otherwise be classified as a current asset to one that included current liabilities, inasmuch as the accrual for a contract loss represents additional contract costs for which the contractor will not be reimbursed by the owner. The following reporting requirements related to the contractor's financial statements should be noted:

- The provision for loss should be accounted for in the income statement as an additional contract cost rather than as a reduction of contract revenues.

- Separate disclosure in the income statement, or in notes to financial statements, is necessary only if the loss provison is material.

- Where the accrual for loss is netted against contract costs that are included as an asset or added to those included in current liabilities, material amounts should be appropriately disclosed parenthetically in the statement caption or in the notes to the financial statements.

JOINT VENTURES

Joint ventures are widely used in the construction industry (see Chapter 2 for a full explanation of the reasons). In addition, since there is little or no authoritative guidance in accounting literature covering the accounting for joint ventures by investors, a wide variety of reporting methods have evolved to accommodate a contractor's joint venture operations. At present, at least five different methods of recording a venturer's interest in joint venture operations are followed, all of which are identified in the *Guide*.

1 *Consolidation.* The venture is fully consolidated, with the other venturers' interests shown as minority interest.

2 *Partial or proportionate consolidation.* The venturer records its proportionate interest in the venture's assets, liabilities, revenues, and expenses on a line-by-line basis and combines the amounts directly with its own assets, liabilities, revenues, and expenses without distinguishing between the amounts related to the venture and those held directly by the venturer.

3 *Expanded equity method.* The venturer presents its proportionate share of the venture's assets and liabilities, separated into current and noncurrent, in capsule form. The elements of the investment are presented separately by including the venturer's equity in the venture's corresponding items under current assets, current liabilities, noncurrent assets, noncurrent liabilities, revenues, and expenses, using a caption such as "investor's share of net current assets of joint ventures."

4 *Equity method.* The equity method is the traditional one-line method prescribed by the Accounting Principles Board (APB) Opinion 18 for investments in corporate joint ventures and for investments in common stock that represent less than a majority interest but that evidence an ability to exercise signficant influence over the investee.

5 *Cost.* The investment is recorded at cost, and income is recognized as distributions are received from earnings accumulated by the venture since the date of acquisition by the venturer.

Figure 2-5 in Chapter 2 summarizes the conditions under which each of the above methods may be most appropriate. It is significant to note here that there is little or no uniformity with respect to the use of these methods. In addition, in the absence of specific guidance in the authoritative literature, the above methods or any combinations thereof have been considered acceptable in the construction industry in accounting for investments in joint ventures. Combinations of these methods, such as using the equity method of reporting the joint venture investment on the balance sheet and using the consolidation of proportional consolidation method on the income statement, are often used in the construction industry (see Chapter 2).

In view of (1) the diversity in practice in the construction industry in accounting for joint ventures and (2) the need to provide further guidance with respect to accounting for equity investments, the agenda of the Financial Accounting Standards Board (FASB) currently includes the subject of reporting the joint venture investment along with the accounting for joint ventures. When this guidance is provided, it is hoped that the unique problems of joint ventures in the construction industry will be recognized.

Depending on the method employed by the contractor in reporting the investment in, and the operations of, joint ventures, the following factors should be considered.

- In those instances where a contractor accounts for the investment in a joint venture on the cost or equity method (and a classified statement is prepared), that investment may be classified as a current asset, unless the venture is of a long-term nature or the operating cycle of the venture is significantly in excess of the operating cycle of the contractor. It is difficult to generalize where classified statements are concerned; relevant facts should be considered for each given situation.

- Where the investment in joint ventures is accounted for on the equity or expanded equity basis, provision should be made for any deferred income taxes resulting from the recognition of joint venture earnings that are not currently subject to taxation. The amount of deferred taxes should be classified as current or noncurrent, consistent with the classification of the joint venture investment as a current or noncurrent asset.

- Where the accounts of the joint venture are fully consolidated with those of the venturer, appropriate provision should be made for deferred income taxes, where applicable. In addition, the liability for amounts due to co-venturers representing their interests in venture assets and earnings (similar to minority interests) should be generally classified as a current liability on a classified balance sheet. If the venture assets include significant noncurrent assets, such as construction equipment, they should be classified as current or noncurrent consistent with the classification of the underlying assets of the joint venture.

- Where joint venture losses are incurred in excess of a venturer's investment, they should be recognized by the venturer if the venture is liable for the obligations of the joint venture or otherwise committed to provide ongoing financial support for the venture. A loss would normally be reflected as a current liability to the extent that an accrual for the loss exceeded the investment in the venture.

- In the event that the accounts of a venture are prepared on a basis that is not in accordance with GAAP, as might be the case in connection with foreign joint ventures, appropriate adjustments should be made prior to applying the equity method of accounting or full or partial consolidation.

Disclosures

The accounting policy followed by the contractor in reporting for joint ventures, along with the contractor's other significant accounting policies, should be indicated in accordance with APB Opinion 22. In addition, the *Guide* recommends that consideration be given by a venturer to the following additional disclosures:

1 The name of each joint venture, the percentage of ownership, and any important provisions of the joint venture agreement.

2 If the joint venture's financial statements are not fully consolidated with those of the venturer, separate or combined financial statements of the ventures in summary form, including disclosure of the accounting principles of the ventures that differ significantly from those of the venturer.

3 Intercompany transactions during the period and the basis of intercompany billings and charges.

4 Liabilities and contingent liabilities arising from the joint venture arrangement.

DEFERRED INCOME TAXES

At present, it is common for contractors to use a method of accounting to determine taxable income that differs from the basis of accounting used for financial-reporting purposes. For example, for income tax purposes, contractors have had available the following methods:

- *Cash method,* whereby gross income (i.e., cash, property, or services) is included in income in the taxable year of receipt (or constructive receipt) and expenditures, including contract costs, are deductible in the year paid.

- *Accrual method,* which reports as revenue amounts billed on contracts, excluding retentions, and deducts contract costs incurred to the date of the most recently rendered contract billing.

These two methods are not in accordance wtih GAAP. The following methods, however, are generally accepted, subject to the guidelines set forth in SOP 81-1:

- Completed-contract method
- Percentage-of-completion method

Inasmuch as contractors have not been required to use the same method of accounting for determining taxable income as that used for financial-reporting purposes, timing differences in the recognition of income frequently occur. In general, APB Opinion 11 requires the recognition of deferred tax effects resulting from timing differences in the determination of income tax provisions in current periods.

In most instances, contractors' financial statements are prepared using the percentage-of-completion method of accounting. Nevertheless, for effective tax and cash management planning purposes, contractors often employ the completed-contract method, or one of the other methods noted above, for income tax reporting purposes. Where this occurs, appropriate provision must be made in income statements for income taxes payable that are deferred until such time when the related contract revenue is recognized for income tax purposes. For contractors, the amount of taxes deferrable is often significant and is, therefore, an important aspect of cash management, particularly when the contractor's profit is represented by contract retentions that are not normally collectible until final completion and acceptance.

The income statement presentation of income tax expense should disclose the composition of income taxes as between amounts currently payable and amounts representing taxes allocable to the period, as follows:

Income before taxes on income		$xxx,xxx
Taxes on income		
Taxes currently payable	$xxx,xxx	
Deferred taxes	xxx,xxx	
		xxx,xxx
Net income		$xxx,xxx

The question of balance sheet classification of deferred taxes arising from timing differences in the recognition of contract revenues is addressed on page 48 of the *Guide*, where it is indicated that contract-related liabilities include:

> Deferred taxes resulting from the use of a method of income recognition for tax purposes different from the method used for financial reporting purposes.

This position is consistent with the guidance set forth in paragraph 22 of SFAS 37, stated below.

> 22. An enterprise reports profits on construction contracts on the completed contract method for tax purposes and the percentage-of-completion method for financial reporting purposes. The deferred income tax credits do not relate to an asset or liability that appears on the enterprise's balance sheet; the timing differences will only reverse when the contracts are completed. Receivables that result from progress billings can be collected with no effect on the timing differences, and the timing differences will reverse when the contracts are deemed to be complete even if there is a waiting period

before retentions will be received. Accordingly, the enterprise would classify the deferred income tax credits based on the estimated reversal of the related timing differences. *Deferred income tax credits related to timing differences that will reverse within the same time period used in classifying other contract-related assets and liabilities as current (for example, an operating cycle) would be classified as current.* [Emphasis added.]

Therefore, it is abundantly clear that when contract-related assets and contract-related liabilities are shown as current on a classified balance sheet, the deferred taxes applicable to contracts in process should be recorded as a current liability.

In the event that the contractor's balance sheet is presented in an unclassified basis, it is recommended that the notes to the financial statements include information as to the periods in which these taxes will ultimately become payable.

There are currently two developments in process that could possibly have a future impact on the accounting for deferred income taxes:

1 The IRS has issued proposed regulations that will restrict the use of the completed-contract method in the future. Although it is anticipated that the implementation of these regulations will narrow the areas of difference between income recognition for tax- and financial-reporting purposes, it is currently believed that these areas of difference will continue in the future and that there will be a continued need to recognize and account for deferred income taxes.

2 The current agenda of the FASB includes reconsideration of the need to account for deferred income taxes as currently set forth in APB Opinion 11. From this review, it is probable that there will be significant revision to the present requirements of APB Opinion 11. However, it is difficult to believe that any future standard will not recognize the unique aspects of deferred taxes for contractors.

Significant Accounting Policies

APB Opinion 22 requires that significant accounting policies should be disclosed in notes to financial statements. The focus here is on those unique disclosure requirements applicable to contractors.

The significant disclosure requirements relating to accounting policies of contracts include the following:

- *Method of reporting affiliated entities.* Information relating to the method of reporting by affiliated entities (e.g., consolidation, combination, or separate statement) should be set forth along with the disclosures set forth under reporting for affiliated entities.

- *Operating cycle.* Where the operating cycle exceeds one year, the normal range of contract duration should be indicated.

- *Revenue recognition.* The method of recognition of revenue, including where applicable the policy with respect to combining or segmenting contracts, should be disclosed. In addition, if the percentage-of-completion method is used, disclosure should be made of the method or methods of computing the percentage-of-completion (e.g., cost-to-cost, labor hours, unit delivery).

  a If the completed-contract method is used, disclosure should be made on the basis of selection of the method and the criteria employed to determine completion.

  b In those instances where a contractor's basic accounting policy is percentage-of-completion, the completed-contract method may be used for a single contract or group of contracts where the criteria for the use of percentage-of-completion are not present. Such a departure from the basic policy should be disclosed.

  c Where there is a question of losses on contracts, disclosure of the amount of the loss provision in the income statement or in the notes to the financial statements is only necessary if the loss provision is material.

Revision of Estimates

Estimating is a continuous and normal process in the construction industry, and changes are not uncommon adjustments in estimated contract profits over the life of a contract. As SOP 81-1 indicates, changes should be accounted for on the cumulative catch-up basis and should be so disclosed. It is not necessary to report the effect of revisions on estimates unless the effect of the change in estimate is material.

Claims

In the event claims are recognized as assets and included in contract revenues, the accounting policy and the amounts, if material, should be disclosed. Similarly, if the contractor follows the policy of recording revenues from claims only when the amounts have been awarded or upon receipt, the policy and the amounts should be disclosed in the notes to the financial statements.

Furthermore, where appropriate, the existence of claims as contingent assets should be disclosed in the notes to the financial statements in accordance with SFAS 5.

Methods of Accounting for Investments in Joint Ventures

The method for accounting for investments in joint ventures should be disclosed in the notes to the financial statements. In addition, other disclosures relating to significant joint ventures are recommended:

- The name of each joint venture, the percentage of ownership, and any important provisions of the joint venture agreement should be disclosed.

- Where the joint venture's financial statements are not fully consolidated with those of the venturer, separate or combined financial statements of the ventures in summary form, including disclosure of accounting principles of the ventures that differ significantly from those of the venturer, are needed.

- Intercompany transactions during the period and the basis of intercompany billings and charges should be disclosed.

- Liabilities and contingent liabilities arising from the joint venture arrangement should be disclosed.

Contract Costs

Information should be provided relating to the aggregate amounts included in contract costs of items subject to uncertainty (e.g., claims together with a description of the nature and status of these items). The amount of progress payments netted against contracts at the statement date should be set forth as well.

Deferred Costs

The contractor's policy with respect to costs deferred in anticipation of future contracts and the amounts involved, if material, should be disclosed.

Other Disclosures

Receivables. The following disclosures are recommended relating to contract receivables:

- A description of the nature and status of uncertain items (e.g., unapproved change orders and claims).

- Disclosure of amounts representing the recognized sales value of performance under contracts if the amounts have not been billed and are not billable to customers at the balance sheet date.

- A description of a contingent asset should be considered if it is improbable that a claim will result in additional contract revenue or if the amount cannot be reasonably estimated. Where disclosing contingencies that may result in gains, in accordance with SFAS 5, care should be exercised to avoid misleading implications concerning the likelihood of realization.

- For amounts maturing after one year, a disclosure of the amounts maturing in each year, if practicable, and the interest rates on major long-term receivable items.

- A description of retainage provisions for amounts billed but not yet paid by customers.

Backlog

One of the most important indexes of a contractor's operations is the amount of backlog of uncompleted contracts at the end of the current period as compared with the amount of backlog at the end of the prior year. Accordingly, the *Guide* strongly encourages the disclosure of this information.

SAMPLE FINANCIAL STATEMENTS

With the approval of the AICPA, the illustrative financial statements that are included in Appendixes 6 and 8 of the *Guide* (1981) are presented as Figures 10-1 and 10-2. These sample financial statements were originally prepared to illustrate the guidance provided by the newly issued AICPA publication. They do not present all the possible alternative presentations and varieties of transactions in the vast and complex construction industry. However, the content of these sample financial statements provides examples of the transactions and disclosures described in this chapter.

**FIG. 10-1   Sample Financial Statements—
Percentage Contractors, Inc.**

Independent Accountants' Report

The Shareholders and Board of Directors
Percentage Contractors, Inc.

We have examined the consolidated balance sheets of Percentage Contractors, Inc., and subsidiaries as of December 31, 19X8 and 19X7, and the related consolidated statements of income and retained earnings and changes in financial position for the years then ended. Our examinations were made in accordance with generally accepted auditing standards and, accordingly, included such tests of the accounting records and such other auditing procedures as we considered necessary in the circumstances.

In our opinion, the financial statements referred to above present fairly the financial position of Percentage Contractors, Inc., and subsidiaries at December 31, 19X8 and 19X7, and the results of their operations and the changes in their financial position for the years then ended, in conformity with generally accepted accounting principles applied on a consistent basis.

(Firm Signature)
Certified Public Accountants

City, State
February 18, 10X0

(continued)

Percentage Contractors, Inc.
Consolidated Balance Sheets
December 31, 19X8 and 19X7

Assets	19X8	19X7
Cash	$ 264,100	$ 221,300
Certificates of deposit	40,300	
Contract receivables (Note 2)	3,789,200	3,334,100
Costs and estimated earnings in excess of billings on uncompleted contracts (Note 3)	80,200	100,600
Inventory, at lower of cost, on a first-in, first-out basis, or market	89,700	99,100
Prepaid charges and other assets	118,400	83,200
Advances to and equity in joint venture (Note 4)	205,600	130,700
Note receivable, related company (Note 5)	175,000	150,000
Property and equipment, net of accumulated depreciation and amortization (Note 6)	976,400	1,019,200
	$5,738,900	$5,138,200

Liabilities and Shareholders' Equity	19X8	19X7
Notes payable (Note 8)	$ 468,100	$ 578,400
Lease obligations payable (Note 9)	197,600	251,300
Accounts payable (Note 7)	2,543,100	2,558,500
Billings in excess of costs and estimated earnings on uncompleted contracts (Note 3)	242,000	221,700
Accrued income taxes payable	52,000	78,600
Other accrued liabilities	36,600	36,000
Due to consolidated joint venture minority interests	154,200	26,200
Deferred income taxes (Note 13)	619,200	408,000
	4,312,800	4,188,700
Contingent liability (Note 10)		
Shareholders' equity		
Common stock—$1 par value, 500,000 authorized shares, 300,000 issued and outstanding shares	300,000	300,000
Retained earnings	1,126,100	649,500
Total shareholders' equity	1,426,100	949,500
	$5,738,900	$5,138,200

The accompanying notes are an integral part of these financial statements.

Percentage Contractors, Inc.
Consolidated Statements of Income and Retained Earnings
Years Ended December 31, 19X8 and 19X7

	19X8	19X7
Contract revenues earned	$22,554,100	$16,225,400
Cost of revenues earned	20,359,400	14,951,300
Gross profit	2,194,700	1,274,100
Selling, general, and administrative expense	895,600	755,600
Income from operations	1,299,100	518,500
Other income (expense)		
Equity in earnings from unconsolidated joint venture	49,900	5,700
Gain on sale of equipment	10,000	2,000
Interest expense (net of interest income of $8,800 in 19X8 and $6,300 in 19X7)	(69,500)	(70,800)
	(9,600)	(63,100)
Income before taxes	1,289,500	455,400
Provision for income taxes (Note 13)	662,900	225,000
Net income (per share, $2.09 (19X8); $.77 (19X7))	626,600	230,400
Retained earnings, beginning of year	649,500	569,100
	1,276,100	799,500
Less: Dividends paid (per share, $.50 (19X8); $.50 (19X7))	150,000	150,000
Retained earnings, end of year	$ 1,126,100	$ 649,500

The accompanying notes are an integral part of these financial statements.

(continued)

Percentage Contractors, Inc.
Consolidated Statements of Changes in Financial Position
Years Ended December 31, 19X8 and 19X7

	19X8	19X7
Source of funds		
From operations		
Net income	$ 626,600	$230,400
Charges (credits) to income not involving cash and cash equivalents		
Depreciation and amortization	167,800	153,500
Deferred income taxes	211,200	(75,900)
Gain on sale of equipment	(10,000)	(2,000)
	995,600	306,000
Proceeds from equipment sold	25,000	5,000
Net increase in billings related to costs and estimated earnings on uncompleted contracts	40,700	10,500
Decrease in inventory	9,400	
Decrease in prepaid charges and other assets		16,100
Increase in accounts payable		113,200
Increase in other accrued liabilities	600	21,200
Increase in amount due to consolidated joint venture minority interests	128,000	26,200
Total	1,199,300	498,200
Use of funds		
Acquisition of equipment		
Shop and construction equipment	100,000	155,000
Automobiles and trucks	40,000	20,000
Dividends paid	150,000	150,000
Increase in contract receivables	455,100	9,100
Increase in inventory		3,600
Increase in advances to and equity in joint venture	74,900	15,400
Increase in note receivable, related company	25,000	50,000
Increase in prepaid charges and other assets	35,200	
Decrease in notes payable	110,300	90,300
Decrease in lease obligations payable	53,700	9,700
Decrease in accounts payable	45,400	
Decrease in accrued income taxes payable	26,600	2,400
Total	1,116,200	505,500
Increase (decrease) in cash and certificates of deposit for year	83,100	(7,300)
Cash and certificates of deposit		
Beginning of year	221,300	228,600
End of year	$ 304,400	$221,300

Percentage Contractors, Inc.
Notes to Consolidated Financial Statements
December 31, 19X8 and 19X7

1. Significant Accounting Policies

Company's activities and operating cycle. The company is engaged in a single industry: the construction of industrial and commercial buildings. The work is performed under cost-plus-fee contracts, fixed-price contracts, and fixed-price contracts modified by incentive and penalty provisions. These contracts are undertaken by the company or its wholly owned subsidiary alone or in partnership with other contractors through joint ventures. The company also manages, for a fee, construction projects of others.

The length of the company's contracts varies but is typically about two years. Therefore, assets and liabilities are not classified as current and noncurrent because the contract-related items in the balance sheet have realization and liquidation periods extending beyond one year.

Principles of consolidation. The consolidated financial statements include the company's majority-owned entities, a wholly owned corporate subsidiary and a 75 percent-owned joint venture (a partnership). All significant intercompany transactions are eliminated. The company has a minority interest in a joint venture (partnership), which is reported on the equity method.

Revenue and cost recognition. Revenues from fixed-price and modified fixed-price construction contracts are recognized on the percentage-of-completion method, measured by the percentage of labor hours incurred to date to estimated total labor hours for each contract.* This method is used because management considers expended labor hours to be the best available measure of progress on these contracts. Revenues from cost-plus-fee contracts are recognized on the basis of costs incurred during the period plus the fee earned, measured by the cost-to-cost method.

Contracts to manage, supervise, or coordinate the construction activity of others are recognized only to the extent of the fee revenue. The revenue earned in a period is based on the ratio of hours incurred to the total estimated hours required by the contract.

Contract costs include all direct material and labor costs and those indirect costs related to contract performance, such as indirect labor, supplies, tools, repairs, and depreciation costs. Selling, general, and administrative costs are charged to expense as incurred. Provisions for estimated losses on uncompleted contracts are made in the period in which such losses are determined. Changes in job performance, job conditions,

---

*There are various other alternatives to the percentage of labor hours method for measuring percentage of completion, which, in many cases, may be more appropriate in measuring the extent of progress toward completion of the contract (labor dollars, units of output, and the cost-to-cost method and its variations).

(continued)

and estimated profitability, including those arising from contract penalty provisions, and final contract settlements may result in revisions to costs and income and are recognized in the period in which the revisions are determined. Profit incentives are included in revenues when their realization is reasonably assured. An amount equal to contract costs attributable to claims is included in revenues when realization is probable and the amount can be reliably estimated.

The asset, "Costs and estimated earnings in excess of billings on uncompleted contracts," represents revenues recognized in excess of amounts billed. The liability, "Billings in excess of costs and estimated earnings on uncompleted contracts," represents billings in excess of revenues recognized.

Property and equipment. Depreciation and amortization are provided principally on the straight-line method over the estimated useful lives of the assets. Amortization of leased equipment under capital leases is included in depreciation and amortization.

Pension plan. The company has a pension plan covering substantially all employees not covered by union-sponsored plans. Pension costs charged to earnings include current-year costs and the amortization of prior-service costs over 30 years. The company's policy is to fund the costs accrued.

Income taxes. Deferred income taxes are provided for differences in timing in reporting income for financial statement and tax purposes arising from differences in the methods of accounting for construction contracts and depreciation.

Construction contracts are reported for tax purposes on the completed-contract method and for financial statement purposes on the percentage-of-completion method. Accelerated depreciation is used for tax reporting, and straight-line depreciation is used for financial statement reporting.

Investment tax credits are applied as a reduction to the current provision for federal income taxes using the flow-through method.

2. Contract Receivables

	December 31, 19X8	December 31, 19X7
Contract receivables		
Billed		
Completed contracts	$ 621,100	$ 500,600
Contracts in progress	2,146,100	1,931,500
Retained	976,300	866,200
Unbilled	121,600	105,400
	3,865,100	3,403,700
Less: Allowances for doubtful collections	75,900	69,600
	$3,789,200	$3,334,100

Contract receivables at December 31, 19X8, include a claim, expected to be collected within one year, for $290,600 arising from a dispute with the owner over design and specification changes in a building currently under construction. The changes were made at the request of the owner to improve the thermal characteristics of the building and, in the opinion of counsel, gave rise to a valid claim against the owner.

The retained and unbilled contract receivables at December 31, 19X8, included $38,600 that was not expected to be collected within one year.

3. Costs and Estimated Earnings on Uncompleted Contracts

	December 31, 19X8	December 31, 19X7
Costs incurred on uncompleted contracts	$15,771,500	$12,165,400
Estimated earnings	1,685,900	1,246,800
	17,457,400	13,412,200
Less: Billings to date	17,619,200	13,533,300
	$ (161,800)	$ (121,100)
Included in accompanying balance sheets under the following captions:		
Costs and estimated earnings in excess of billings on uncompleted contracts	$ 80,200	$ 100,600
Billings in excess of costs and estimated earnings on uncompleted contracts	(242,000)	(221,700)
	$ (161,800)	$ (121,100)

4. Advances to and Equity in Joint Venture

The company has a minority interest (one-third) in a general partnership joint venture formed to construct an office building. All of the

(continued)

partners participate in construction, which is under the general management of the company. Summary information on the joint venture follows:

	December 31, 19X8	December 31, 19X7
Current assets	$ 483,100	$280,300
Construction and other assets	220,500	190,800
	703,600	471,100
Less: Liabilities	236,800	154,000
Net assets	$ 466,800	$317,100
Revenue	$3,442,700	$299,400
Net income	$ 149,700	$ 17,100
Company's interest		
Share of net income	$ 49,900	$ 5,700
Advances to joint venture	$ 50,000	$ 25,000
Equity in net assets	155,600	105,700
Total advances and equity	$ 205,600	$130,700

(For the purposes of illustrative financial statements, the one-line equity method of presentation is used in both the balance sheet and the income statement. However, the pro rata consolidation method is acceptable if the investment is deemed to represent an undivided interest.)

5. Transactions With Related Party

The note receivable, related company, is an installment note bearing annual interest at 9¼%, payable quarterly, with the principal payable in annual installments of $25,000, commencing October 1, 19Y0.

The major stockholder of Percentage Contractors, Inc. owns the majority of the outstanding common stock of this related company, whose principal activity is leasing land and buildings. Percentage Contractors, Inc., rents land and office facilities from the related company on a ten-year lease ending September 30, 19Y6, for an annual rental of $19,000.

6. Property and Equipment

	December 31, 19X8	December 31, 19X7
Assets		
Land	$ 57,500	$ 57,500
Buildings	262,500	262,500
Shop and construction equipment	827,600	727,600
Automobiles and trucks	104,400	89,100
Leased equipment under capital leases	300,000	300,000
	1,552,000	1,436,700
Accumulated depreciation and amortization		
Buildings	140,000	130,000
Shop and construction equipment	265,600	195,500
Automobiles and trucks	70,000	42,000
Leased equipment under capital leases	100,000	50,000
	575,600	417,500
Net property and equipment	$ 976,400	$1,019,200

7. Accounts Payable

Accounts payable include amounts due to subcontractors, totaling $634,900 at December 31, 19X8, and $560,400 at December 31, 19X7, which have been retained pending completion and customer acceptance of jobs. Accounts payable at December 31, 19X8, include $6,500 that are not expected to be paid within one year.

8. Notes Payable

	December 31, 19X8	December 31, 19X7
Unsecured note payable to bank, due in quarterly installments of $22,575 plus interest at 1% over prime	$388,100	$478,400
Note payable to bank, collateralized by equipment, due in monthly installments of $1,667 plus interest at 10% through January, 19Y3	80,000	100,000
	$468,100	$578,400

At December 31, 19X8, the payments due within one year totaled $110,300.

(continued)

9. Lease Obligations Payable

The company leases certain specialized construction equipment under leases classified as capital leases. The following is a schedule showing the future minimum lease payments under capital leases by years and the present value of the minimum lease payments as of December 31, 19X8:

Year ending December 31	
19X9	$ 76,500
19Y0	76,500
19Y1	76,500
Total minimum lease payments	229,500
Less: Amount representing interest	31,900
Present value of minimum lease payments	$197,600

At December 31, 19X8, the present value of minimum lease payments due within one year is $92,250.

Total rental expense, excluding payments on capital leases, totaled $86,300 in 19X8 and $74,400 in 19X7.

10. Contingent Liability

A claim for $180,000 has been filed against the company and its bonding company arising out of the failure of a subcontractor of the company to pay its suppliers. In the opinion of counsel and management, the outcome of this claim will not have a material effect on the company's financial position or results of operations.

11. Pension Plan

Pension costs charged to earnings were $61,400 in 19X8 and $57,300 in 19X7. At December 31, 19X8, the estimated actuarial value of vested benefits exceeded the fund assets (at market) and contribution accruals by $197,600.

12. Management Contracts

The company manages or supervises commercial and industrial building contracts of others for a fee. These fees totaled $121,600 in 19X8 and $1,700 in 19X7 and are included in contract revenues earned.

13. Income Taxes and Deferred Income Taxes

The provision for taxes on income consists of the following:

	December 31, 19X8	December 31, 19X7
Currently payable, net of investment credits of $9,400 and $13,800	$451,700	$300,900
Deferred		
Contract related	204,200	(80,900)
Property and equipment related	7,000	5,000
	$662,900	$225,000

At December 31 of the respective years, the components of the balance of deferred income taxes were:

	December 31, 19X8	December 31, 19X7
Contract related	$594,000	$389,800
Property and equipment related	25,200	18,200
	$619,200	$408,000

14. Backlog

The following schedule shows a reconciliation of backlog representing signed contracts, excluding fees from management contracts, in existence at December 31, 19X7 and 19X8:*

Balance, December 31, 19X7	$24,142,600
Contract adjustments	1,067,100
New contracts, 19X8	3,690,600
	28,900,300
Less: Contract revenue earned, 19X8	22,432,500
Balance, December 31, 19X8	$ 6,467,800

In addition, between January 1, 19X9 and February 18, 19X9, the company entered into additional construction contracts with revenues of $5,332,800.

---

*The presentation of backlog information, although encouraged, is not a required disclosure.

**FIG. 10-2   Sample Financial Statement—Completed
Contractors, Inc.**

Independent Accountants' Report

The Stockholders and Board of Directors
Completed Contractors, Inc.

We have examined the balance sheets of Completed Contractors, Inc.,
as of December 31, 19X8 and 19X7, and the related statements of income
and retained earnings and changes in financial position for the years then
ended. Our examinations were made in accordance with generally accepted auditing standards and, accordingly, included such tests of the
accounting records and such other auditing procedures as we considered
necessary in the circumstances.

In our opinion, the financial statements referred to above present fairly
the financial position of Completed Contractors, Inc., at December 31,
19X8 and 19X7, and the results of its operations and the changes in its
financial position for the years then ended, in conformity with generally
accepted accounting principles applied on a consistent basis.

(Firm Signature)
Certified Public Accountants

City, State
February 18, 19X9

Completed Contractors, Inc.
Balance Sheets
December 31, 19X8 and 19X7

Assets	19X8	19X7
Current assets		
Cash	$ 242,700	$ 185,300
Contract receivables (less allowance for doubtful accounts of $10,000 and $8,000) (Note 2)	893,900	723,600
Costs in excess of billings on uncompleted contracts (Note 3)	418,700	437,100
Inventories, at lower of cost or realizable value on first-in, first-out basis (Note 4)	463,600	491,300
Prepaid expenses	89,900	53,900
Total current assets	2,108,600	1,891,200
Cash value of life insurance	35,800	32,900
Property and equipment, at cost		
Building	110,000	110,000
Equipment	178,000	163,000
Trucks and autos	220,000	200,000
	508,000	473,000
Less: Accumulated depreciation	218,000	203,200
	290,000	269,800
Land	21,500	21,500
	311,500	291,300
	$2,456,100	$2,215,400

Liabilities and Stockholders' Equity	19X8	19X7
Current liabilities		
Current maturities, long-term debt (Note 5)	$ 37,000	$ 30,600
Accounts payable	904,900	821,200
Accrued salaries and wages	138,300	155,100
Accrued income taxes	53,000	36,200
Accrued and other liabilities	116,400	55,550
Billings in excess of costs on uncompleted contracts (Note 3)	34,500	43,700
Total current liabilities	1,284,100	1,142,350
Long-term debt, less current maturities (Note 5)	245,000	241,000
	1,529,100	1,383,350
Stockholders' equity		
Common stock—$10 par value; 50,000 authorized shares, 23,500 issued and outstanding shares	235,000	235,000
Additional paid-in capital	65,000	65,000
Retained earnings	627,000	532,050
	927,000	832,050
	$2,456,100	$2,215,400

The accompanying notes are an integral part of these financial statements.

(continued)

Completed Contractors, Inc.
Statements of Income and Retained Earnings
Years Ended December 31, 19X8 and 19X7

	19X8	19X7
Contract revenues	$9,487,000	$8,123,400
Costs and expenses		
Cost of contracts completed	8,458,500	7,392,300
General and administrative	684,300	588,900
Interest expense	26,500	23,000
	9,169,300	8,004,200
Income before income taxes	317,700	119,200
Income taxes	164,000	54,200
Net income ($6.54 and $2.77 per share)	153,700	65,000
Retained earnings		
Balance, beginning of year	532,050	525,800
	685,750	590,800
Dividends paid ($2.50 per share)	58,750	58,750
Balance, end of year	$ 627,000	$ 532,050

The accompanying notes are an integral part of these financial statements.

Completed Contractors, Inc.
Statements of Changes in Financial Position
Years Ended December 31, 19X8 and 19X7

	19X8	19X7
Source of working capital		
Net income	$153,700	$ 65,000
Charge to income not requiring outlay		
of working capital—depreciation	54,800	50,300
Working capital from operations	208,500	115,300
Proceeds of notes payable	44,000	68,000
	252,500	183,300
Use of working capital		
Purchase of property and equipment	75,000	53,500
Reduction of long-term debt	40,000	28,000
Payment of dividends	58,750	58,750
Increase in cash value of life insurance	2,900	2,685
	176,650	142,935
Increase in working capital	$ 75,850	$ 40,365
Changes in components of working capital		
Increase (decrease) in current assets		
Cash	$ 57,400	$ (26,435)
Contract receivables	170,300	36,500
Costs in excess of billings on		
uncompleted contracts	(18,400)	49,100
Inventories	(27,700)	3,400
Prepaid expenses	36,000	(16,500)
	217,600	46,065
Decrease (increase) in current liabilities		
Current maturities, long-term debt		
Notes payable, bank	(6,000)	(12,000)
Mortgage payable	(400)	(500)
Accounts payable	(83,700)	(24,600)
Accrued salaries and wages	16,800	(24,300)
Accrued income taxes	(16,800)	6,300
Accrued and other liabilities	(60,850)	33,100
Billings in excess of costs on		
uncompleted contracts	9,200	16,300
	(141,750)	(5,700)
Increase in working capital	$ 75,850	$ 40,365

The accompanying notes are an integral part of these financial statements.

(continued)

Completed Contractors, Inc.
Notes to Financial Statements
December 31, 19X8 and 19X7

1. Significant Accounting Policies

Company's activities. The company is a heating and air-conditioning contractor for residential and commercial properties. Work on new structures is performed primarily under fixed-price contracts. Work on existing structures is performed under fixed-price or time-and-material contracts.

Revenue and cost recognition. Revenues from fixed-price construction contracts are recognized on the completed-contract method. This method is used because the typical contract is completed in two months or less and financial position and results of operations do not vary significantly from those which would result from use of the percentage-of-completion method. A contract is considered complete when all costs except insignificant items have been incurred and the installation is operating according to specifications or has been accepted by the customer.

Revenues from time-and-material contracts are recognized currently as the work is performed.

Contract costs include all direct material and labor costs and those indirect costs related to contract performance, such as indirect labor, supplies, tools, repairs, and depreciation costs. General and administrative costs are charged to expense as incurred. Provisions for estimated losses on uncompleted contracts are made in the period in which such losses are determined. Claims are included in revenues when received.

Costs in excess of amounts billed are classified as current assets under costs in excess of billings on uncompleted contracts. Billings in excess of costs are classified under current liabilities as billings in excess of costs on uncompleted contracts. Contract retentions are included in accounts receivable.

Inventories. Inventories are stated at cost on the first-in, first-out basis using unit cost for furnace and air-conditioning components and average cost for parts and supplies. The carrying value of furnace and air-conditioning component units is reduced to realizable value when such values are less than cost.

Property and equipment. Depreciation is provided over the estimated lives of the assets principally on the declining-balance method, except on the building where the straight-line method is used.

Pension plan. The company has a pension plan covering all employees not covered by union-sponsored plans. Pension costs charged to income include current-year costs and the amortization of prior-service costs over 30 years. The company's policy is to fund the costs accrued.

Investment tax credit. Investment tax credits are applied as a reduction to the current provision for federal income taxes using the flow-through method.

2. Contract Receivables

	December 31, 19X8	December 31, 19X7
Completed contracts, including retentions	$438,300	$408,600
Contracts in progress		
Current accounts	386,900	276,400
Retentions	78,700	46,600
	903,900	731,600
Less: Allowance for doubtful accounts	10,000	8,000
	$893,900	$723,600

Retentions include $10,300 in 19X8, which are expected to be collected after 12 months.

3. Costs and Billings on Uncompleted Contracts

	December 31, 19X8	December 31, 19X7
Costs incurred on uncompleted contracts	$2,140,400	$1,966,900
Billings on uncompleted contracts	1,756,200	1,573,500
	$ 384,200	$ 393,400
Included in accompanying balance sheets under the following captions:		
Costs in excess of billings on uncompleted contracts	$ 418,700	$ 437,100
Billings in excess of costs on uncompleted contracts	(34,500)	(43,700)
	$ 384,200	$ 393,400

4. Inventories

	December 31, 19X8	December 31, 19X7
Furnace and air-conditioning components	$303,200	$308,700
Parts and supplies	160,400	182,600
	$463,600	$491,300

Furnace and air-conditioning components include used items of $78,400 in 19X8 and $71,900 in 19X7 that are carried at the lower of cost or realizable value.

(continued)

5. Long-Term Debt

	December 31, 19X8	December 31, 19X7
Notes payable, bank		
Notes due in quarterly installments of $2,500, plus interest at 8%	$140,000	$150,000
Notes due in monthly installments of $1,500, plus interest at prime plus 1½%	87,000	58,000
Mortgage payable		
Due in quarterly payments of $3,500, including interest at 9%	55,000	63,600
	282,000	271,600
Less: Current maturities	37,000	30,600
	$245,000	$241,000

6. Pension Plans

The total pension expenses for the years 19X8 and 19X7 were $31,200 and $27,300, including contributions to union-sponsored plans.

At December 31, 19X8, the estimated actuarial value of vested benefits of the company's plan exceeded the fund assets (at market) and contribution accruals by $48,000.

7. Backlog

The estimated gross revenue on work to be performed on signed contracts was $4,691,000 at December 31, 19X8, and $3,617,400 at December 31, 19X7. In addition to the backlog of work to be performed, there was gross revenue, to be reported in future periods under the completed-contract method used by the company, of $2,460,000 at December 31, 19X8, and $2,170,000 at December 31, 19X7.*

---

*The presentation of backlog information, although encouraged, is not a required disclosure.

CHAPTER **11**

Cash Management

Richard S. Hickok
Elliott C. Robbins

IMPORTANCE OF CASH MANAGEMENT

The importance of a sound cash management system for contractors cannot be overemphasized, as it is one of the keys to a successful construction operation. Construction projects have inherent cash management problems. One of the major problems is the range in size of projects, which can vary in cost from several thousand to many millions of dollars. Cash management is further affected by the extended periods of time over which projects are undertaken; this problem is most visible when the timing is such that large cash outflows take place during periods when cash inflows do not. Cash management problems are often exacerbated by other factors, including the following:

- Sensitivity to current economic trends
- Inadequate capitalization
- Lack of sufficient bank credit facilities
- Low industry profitability
- Inherent high-risk nature of the business
- Retainages on contracts in process
- Inadequate tax planning

Need for Planning at All Project Stages

Cash management can only be achieved through appropriate planning, budgeting, and forecasting. Construction contractors, perhaps more than most other business professionals, are constantly involved in planning. Planning begins when contractors receive requests for proposals. Bids are then prepared, and contracts are awarded. Contractors continue to plan as they purchase materials, schedule and purchase required equipment, hire personnel, and schedule workloads.

Planning and Overall Profitability

Proper planning techniques should be conducive to profitable operations or should reduce losses on potentially unprofitable projects. Conversely, poor or inadequate planning may contribute to less than optimal contract profits or even result in losses. Lack of planning can contribute to:

- Inefficient contract management techniques
- Inadequate field supervision to complete contracts on schedule
- Poor equipment management (such as idle equipment or the purchase of excessive equipment)

- Panic buying to meet project requirements
- Excessive labor costs
- Unprofitable operations and resultant cash problems

The greatest risk factor in the construction industry stems from the method of pricing. Pricing and, therefore, overall profitability are functions of planning. If the items listed above are not considered in the pricing structure, cash flow problems may result. A contractor who is selling his services must set his prices in the bidding or negotiating process before absolute total costs are determined. Cash management techniques are a vital link to contract management planning; profitable operations and maintenance of positive cash flow are the goals of these techniques.

PROJECT CASH FLOW

Typical Project Cash Flow

Each construction project is an individual profit center with its own cash cycle based on the cost of the activities involved in the project and on payments from the owner, which are determined by the contract terms. Typical cash flow on a construction project includes the following components listed under cash disbursements and cash receipts:

PROJECT CASH FLOW

Cash Disbursements (Outflow)

• Bid costs	xx
• Preconstruction costs (engineering, design, mobilization, etc.)	xx
• Materials and supplies	xx
• Equipment and/or equipment rental	xx
• Payments to subcontractors	xx
• Labor	xx
• Overhead	xx
Total	xxx

Cash Receipts (Income)

• Billings (less retentions)	xx
• Retentions (usually paid on completion and acceptance)	xx
• Claims and change orders (frequently delayed payment)	xx
Total	xxx
NET CASH FLOW	xxx

If a contractor understands the basic concept of the cash flow of a project, then it is possible for him to form an effective cash management program. The goal in measuring project cash flow is to accelerate cash receipts so that there is a positive cash flow from each contract or, alternatively, a minimization of negative cash flow. A contractor's primary goal is to receive payments from the owner before the cash payments to subcontractors, material suppliers, employees, and others are due. This goal is frequently difficult to achieve, due to the normal owner retention clauses in construction contracts and the delays attendant in approving change orders and resolving claims.

Cash Flow Analysis

Job-Related Cash Flows. From his knowledge of the basic contract cycle, every contractor should have the ability to prepare a cash flow analysis, that is, a cash forecast for each separate project and a total forecast for all jobs in progress.

The cash flow analysis should be based on time periods (e.g., monthly, quarterly) for each contract or for each segment or activity of each contract in process. Therefore, the contractor must schedule the time frame within which various aspects of a project will be performed, planning when the payment for these activities will be disbursed and how the terms of progress billing to the owner will provide actual cash flow.

The cash flow analysis by time period indicates the positive or negative cash flow of projects in process. Although a summary such as this never completely eliminates all cash management problems, it should significantly reduce these problems and identify cash needs in sufficient time for the contractor to take appropriate corrective measures. Figure 11-1 provides a typical projected cash flow by period for projects in process.

Non-Job-Related Cash Flows. In order to complete the cash-forecasting function, the projected cash flow from jobs in process should be incorporated into the contractor's total plan for the entire organization. The cash flow analysis for projects in progress should be combined with non-job-related cash flows, such as:

- General and administrative expenses
- Interest
- Fixed assets expenditures not related to jobs, such as office furniture and fixtures
- Taxes based on income and dividends to shareholders

FIG. 11-1   Projected Cash Flow for Projects in Process

MH CONTRACTORS
PROJECTED CASH FLOW
Year Ending December 31, 19XX

	Current Quarter						
	January	February	March	2nd Qtr	3rd Qtr	4th Qtr	Total
CONTRACTS							
Job #101							
Cash Disbursements for Job Costs							
Materials	$20,000	$ 53,000	$ 30,000	$ 50,000	$ 70,000	—	$ 220,000
Subcontractors	—	—	80,000	120,000	—	—	200,000
Labor	—	33,000	70,000	150,000	180,000	$ 60,000	490,000
Overhead (actual)	10,000	22,000	30,000	80,000	70,000	30,000	240,000
TOTAL	$30,000	$100,000	$210,000	$400,000	$320,000	$ 90,000	$1,150,000
Billings							
TOTAL	$40,000	$120,000	$200,000	$450,000	$300,000	$190,000	$1,300,000
Less: Retainages withheld	4,000	12,000	20,000	45,000	30,000	19,000	130,000
Add: Retainages due	—	—	—	—	—	130,000	130,000
Cash receipts	36,000	108,000	180,000	405,000	270,000	301,000	1,300,000
Net cash flow	$ 6,000	$ 8,000	$(30,000)	$ 5,000	$ (50,000)	$211,000	$ 150,000
Cumulative cash flow	$ 6,000	$ 14,000	$(16,000)	$ (11,000)	$ (61,000)	$150,000	$ 150,000
TOTAL CASH FLOW— PROJECTS IN PROCESS	$48,000	$(35,000)	$150,000	$(30,000)	$(120,000)	$240,000	

FIG. 11-2 Projected Total Cash Flow

MH CONTRACTORS

PROJECTED CASH FLOW

Year Ending December 31, 19XX

	Current Quarter			2nd Qtr	3rd Qtr	4th Qtr	Total
	January	February	March				
Cash flow from projects in process	$48,000	$(35,000)	$150,000	$ (30,000)	$(120,000)	$240,000	$ 253,000
Less Non-Job Expenditures:							
Selling expenses[1]	22,000	16,000	6,000	31,000	30,000	45,000	150,000
Administrative expenses[1]	12,000	12,000	12,000	36,000	36,000	36,000	134,000
Interest	1,000	1,000	1,000	4,000	5,000	5,000	17,000
Federal and state income taxes	6,000	–	–	3,000	3,000	3,000	15,000
Dividends to stockholders	–	–	–	–	10,000	–	10,000
Expenditures for property, plant, and equipment	–	22,000	–	–	–	8,000	30,000
Other (identify)	–	–	–	–	–	–	–
	41,000	51,000	19,000	74,000	84,000	107,000	376,000
Add Non-Job Receipts:							
Interest	1,000	1,000	–	–	–	–	2,000
Other (identify)	–	–	2,000	–	2,000	–	4,000
	1,000	1,000	2,000	–	2,000	–	4,000
CASH INCREASE (DECREASE)	$ 8,000	$(85,000)	$133,000	$(104,000)	$(202,000)	$133,000	$(117,000)

(1) These items can be set forth in total or by detailed expense classifications.

Figure 11-2 illustrates the incorporation of job-related and non-job-related cash flows. It is preferable that these two cash flows be combined on a regular basis so that future cash flow problems can be identified and, thus, their impact can be buffered. This combined approach should significantly improve cash management and also enable the contractor to plan effectively for future cash requirements.

INTEGRATION OF CASH AND PROJECT MANAGEMENT

Monitoring and Updating

In order to maximize income and cash flow, the data required to implement and monitor an appropriate cash management system should ideally be integrated with a contractor's procedures for estimating and bidding projects and for scheduling and monitoring performance on contracts in process. This data can be maintained by one of the many integrated computer systems available. The recent development of microcomputer terminology and the corresponding cost reduction thereof, leaves little, if any, reason for the lack of project planning, scheduling, and related cash flow forecasting that contributed to many contractor failures in the past. (See also Chapter 5, which is concerned with management information systems (MIS).)

A project-planning and cash analysis system must be constantly monitored and updated, so that changes in scheduling and cash flow reflect actual, current activity. For example, delays caused by weather, strikes, or similar factors must be factored into the system; otherwise, the output will be meaningless. Further, a cash project management system of the type described herein must be updated at least monthly and must always be projected forward for a period of one year or more.

Internal and External Benefits

A major factor contributing to the high rate of insolvencies in the construction industry for years has been the failure by contractors to plan and control cash requirements. Cash management techniques have been used for some time by companies in other industries and are now available to, and should be used by, all contractors.

In addition to the benefits inherent in proper cash management, a sound integrated system of project planning is beneficial in these other areas related to construction management:

- Supervision and management of jobs in process

- Allocation of human resources to contracts in process
- Management of equipment on jobs in process
- Financial accounting for the contractor
- Constant monitoring of job costs, comparison with original estimated costs, and projected job profit for income-recognition purposes

Furthermore, an effective project-planning system should assist contractors in their relationships with surety companies and with the banking industry, since both cash and financing requirements will be identified in sufficient time for necessary planning or financing arrangements to be made long in advance of actual needs.

BILLING PRACTICES

In addition to an effective cash analysis and planning system, billing practices and the controlling of job expenditures are two important areas that further enhance a contractor's cash flow. In the construction industry, billing practices are influenced by two major objectives: (1) to generate sufficient cash flow to finance the ongoing progress of the project and thus minimize the investment of cash resources by the contractor; and (2) to aid the owner's financing requirements while ensuring him suitable protection that the contract will complete the project satisfactorily.

Consequently, the billing practices in the industry provide for (1) progress billing over the duration of the project and (2) retention by the owner, which is usually 10 percent of the progress billing of a portion of the cash due on the contract. This is payable upon completion and acceptance of the project or in any other manner provided in the contract.

Billing Terms

Billing terms are customarily set forth in the contract between the owner and the contractor. These terms are of considerable importance to the contractor for cash flow purposes. The contractor is able to accelerate the cash flow from the project and minimize internal and external financing costs if favorable terms can be negotiated. The following factors usually affect the billing arrangements provided in the contract:

- Completion of certain stages of work as provided in the contract, such as grading, excavation, and foundations
- Delivery and/or installation of material

- Costs incurred on cost-plus type contracts
- Architect's or engineer's estimates of completion
- Specified time schedule
- Quantity measurements, such as cubic yards excavated or yards of concrete poured

In addition, the contract may also provide for an advance to the contractor by the owner to cover costs incurred by the contractor during the start-up phase of the project — such as engineering, project planning, and mobilization at the project location — or to provide project working capital for the contractor.

Although billing terms vary from one contract specialty (e.g., general building, highway, and subcontractor specialties) to another, they are, consistently, of critical significance to the contractor. The contractor must give consideration to billing terms in the contract as well as in his negotiations with the owner and, where applicable, with the subcontractor. To the extent possible, billing terms in the contract should provide for a billing schedule correlated directly with the time schedule as set forth in the project plan. The purpose of this is to ensure that the cash flow on each job, if properly planned, will provide the funds necessary to meet the cash requirements of the project without having to go beyond specified contract limits for financing.

Front-End Loading

Use by Contractors. Many contractors employ the technique described as front-end loading to improve project cash flow. Using this technique, the contractor assigns in the billing clauses of the contract a higher percentage of the contract price to those segments or activities normally completed in the early stages of the project (e.g., grading, excavation, laying the foundation, and so on). This allows the contractor to bill and collect amounts in excess of costs incurred to that point and to increase the cash flow prior to release of the retention by the owner. This practice assists the contractor in financing jobs in process, accelerates cash flow, and serves as a significant cash management strategy for many successful contractors. In addition to providing adequate financing of jobs in process, the contractor may benefit from the investment of excess cash, the reduction of external financing, and, at times, the reduction of materials costs from improved purchasing practices.

An illustrative example of the principle of front-end loading is shown in Figure 11-3. The positive and negative cash flows caused by

FIG. 11-3   Billing Schedule Illustrating Front-End Loading

BIPLEX CONSTRUCTION COMPANY

PROJECT: TWO-STORY PROFESSIONAL BUILDING

Job Activity	Estimated Cost (1)	Billing Schedule Total (2)	10% Retention (3)	Net Payment (4)	Positive (Negative) Cash Flow (5) (4 − 1)
Land Development					
Grading (40,000 sq. ft.)	$ 2,500	$ 5,000	$ 500	$ 4,500	$ 2,000
Excavation	10,000	15,000	1,500	13,500	3,500
Foundation					
Forms	20,000	25,000	2,500	22,500	2,500
Steel (rods and mesh)	8,000	10,000	1,000	9,000	1,000
Concrete	25,000	30,000	3,000	27,000	2,000
Steel					
Delivery	50,000	75,000	7,500	67,500	17,500
Erection	20,000	20,000	2,000	18,000	(2,000)
Other					
Roof	25,000	30,000	3,000	27,000	2,000
Siding and windows	25,000	30,000	3,000	27,000	2,000
Insulation	10,000	12,000	1,200	10,800	800
Heat, air conditioning, ventilation, etc.*	40,000	40,000	4,000	36,000	(4,000)
Interiors (dry wall, etc.)*	25,000	20,000	2,000	18,000	(7,000)
Plumbing, electrical*	20,000	20,000	2,000	18,000	(2,000)
Painting*	15,000	15,000	1,500	13,500	(1,500)
Cleanup (punch list, etc.)	5,000	2,500	250	2,250	(2,750)
Parking lot	20,000	20,000	2,000	18,000	(2,000)
Landscaping	10,000	5,000	500	4,500	(5,500)
	$330,500	$374,500	$37,450	$337,050	$ 6,550
Estimated profit and contingencies	44,000				
TOTAL	$374,500				
Collection of retention					37,450
CASH FLOW FROM PROJECT					$44,000

* Subcontracted—total $100,000

front-end loading are shown in Column (5). In this illustration, the cash flow created by front-end loading the project — after completion of all job activities and payment of related costs but prior to completion, acceptance, and release by the owner of the $37,450 retention — is $6,550. There is also a significant positive cash flow during the early project activities.

Retentions to Subcontractors. Contractors can further enhance their cash flow by providing similar 10 percent retention schedules in the contracts with subcontractors, whereby the retention to subcontractors would be released when the building is completed, accepted, and the owner releases the retention of $37,450 (using Figure 11-3) to the general contractor. Since subcontractor costs of $100,000 are provided for, the indicated negative flow related to billing those segments of the job would be reduced by $10,000 and positive cash flow prior to release of the retention would be $16,550.

Proper Use of Front-End Loading. Although front-end loading is a significant component of contractor cash management, it must be properly used. Contractors should be cognizant of the technique's negative aspects.

- Most owners are aware of the practice of front-end loading, so it is important that contractors not make unreasonable assessments of costs and billing in the early stages of the project.

- The positive cash flow in the early stages of the project must be available for the eventual negative cash flow later on or whenever the cash flow reversal, the point at which the flow goes from positive to negative, occurs. A frequent abuse of this management technique occurs when the positive cash flow, which is generated on a current job, is used to complete a prior job or jobs that are currently in final stages of completion and in a negative cash flow position. This abuse has occurred more frequently in the past than it does now, as sureties and other credit grantors have become watchful for it.

- Front-end loading can engender a false sense of confidence about the ultimate profitability of projects in process. Profits cannot be determined by measuring billing on jobs as compared with costs incurred to date. If the contractor uses billings as the basis upon which the percentage of completion is measured and

FIG. 11-4   Illustration of Incorrect Revenue Calculation Caused by Front-End Loading

Windy Corporation signs two contracts to complete the excavation of two separate building sites (Windward and Leeward). The Windward contract is negotiated so that Windy is able to front-end load. Each contract is for $300,000. Each contract has three phases: (1) planning and delivery of equipment on site, (2) excavation, (3) cleanup and grading. The total cost of each contract is $270,000.

| | Windward | | Leeward | |
| | Estimated Cost | Billing per Contract | Estimated Cost | Billing per Contract |
Phase				
1	$ 60,000	$100,000	$ 60,000	$ 70,000
2	150,000	100,000	150,000	160,000
3	60,000	100,000	60,000	70,000
	$270,000	$300,000	$270,000	$300,000

At year-end, Phase 1 on each contract has been completed. If Windy were to use billing as the percentage of completion, Windward would be 33⅓ percent complete, while Leeward would be only 23⅓ percent complete. Therefore, even if the correct cost figures were used, the profit recognized would be inaccurate.

In summary, front-end loading is a useful strategy for contractors to employ as part of their overall cash management scheme; it must, however, be used realistically and prudently.

the estimated profit earned on a project computed, then the recognition of revenues will be inaccurate. This is true even if the correct total cost and estimated cost to complete are factored into the calculations. A simple illustration of this is shown in Figure 11-4.

PROGRESS BILLINGS

Progress billings are amounts billed in accordance with the provisions of the contract, based on progress to date under the contract. A contractor should have an established system for rendering progress billings on jobs and for assuring the prompt collection of these billings. Those contractors who perform a single construction activity (e.g., paving, plumbing, or electrical installation) can establish an effective system for billing on all contracts in process because of the relative uniformity of construction activities and standard contract billing terms with owners. Contractors engaged in a wide range of construction activities have billing terms tailored to each contract. Out of necessity, procedures for billing progress payment requests must be established for each individual contract.

Requirements for Timely Billing Practices

Certain activities in each contract trigger a billing for the contractor. These activities are generally related to costs incurred on a contract, the percentage of completion of various segments and activities of the entire contract, or quantity measurements of performance.

A good program of billing and collecting receivables requires that:

- Invoices be rendered by the contractor on a timely basis,
- Billings be processed for payment, and
- Payment by the owner meets contract guidelines.

Timely billing requires that invoices be prepared and submitted for payment in accordance with the terms of the contract as soon as the event occurs; this enables the contractor to bill the owner. Therefore, there must be a regular and timely flow of information from project management to the administrative personnel responsible for billing owners.

Guidelines on How to Accelerate Billings

In view of both a contractor's need for cash to finance projects and the interest cost involved in securing external funds, it is undesirable to defer billings for any period of time, such as until after the closing of the accounting records for the prior month. Billings should be made as often as prescribed by the terms of the contract. The diversity of construction operations and contract terms makes the variables used to accelerate billings limitless. Following are several practical examples, which the contractor can employ as guidelines:

☐ *Basis of billing:*
  - If the progress billing is based on the estimated percentage of completion of individual segments or activities, then the estimated percent complete is available at any time and a progress billing can therefore be rendered on the first working day of a given month.

- If progress billings are based on achieving certain specified percentage-of-completion levels on various segments of a project (i.e., 50 percent of excavation) then these events should be reported by project management so that billings can be rendered as soon as the condition for billing is satisfied.

- If progress billings are based on costs incurred, either to measure percentage of completion or to be used as a basis for billing on cost-plus type contracts, then thought should be given to an early cutoff on field project costs to accelerate the billing date.

- For cost-plus type contracts, consideration should be given to billing estimated costs during the period and making a subsequent adjustment to actual cost later on. In this manner, estimated costs (plus fees) could be billed on the first of the month and, after job costs are accumulated and finalized, an additional billing (or credit) could be submitted to the owner.

☐ *Event or action triggers billing:*

- Projects with significant material requirements should provide for billing on delivery to the job-site rather than when material is installed or fabricated.

- In those instances where the owner's architect, consulting engineers, or engineering staff must attest to percentage-of-completion levels as determined by the contractor, it is desirable to establish an appropriate liaison with these individuals so that the invoicing and collection of progress billing is not unduly delayed.

☐ *Coordination with owner's accounts payable system:*

- The contractor must be familiar with the owner's accounts payable system and payment cycle. This is relevant in that payment requests must be directed to appropriate personnel in the owner's organization, thus avoiding any undue delay in obtaining required payment authorization. It should be noted on invoices that payment is due on receipt, as internal procedures in many entities provide for payment in 30 days unless otherwise indicated or unless discounts for early payment are available. It is valuable to know the owner's payment cycle so that billings can be submitted in time for prompt payment. Under most plans, invoices due and approved for payment are paid on predetermined dates, such as the 15th and 30th of every month. Familiarity with the owner's internal procedures can expedite the processing of billings for payment.

A careful review of contract terms and contractor billing procedures allows most organizations to devise appropriate internal billing systems to ensure timely submission of progress billings.

Collections

Since cash is not collected until after progress billings are submitted, appropriate attention must always be given to collection as an aspect of contractor operations.

Creditworthiness of the Owner. One of the first elements of an effective collection program is the assessment of the creditworthiness of the owner. This is a significant factor in the management decision-making process at the proposal and bidding stage. Unfortunately, because of the competitive constraints of the industry and the need for new work in periods of slow economic activity, creditworthiness does not always receive appropriate attention at the bidding stage. If there is a significant credit risk, outstanding balances must be carefully monitored to minimize losses from bad debts. In extreme cases, it may even be necessary for the contractor to shut the job down or halt work to minimize possible future losses. Generally, contractors' losses from bad debts should be minimal, provided that there are appropriate review procedures at the bidding stage and a careful monitoring of accounts receivable in the production stages.

Follow-Up System. The timely submission of progress billings is only the first part of the billing and collection cycle. This procedure must be carried through with an efficient and effective follow-up system that encourages prompt collection of amounts due from owners. It is not enough to bill promptly and then have to wait to receive payment for costs incurred 60 to 90 days previously. Assertiveness is necessary; without it, a construction contractor may find that his business has become a financing business.

Aged Analysis of Accounts Receivable. The contractor's accounting system should provide an aged analysis of accounts receivable balances on a periodic basis of at least once a month. Each contractor should

FIG. 11-5 Accounts Receivable Analysis

KMG CONSTRUCTION COMPANY
ACCOUNTS RECEIVABLE ANALYSIS
December 31, 19XX

Job No.	Project Manager	Invoice Date	Amount Original Billing	Balance Outstanding	Progress Billing Balance			Retentions
					Current	30–60 Days	Over 60 Days	
Due from								
Waste Treatment Authority #1								
WT 101	Cranstoun	9/1/XX	$100,000	$ 10,000	$ —	$ —	$ —	$10,000
WT 101	Cranstoun	10/1/XX	75,000	10,000	—	—	2,500*	7,500
WT 101	Cranstoun	11/1/XX	65,000	6,500	—	—	—	6,500
WT 101	Cranstoun	12/1/XX	100,000	100,000	90,000	—	—	10,000
TOTAL WT 101			$340,000	126,500	90,000	—	2,500	34,000
Waste Treatment Authority #1								
WT 102	Cranstoun	10/1/XX	$ 75,000	7,500	—	—	—	7,500
WT 102	Cranstoun	11/1/XX	125,000	125,000	—	112,500	—	12,500
TOTAL WT 102			$200,000	132,500	—	112,500	—	20,000
TOTAL WASTE TREATMENT Authority #1				$259,000	$90,000	$112,500	$2,500	$54,000

* Disputed

design these analyses to meet his own particular requirements. The following additional information, however, is necessary to assist in the control and collection of balances due:

- Name of owner
- List of invoices with open balances by project (subtotals by project and total by owner)
- Name of project manager
- Invoice date
- Original invoice amount
- Balance due on invoice
- Aging of progress billing—current, 30 to 60 days, over 60 days
- Analysis of retentions

A typical accounts receivable report setting forth this data is shown in Figure 11-5.

By utilizing the data set forth in this type of analysis, management can identify; (1) the amounts due on each contract as well as the total amount due from each owner; (2) the project manager responsible; and (3) the overdue balances for both progress billings and retentions.

In addition to an adequate accounts receivable analysis, an appropriate collection system must be in place to monitor and follow up on noncurrent balances on progress billings (see the section on retentions that follows). Noncurrent balances should be carefully investigated and appropriate collection procedures should be employed. Collection procedures might include monthly statements, dunning letters, telephone calls, and personal contact with owners.

Lengthy collection delays or partial payment of amounts due on progress billings may be indicative of a misunderstanding with owners on payment terms or of deductions for disputed billings. Overdue balances should be reviewed with project management to avoid alienating the owner by making inappropriate or excessive collection efforts.

In the event excessive delays on currently due progress billings continue to occur on specific projects or with individual owners, the contractor should consider the feasibility of assessing interest on overdue balances. This is a reasonable business practice, particularly in the current environment of high carrying costs associated with financing business activities.

The combination of timely billing practices and effective management of collection procedures significantly enhances the contractor's cash management program.

Retentions

The management of retentions and the management of progress billings are of equal importance to the contractor's financial health, especially because retentions (typically 10 percent) often represent either the entire gross profit on the project or at least a significant portion of it.

Retentions are frequently not fully released until completion and acceptance by the owner; thus, it is critical that every effort be made to accelerate completion and acceptance so that retentions are released on schedule.

As projects near completion, construction firms all too often concentrate their efforts on starting up new projects and searching for new work. Despite the importance of concentrating efforts on new projects, sufficient attention must be devoted to completing projects in process so that the owner's approval for the release of retention is received. Furthermore, from the point of view of client relations, an otherwise satisfied owner can be alienated by relatively minor problems in the final completion of the project.

The project punch list (a "to do" list) should be agreed upon with the owner and promptly completed to accelerate the release of the retention. In the event that there are minor items on this list that are delayed for abnormal reasons (e.g., out-of-stock materials), the contractor should consider entering into negotiations with the owner to obtain at least a partial release of retainages and thereby reduce the uncollected balance of the total retention.

In the event of owner claims or back charges that may delay release of retentions, the contractor should accelerate resolution of these open items. If excessive delay is contemplated, efforts should be made to obtain partial release of retentions. As noted earlier, the collection of normal retentions—those that are due based on projects completed and accepted—should be subject to normal collection procedures to accelerate cash flow.

ADJUSTMENTS IN CONTRACT PRICE AND BILLING ARRANGEMENTS

In addition to progress and retention billings set forth in the original contract between owner and contractor, adjustments to the contract price and billing arrangement are frequently necessary to cover events that occur during the period of contract performance or at the completion of the contract. These modifications—such as those arising from change orders, back charges, escalation, performance incentives, or claims—have an impact on a contractor's cash flow and must be analyzed not only to maximize cash flow but, more importantly, to control project profitability as well.

Change Orders

On most projects, change orders are a common occurrence. They cover changes in project specifications or magnitude that require an adjustment of the total contract price. The magnitude, extent, and control of change orders frequently represent the difference between profit and loss on a project. Therefore, it is critical that these changes be properly authorized and put in writing by the owner *as they occur*. A proper system of handling change orders minimizes billing and collection difficulties and, more importantly, reduces the impact of claims arising from unapproved change orders. The appropriate format for a change order billing is shown in Figure 11-6.

Steps for Adjusting the Billings. Change orders can significantly affect total contract revenues. Thus, current progress billings must be adjusted to reflect increases in amounts billable as a result of these changes. If contracts are structured to prohibit additional billings outside of the billing schedule, then the contractor necessarily incurs additional costs, with collection deferred until the project is completed and accepted. The frequent occurrence of change orders on construction projects mandates that where possible the billing for change orders be issued and collected as incurred during the life of the project.

The normal format for the billing of change orders during the contracting cycle should include the following steps.

1 *Appropriate authorization.* Authorization for the additional work, including approval by the owner or the owner's representative. The contractor must make certain that the individual suggesting or directing the change has the authority to make the change and that the additional cost to the owner is clearly understood.

2 *Documentation.* Written documentation of the change is required; most contracts do not allow oral modifications. The contractor must provide, or be provided with, the appropriate writing in advance. The nature of the change order sometimes makes it impractical to obtain the writing in advance, but this

---

FIG. 11-6   Change Order Billing

[Letterhead]

[Date]:

Alteration Corp.
One Boston Place
Boston, MA 02102

Re: DODGE MEMORIAL LIBRARY PROJECT
     SPECIAL BILLING CHANGE ORDER #4

Original contract		$2,150,000
Change orders authorized		
Prior to this billing	$67,300	
Change order #4	9,100	
Total changes		76,400
Total contract		$2,226,400
Original contract completed to date		835,000
Change orders—all completed		76,400
		911,400
Total previously billed		902,300
Due this billing		$ 9,100

---

should remain the ideal. At the very least, a change order must be followed up quickly so that, at some later date, the parties do not allow the change to become a disputed item.

3 *Billing of the change.* The contractor's information system must be organized so that the billing function is notified and reacts promptly to the new change in the contract.

4 *Collection.* When a properly authorized, documented, completed, and billed change order becomes an integral part of the contract, standard assertive collection methods must be applied to the change just as they are to the base contract. If the contractor exercises due care over the change order, it should result in a minimum period cash outflow. If there is no effective system, a negative cash flow may cause a significant cash drain.

Escalations, Incentives, and Penalties. The long-term nature of many construction projects often requires that contracts have escalation features to provide for the impact of inflation on material or labor costs. In some contracts, incentives are given for early completion, acceptance, or levels of performance; conversely, penalties are set for delay or lack of performance. The contractor should establish suitable controls to bill all these adjustments of total contract price on a timely basis.

Back Charges

Back charges are billings for work performed or costs incurred by one party that, by contract, should have been performed or incurred by the party billed. Back charges are often not recorded as accounts receivable or payable but merely as received or paid by the construction firm. Although back charges seldom represent significant amounts in relation to total project cost, they can be material relative to the ultimate contract profitability. Furthermore, if back charges are not processed on a current basis, the amount of the charge may be disputed and therefore result in a claim. Back charges should be processed and billed on as current a basis as possible.

The most important controls exercised over back charges include:

- *Documentation.* The item causing the back charge should be verified in writing and then referenced to the original contract. The responsibility for the charge can thus be properly identified and the authorization validated.

- *Accounting.* The increase or decrease to the contract should be reflected in the next standard billing, so that the parties are all formally apprised of the change in contract terms created by the back charge.

Claims and Litigation

Claims are amounts in excess of the agreed-upon contract price that a contractor seeks to collect for customer-caused delays, errors, and so forth. There is an ever increasing quantity and dollar amount of claims relative to the number of projects in process, possibly because of current economic conditions and the litigious nature of our society. From the point of view of both the contractor and owner, delays in the settlement of claims and the attendant legal and related costs affect all parties.

Claims and related litigation or arbitration result in significant delays in releasing contract retentions and determining the ultimate resolution of the disposed amount. In addition to the interest and related costs, there are significant legal and internal costs involved in resolving claims. Besides procedures designed to minimize claims—such as obtaining approval for all change orders and properly pricing such change orders to reflect all the costs associated with a change in specifications or scope—the contractor must accumulate adequate cost information and supporting documentation in order to be properly compensated for the work performed. A failure to accumulate and document these costs properly will adversely affect both the ultimate resolution of the claim and the period of time required to collect the claim. Claims management and contract litigation are important topics that are discussed in depth in Chapters 13 and 17, respectively.

INVENTORY MANAGEMENT

Sound inventory management is another important aspect of cash management. One of the unique aspects of the construction industry is that many of the small contractors do not find it necessary to maintain inventories of materials and supplies to support their operations. Therefore, they purchase materials and supplies only to meet the requirements of the project and for delivery to the job-site. On the other hand, larger contractors maintain substantial inventory levels of material and supplies. Regardless of the contractor's policy on maintaining inventories, purchasing practices and control of materials can have a significant impact on cash flow and job profitability.

Purchasing Inventory

The purchasing requirements of all jobs in process should be considered in order to obtain any available price advantage from volume purchasing. For example, if two or more jobs are in process that require materials such as structural steel, reinforcing rods, or the like, the total requirements should be consolidated to obtain volume discounts. Scheduling of deliveries of these materials to job-sites must be carefully coordinated with the scheduling plan of the job to ensure that the materials are on hand when required. Again, proper planning allows the contractor to match his payable cycle with the expected collection from the owner. If it is not feasible to coordinate delivery of materials to job-sites directly with job scheduling requirements and deliveries are therefore made earlier than is necessary, then the contractor must compare the time value of money with the savings accrued from volume purchasing and the impact on his cash flow.

Maintaining Inventory

Some contractors maintain an inventory of materials and supplies to meet the requirements of commonly used items for jobs in process. The carrying of such inventories can be justified for the following reasons:

- The maintenance of inventories enables a contractor to estimate some portion of job material costs with a high degree of certainty and with less use of estimated cost of purchased materials.

- In view of continually escalating raw material costs, the cost of materials and supplies held in inventory may be less than the replacement cost.

- The maintenance of inventory stocks permits a contractor to acquire raw material in economical order quantities.

- The contractor is less subject to the hazards related to reliance on suppliers in meeting delivery schedules required for each project. Delays in the job-site receipt of critical materials due to supplier or transport strikes can be alleviated if the materials are in inventory.

- Excess materials and supplies at the completion of a project can be returned to inventory and then used on subsequent projects. The contractor can thus avoid the necessity of disposing of this material at distress prices.

There are many advantages inherent in maintaining an inventory of commonly used material and supplies; however, to do so often requires a substantial investment of cash. Accordingly, contractors should implement sound inventory control procedures to avoid the problems related to excess inventory levels, obsolescence, and high carrying costs.

EQUIPMENT

The acquisition of construction equipment is normally perceived as a negative cash flow item. Thus, when the decision to acquire or not to acquire equipment is being made, consideration must be given to some basic cash flow decision-making models. Any analysis should include items of input, such as payback period measurement, the internal rate of return, and a simple present value model. Construction equipment acquisitions should be based on the economic fundamentals of need and the ability to finance the cost. Too often, companies that are undercapitalized, have inadequate cash resources, or lack financing alternatives purchase heavy equipment on a long-term basis. This action could result in excessive carrying costs and have an adverse impact on working capital. Such purchases can result in reduced profits, which may consequently hamper the contractor's capability to bid on new work.

Cash management of fixed asset needs is both manageable and critical. Options available to the contractor are varied, and a poorly planned execution of the meeting of the equipment requirement can bring about predictable and troublesome effects. Accordingly, these alternative methods should be considered by the contractor prior to making the capital investment decision related to equipment purchases:

- *Leasing* – a favorable option where equipment is needed on a project for short duration, or future utilization of the equipment on other jobs is not assured.

- *Rental with option to buy* – a common practice, whereby equipment is leased on a long-term basis with the option to purchase–using all or part of the lease payments to apply against the purchase price. The federal income tax regulations related to leases with purchase options are highly complex and, therefore, must be considered when using this alternative.

- *Equipment purchase* – generally the least costly option if the equipment has a high degree of utilization. In purchasing equipment, consideration must also be given to the income tax regulations on investment credits and depreciation (including the accelerated cost recovery system (ACRS) under the Economic Recovery Tax Act of 1981 (ERTA)), which are discussed in Chapter 18.

In considering these and other alternatives, the contractor must be continually aware of the impact of his decision on the firm's cash flow and financial capacity.

ACCOUNTS PAYABLE

Another area related to the management of cash resources concerns the payment of invoices from suppliers and the amounts due to subcontractors. The contractor must meet obligations to suppliers as they are due. This practice ensures a favorable credit rating, which in turn aids the future availability of credit from both suppliers and credit grantors. While it is important to pay suppliers' bills on a timely basis, the payments must also be managed.

Vendors' billings frequently permit cash discounts for early payment of invoices. These discount terms may permit a deduction of 1–2 percent of the invoice price for payment within the specified time period. Accounts payable procedures should take advantage of such discounts to reduce costs of material, supplies, and operating expenses.

Progress billings for subcontractors should also be processed for payment on a timely basis. Contractual arrangements with subcontractors should provide for retentions comparable with the retention provisions in the master contract between the owner and the general contractor. It is a simple planning rule that the release of retentions to subcontractors be coordinated with the release of retentions by the owner to the general contractor.

FEDERAL INCOME TAXES

One of the major planning areas with significant cash flow implications relates to the alternatives available to contractors with respect to federal income taxes. While this is also true of most business entities today, the planning alternatives available to contractors and the complexities of income tax accounting for the industry present unique tax opportunities. The selection of alternatives and the use of advantageous income tax accounting methods can provide significant cash flow opportunities for contractors. Chapter 18, which concerns taxation and tax planning, presents a discussion of the available options, which contractors and their advisers should study with care.

CONCLUSION

A successful, profitable construction contractor must have a working program of cash management. Financial capacity, as represented by effective cash flow and the availability of financing alternatives, is critical to a contractor's success. Figure 11-7 depicts the warning signs and possible causes of poor cash management planning. The need for cash budgeting and forecasting, and the integration of these processes with job scheduling, are specific areas that the contractor should consider in the cash

---

**FIG. 11-7   Poor Cash Management Planning — Warning
Signs and Possible Causes**

Warning Signs

- Cash balances are lower than usual.
- Management's major priority is pursuing dollars, not running the company.
- Payables are higher than normal without a corresponding increase in volume.
- Creditors have imposed credit restrictions such as "Cash Only" terms.
- Cash discounts are no longer taken.
- Loan payments are made late.
- Penalties are assessed on payroll taxes.
- Ratios, such as deteriorating-current ratio, acid-test ratio, prove inadequate.
- Long-term debt is used to finance normal operations rather than for expansion or fixed-asset acquisition.

Possible Causes

- Operating losses.
- Extraordinary expenditures (e.g., lawsuit settlements, fire loss, etc.).
- Inefficient bidding procedure—timing of billing cycle not appropriate.
- Receivables collection periods lengthened.
- Fixed assets financed through current assets.
- Change orders not processed properly.
- Increased noncontract expenses.

---

management process; these are the survival links for the contractor. Constant vigilance over all aspects of cash management can only enhance the contractor's profitability and financial capacity.

CHAPTER 12

Financing

Robert Gardella

INTRODUCTION

Expressed or implied, written or oral, contracts form the structure of our personal and business existence, promising fulfillment of goals through mutual understanding, agreement, and commitment. Contracts form the essence of the operations of the economy and are also vital to those professionals referred to as contractors.

CHARACTERISTICS AND RISKS

The business of contractors is subject to one or more of the following characteristics:

- An extended period of time (often measured in years) needed to complete a job

- A large investment accrued in contract assets (materials, supplies, construction-in-process, receivables including retentions)

- A substantial employment of capital assets (heavy equipment) and/or labor

- A custom-made, special, or limited-use product, built to the owner's specifications

Any one of these characteristics limits acceptability of the credit risk for normal commercial banking purposes. The contract is a set of conditions agreed to by both the seller (contractor) and the purchaser (owner) that, in effect, assures a sale in advance of production. This is the principal benefit of the contract to the contractor. The owner is also protected in the contract by the pronouncement of exacting specifications of performance, time and place of delivery, and price and terms of payment. These pronouncements, in the foreground of the characteristics inherent in the business, suggest significantly increased business risks. These include the following requirements:

- Delivery of the project on time and according to specifications

- Completion of the project at a cost within the contract price

- Achievement of the financing needed for labor and materials

Fundamentally, these risks are borne by the contractor. However, in an atmosphere where unusual events may occur, terms are often included in the contract to alleviate the effects of external uncontrollable events.

- Time frames can be made subject to adjustment in the event of *force majeure* (acts of God), war, strikes, or events that cannot be reasonably anticipated.

- Hazards of cost overrun deriving from inflation can be moderated by escalators based on independent published indices (e.g., the Consumer Price Index). Research and development contracts can be negotiated on a cost-plus basis assuring reimbursement of costs plus a preset markup or incentive fee.

- Financing can be prearranged by a provision in the contract for advance or progress payments during the course of performance and before completion.

MANAGING THE RISKS

Contracts are legal commitments, enforceable by the parties in a court, that can order performance as specified or award money damages in lieu of specific performance. Bankers should be familiar with the basic ingredients, operation, and terminology of contracts and should always consult with an attorney to be sure that any contract essential for the protection of the bank or a credit grantor is reviewed for validity and enforceability of rights in the best interests of that lender. It must always be assumed that the rights of parties against the lender or borrower are fully enforceable.

Because of the seriousness of contracts and the fundamental risks borne by the contractor, contracts should be entered into only after very careful study and analysis. Performance under contracts must be conducted under very stringent control.

Job Performance

The first of the purchaser's considerations prior to entering into a contractual agreement is the evaluation of the contractor's projected ability to perform the job according to specifications. This includes a determination of:

- Technological and facilities requirements compared with contractor's abilities, and
- Manpower and management required compared to capacity.

Contract Price

The second consideration is more difficult—arriving at a contract price. The basis of the contract price is the estimated cost to the contractor to deliver the job. If the bid based on the estimate is too high, the contract will not be awarded. Alternatively, if the bid proves to be too low, the contractor may end up with a nonprofitable job.

Job Progress

Because the potential for cost overrun exists on all but cost-plus type contracts, it is critical that contractor management pay very close attention to job progress to assure that it follows a precise schedule of cost and performance accomplishment. In that way, any departures can be corrected immediately so that the economic viability of the job or even the entire company need not be put in jeopardy.

Protection From Risk of Failure

The purchaser also seeks protection from the risk of contractor failure and the consequence of facing up to having invested (via advance or progress payments) in an incomplete job of dubious value. Protection usually takes the form of a surety bond issued for the account of the contractor and the benefit of the purchaser. The bond promises completion of the job as specified (completion bond) or the payment of money (payment bond).

CONTRACTOR FINANCING

The financing needs of contractors, like those of most business professionals, include both capital asset investments and working capital. The sources who need cash are usually the project owner, through part payments on contracts; banking sources, through equipment or working capital loans; and the contractor himself, through his own working capital.

Capital Asset Borrowing

Capital assets (plant and equipment) are usually financed in a conventional manner. The contractor initiates an equity investment for a portion of the cost of a plant or piece of equipment and, using that investment as a base, debt-finances the balance with a security on the remainder. Ordinarily, these are straight asset-based loans made by:

- Banks or finance companies,
- Purchase money mortgages arranged directly with the equipment manufacturer, or
- Finance or operating leases with the manufacturer or third-party lessor.

Working Capital Loans

The working capital needs of contractors are financed differently than they are for other businesses in that they are most often centered on specific contracts. There are also differences between the two broad categories of contractors—construction contractors and manufacturing contractors.

Construction Contractors. The category of construction contractor refers to those contractors engaged to erect structures (buildings) or infrastructures (bridges, viaducts, tunnels, roads). Implications are that the job will have a high unit cost and require a long term for completion. These features, in turn, imply high investment in materials, work-in-progress, value added (labor) inventories, and a scheduled turnover period considerably longer than one year. The high working investment in each job, the slow liquidation rate, and the high risks inherent in the business combine to narrow and outpace the borrowing capacity of most contractors; this is especially true in consideration of the reality that a contractor normally performs on more than one job at a time. The financing of working capital needs is project oriented. The financing of the project by the general contractor includes a certain amount of downstreaming of part of the load via subcontracting. However, the major source of funds is derived from advance and progress payments from the purchaser/developer of the property. Banks are very active in providing for this indirect source of working capital through the developer, the other party to the construction contract, for the following reasons:

- The bank is insulated from direct risks associated with the contractors.

- The developer provides equity as a cushion to absorb loss and often furnishes substantiated guarantees of payment of the loan.

- The developer-borrower is the beneficiary of the surety bond, which helps ensure completion.

- The loan is secured by a mortgage on the real estate and/or the improvement.

- The bank can better ensure that the proceeds of each borrowing are invested in the collateral securing its payment.

- The permanent financing can be prearranged before the first borrowing.

Consensus among bankers is that lending directly to construction contractors is, categorically, the most hazardous of bank-lending activities. The characteristics, inherent risks, sizes of exposure relative to capital in an atmosphere of time and unpredictability, and the absence of a proprietary interest in the product are the main reasons for this prevailing opinion. Notwithstanding this assessment, banks do occasionally make available to the contractor lines of credit for general needs, with the stipulation that any borrowing be explained as to exact use, time, and source of repayment. General lenders have an added risk if the proceeds of a loan are invested in project assets against which the developer or surety company might have first lien or claim. The problem might be compounded if the proceeds are invested in a contract different from the one providing the repayment source, in which case the credit exposure might, in fairness to the surety companies, be held to be the sum of the debits to the loan account rather than the balance outstanding.

Manufacturing Contractors. The manufacturing contractor presents a significantly different picture with a different set of circumstances. While the length of the contract can span several years, the production time for each unit is usually much shorter, and the value of each unit is usually much smaller than the total contract value. Whereas the values of real estate and improvements for the construction segment are "in-place values" (unique location and limited purpose product), the manufactured good is portable and can be of wider use. Delivery is usually geared to rate of production and payment is usually arranged accordingly. Thus, the investment in working assets is small relative to the total contract value, and turnover is close to the normal standard for manufacturers. In fact, many businesses produce only against confirmed orders and are true manufacturing contractors. For the larger enterprises or the job order shop, this concern is more or less academic. Awareness heightens where a business with modest equity capital entertains a contract of inordinate total value. This imbalance of relationships appears regularly among United States government Department of Defense contractors. High technology products, know-how, and high-precision manufacturing techniques are often present in the entrepreneurial business of unusual capability but limited means. With a demonstrated ability to produce and a customer of unquestioned credit standing, direct bank loans for working capital needed over the entire production cycle can be arranged. Usually these loans are made against assignment of monies due, or that will become due under the contract, or are at least regulated by a borrowing base including billings and inventories. (Assignments of United States government receivables must be perfected in accordance with the Assignment of Claims Act (1940) .) Neither of these methods secures loans in a traditional sense of independent value, since the collateral consists of contract assets having questionable value in the event of default. However, either arrangement serves as a self-leveling governor on the activity/debt relationship and identifies the expected source of repayment.

CREDIT ANALYSIS

Analyzing the credit of a manufacturer generating a regular and steady output of standard goods or services does not differ from analyzing any industrial company. Construction contractors and manufacturing contractors producing very large units requiring long production times, such as ships and printing presses, present more of a challenge to the analyst. Financial statements, the main documents evidencing financial strength and operating performance, are the product of analysis.

If financial statements are prepared on the percentage-of-completion basis, income recognized for any period is based on the estimated gross profit on each job in process and a determination of the percentage of the job completed in the respective period. The former is tenuous; the latter is often arbitrary and inconsistent and is seldom accurate. Tax accounting for that same company is usually on the completed-contract method. The statements offered as a fair representation of the operations for the period and the financial condition at the close of that period are a composite of these fluid data. Supporting schedules, although they are not ordinarily furnished with financial statements, need to be examined periodically. Backlog and work schedules should be reviewed to determine that commitments are within the capacity of the company to perform. A work-in-progress statement should be studied and compared from period to period. All significant revisions of estimated gross profit on jobs should be investigated. The relationship of estimated costs to complete projects and billings to date in excess of costs should be carefully watched. An assessment of the degree of danger facing the contractor who has to complete contracts with borrowed money but has insufficient latitude to bill for repayment sources must be a continuing part of the loan evaluation.

Generally, banks do not lend to construction contractors. Some banks entertain small lines of credit for incidental short term needs of construction contractors, but recognize that such lending activity is at high risk, takes great care, and requires diligent study. Although banks are a source of funds used in construction, the preferable route for many banks is through the developer in the form of construction loans, which, in turn, enable the developer to make progress payments to the construction contractor.

The next section of this chapter describes normal construction project-lending practices.

CONSTRUCTION PROJECT LENDING

Commercial banks are a major source of funds used during construction of residential and commercial properties. For many banks, construction loans are an important segment of their earnings assets in terms of both size and yield. On December 31, 1983 construction and land development loans amounted to more than $60 billion or 5 percent of the total of loans outstanding at all insured domestic banks. For large banks, the percentage was undoubtedly higher. During the week of August 6, 1984, commercial banks made 33,300 such loans aggregating $1.3 billion, having a weighted-average maturity of 9.2 months, and bearing a weighted-average interest rate of approximately 14.56 percent. This compared with a like number of term commercial and industrial loans made in that same week aggregating $4 billion, having a weighted-average maturity of 49.4 months, bearing a weighted-average interest rate of 13.81 percent (Federal Reserve Bulletin, February 1985). The apparent contradiction of higher rate on shorter maturity is actually a reflection of the greater credit risk associated with the construction industry. A ranking of high-loss industries nationwide includes members of the construction industry among the top 25. First were investors, including personal borrowers; second were eating places; third were single-family-housing construction loans; oil and gas field services were fourth; and real estate agents and managers were fifth in the frequency of reported charge-offs. (Robert Morris Associates, Report on Domestic and International Loan Charge-Offs for the year ended December 31, 1981.)

UNDERSTANDING THE RISK

Because of its importance to both borrowers and lenders, and in light of its high-risk nature, construction lending requires specialized handling. Regardless of the organizational lines of a bank and its credit approval/ administration procedures, its construction loan business is almost always conducted by a separate unit in the bank staffed by professionals. By careful practice, the degree of risk can be reduced to a point where the nature of the risk can be accepted and the business can be made not only bankable but attractive. Individual members of a construction-lending unit must be well grounded in the fundamentals of bank credit evaluation and technically trained in the field of construction with: an ability to understand such things as blueprints, engineering reports, and building codes; a capability to assess project economics and local markets; and a familiarity with real estate law, documents, and regulations.

PRINCIPLES OF CREDIT ADMINISTRATION

"Know your borrower." "Mind your business." These phrases summarize the general principles of prudent credit administration. These principles pertain even more directly to construction lending, indeed demanding more than the usual vigilance in their application, specifically with regard to communication with the borrower and an understanding of the project.

The lender must thoroughly investigate the borrower's situation at the outset. Knowing the borrower requires, from the beginning, a thorough investigation of the people involved and the establishment of positive evidence that they are of sound character and integrity. This process includes the following basic steps:

1   Checking with references, reporting agencies, trade sources, and financial institutions. (Any material doubt at this juncture ordinarily precludes further consideration of lending to a subject on any basis).

2   Understanding the physical project and determining the ability, capacity, and experience of the developer and the contractor to manage and build the project.

3   Studying the underlying market concepts and the projected economics of the property and assessing these for reasonableness.

4   Visiting the proposed building-site, visualizing the planned improvement, and analyzing the feasibility study, including management's projection of income to be derived from the completed project and their underlying assumptions.

5   Investigating local circumstances for environmental impact, building codes, laws, and regulatory reputation.

Financial and Economic Evaluation

If the project passes these subjective assessments, it is ready for financial and economic evaluation. This is best initiated by comparing the estimated total costs to complete the project with an appraisal of the in-place value at completion prepared by a qualified independent appraiser of recognized stature. The margin of appraised value over estimated cost indicates the developer's economic motivation for the project and sheds light on the potential cushion of safety for a lender contemplating the construction load. The adequacy of that margin depends upon the particulars of each project. The lender should keep in mind that the appraisal is not at all concerned with the risks of bringing the project in as specified—on budget and on time; these are the owner's and contractor's risks and do not affect the evaluation of the lender's safety cushion.

Review of Total Cost Estimates

Examination of the components of total cost estimates must be reviewed. These include: the cost of the raw land improvements (so-called hard costs); the cost of engineering, design permits, feasibility studies, and other paperwork necessary to initiate and maintain the project (so-called soft costs); and the estimated financing costs. The cost/value relationship of the land is generally straightforward and the soft costs are largely a matter of actual history. The costs of the improvement, the hard costs, are the most difficult to assess. There are no exact formulae available for analysis; there are only judgments made by informed lending. A bonded fixed-price contract will, of course, alleviate much of this risk. Otherwise, assessment is a matter of professional judgment of the degree of difficulty in the project and the types of departures, if any, from the experience of the developer and the contractor. Financing costs are usually pegged at, or adjusted to, the upper range of the consensus of the experts for interest rates over the 18–36-month horizon of the typical construction loan.

Assessment of the Financing Side

As with all balance sheets, assessment of the financing side (liabilities) is as important as appraisal of the investment side (assets) in terms of both sufficiency (Has enough money been lined up?) and maturity schedule (Does maturity of loans and expiration of commitments allow enough time?) of all commitments. A margin for error, beginning at a minimum of about 10 percent for the simple project and scaled upward as project complexity increases, should be built in to allow for cost-overruns. The equity on the project should: be paid in full before the first loan advance; be ample enough to provide an additional cushion for contingencies; assure the maintenance of a safe value/loan relationship; and preserve the cost/in-place value upon completion.

Looking beyond the project's balance sheet is part of the assessment too, since a construction lender seeks the comfort of having additional capital available if needed in the project. That availability is a function of the financial resources of the developer and the developer's commitment of those resources (dilution of capital).

Construction loans are intended to apply only to the actual construction period. Every lender expects that on completion of the project the funds for repayment of the construction monies will be available. Very often the source of the funds is the owner's permanent financing, or take-out, mortgage. Since the permanent mortgage is significant to the construction lender's ability to be repaid, the terms of the takeout commitment and conditions that have to be met before it is effective must be fully understood. The financial resources and general credit standing of the takeout source (for the most part, large insurance companies, thrifts, pension funds, or industrial companies) are equally important.

In summary, the principles of credit administration require that the construction lender do, at the very least, the following:

- Create evidence of the sound corporate character of the borrower.

- Evaluate the financial and economic sense of the project based primarily upon profitability analysis.

- Make a specific review and assessment of the project's cost estimates.

- Measure the contractor's and the project's total financing structure including equity sufficiency, capital availability, and the terms and conditions of permanent financing.

STRUCTURE OF THE CONSTRUCTION LOAN

Each construction loan is project oriented, that is, isolated, insulated, assessed, and documented as an individual entity. The project stands on contracts and is held together by the underlying commitments. All arrangements are formalized, binding, perfected, and enforceable, including those for the construction loan, which is controlled by a commitment entered into after credit approval and satisfactory review by account officers and legal counsel of all details. The commitment is a comprehensive document that not only describes in detail the parameters of the project but also prescribes the requirements and method of each loan advance. Each loan is very much a "hands-on" arrangement. Advances are made as the job progresses, and each advance is measured against the project budgets for actual performance. Repayment is not expected before completion of the project so that the lender must follow its development, make periodic on-site inspections, review engineering progress reports, and interview contractors and foremen.

Closing documents in addition to the loan agreement include:

- A note for the full amount of the commitment. A grid is used to record each advance.

- A security agreement granting first mortgage lien on the land and all improvements. The security agreement should normally prohibit secondary liens. These first two documents are commonly referred to as bond and mortgage.

- Title insurance on the land and improvements.

- Assignment of contractors' and subcontractors' surety bonds and all other contract rights of those parties.

- Assignment of permanent financing commitments, purchase commitments, standby commitments, and all leases. Preferably, the end loan or purchase should be preclosed with all conditions, but actual construction should be fulfilled before the first loan advance.

  There is some room here for variance and negotiation depending upon the nature of the property and the developer. Latitude ranges from a requirement of no more than passage of time, to complete construction, to the achievement of a certificate of occupancy and a fully leased project with documented minimum rent-roll. To the extent that loan advances are contemplated before fulfillment of any condition other than construction, the lender adds marketing risk to the construction risk.

- Recording and other documents required for perfection of all liens and assignments.

- Guarantees, if any.

OPERATION OF THE CONSTRUCTION LOAN

Documentation

Documentation does not end with the first closing. Each draw-down under the commitment is fully documented, from the borrowers' certification of compliance with all terms and conditions of all contracts, to the submission of invoices substantiating additional investment and value added to the project, to the contractors' and subcontractors' affidavits of the amounts due them. Disbursement is best made through the title company to ensure that no encumbrance has arisen that might compromise the bank's collateral position. If disbursement is made directly, a title check should be made and documented. The bank should have its consultant engineer review invoices and make periodic on-site inspections to ensure that physical progress approximates that indicated by documents. Visits by the account officer to the job-site and interviews of project managers are essential. Documentation holds a deal together and is vital to the proper maintenance of the loan. Because the documentation is almost as important as the econom-

ics of the project itself, construction loans also require greater legal assistance than ordinary commercial loans and are seldom acquired via remote
participation from other lenders. Occasionally, a participation in a loan is
considered for purchase, but this should only occur when the original bank
has a construction lending unit of known and proven ability and, even
then, only after a careful review of the loan is made.

It is good practice to include a local bank in a construction loan for
several reasons:

- It can keep an almost day-to-day communication with the
  project and immediately notice a significant event.

- It is in the news area of the project and is likely to hear about
  things going on in the community that can affect the project.

- It is a good device for developing or cultivating a correspondent
  relationship, which can lead to a continuing relationship with
  that bank.

Fronting a Construction Loan

In some instances, it is necessary to front a construction loan through a
local bank because of business problems in the area of the project. Due to
the immobile nature of the basic collateral (real estate), the lender must be
in a position to act (foreclose or else take title) in the jurisdiction of the
property. Some jurisdictions to not recognize "foreign" (out of state) corporations unless those corporations are licensed to do business in that
state, which usually requires payment of substantial taxes and fees. A
method used to clear this obstacle is to make the loan through a local bank
and immediately purchase a 100 percent participation in the loan. Risk
passes through the lender of record. A small participation in the loan or
just a fee is often sufficient inducement for a bank to front the deal.

Project-oriented construction lending can be a safe and profitable
business opportunity, but it requires commitment to specialization and
demands both time and diligence. Compensation in the form of interest
is usually higher than on ordinary loans, but overhead for handling is
higher as well.

PROFILE OF A CONTRACTOR SUITABLE FOR FINANCING

The selection of the word "suitable" in the heading of this section represents the triumph of reason over inclination. At this point, the ten-

dency is to use an adjective such as "ideal," "typical," or "model," but none of these would be logically derived from the premise of the earlier discussions in this chapter. In fact, these terms would describe a utopian situation in almost any industry. Risk appraisal, however, usually includes the practice of measuring departures from a recognized or perceived standard and then weighing them on a balance countered by the potential benefits. This invariably leads to the objective study of relationships among and between balance sheet and income statement accounts and a comparison with composite data for the industry or with those of a subject recognized or considered as an industry leader. This study helps the analyst detect quantitative changes, trends, and departures from the norm. Analysis is concerned with the qualitative interpretation of these changes, trends, and departures, since there is no list of absolute criteria.

Financial statements are the product of analyses, estimates, and judgments. The production cycle for a contractor seldom coincides with the conventional accounting period. Therefore, any attempt to design a profile of a contractor suitable for financing becomes dependent on subjective determinations. This status suggests that the analyst regress and begin the appraisal process by evaluating the fundamental processes supporting the financial statements.

Understanding that by definition the ideal does not exist and recognizing that a standard is used only as a yardstick, one might think of the suitable contractor candidate for credit from the creditor's point of view, documenting a request as follows.

- Financial statements are audited annually, without qualification, by a CPA firm with a proven record of quality and competence in the construction business.

- Schedules of backlog and work-in-process are submitted in detail adequate to track major contracts—both those completed and in process—over a representative number of periods in order to assess the accuracy of estimates. Any significant revisions noted and a major portion of the gross profits recognized in each period are also realized in those periods.

- Assurance by the CPA firm as to the adequacy of controls and personnel. This is to ensure responsibility for accurate cost estimates and early detection for remedial action of departures from budget, while also ensuring that administrative procedures are adequate to assure realization of transactions recognized.

- Opinion of responsible legal counsel in relation to status, merit, and likely consequence's of all significant litigation pending against the contractor.

- Schedules of cash receipts and cash disbursements submitted showing, in adequate amount and detail, time and source of receipts and required or planned disbursements.

A verifiable and working request can be enhanced by good and complete documentation. Aside from speeding the approval process, this documentation can facilitate review of the credit file by examining authorities and ensuring that the name will not be listed in a deficiency report for reason of inadequate information. The purpose of examining the supporting schedules is to understand, not assess, the auditors. Bankers and other creditors rely heavily on the opinion of auditors in performing their evaluation and in making a decision.

CHAPTER 13

Contract Change Management

Francis J. Callahan

INTRODUCTION

Historically, the contract claim or change order was viewed as evidence of the contractor's mismanagement and inefficiency. The construction process was by and large conducted in a nonlitigious environment. In recent years, however, all of this has materially changed. The introduction of design/build, fast track, and other new contract techniques; the availability of more sophisticated construction techniques; and the general economics of the construction industry have made proper management of construction contract changes as important as any other phase of construction management.

The era of the "good old boy" owner, who understood the contractor's problems as they arose and who had a long history with the contractor, has passed. The time when a contractor was not involved in construction litigation has passed as well. Today, in the era of more complex and sophisticated projects, of more competitive pricing of projects, and of greater economic pressures, there is the logical evolution of a much tougher owner with whom the contractor must work. Once signed, the construction contract does not allow for unanticipated cost increases, poor scheduling and management, and general inefficiency. More and more, contracts are written to transfer as much risk to the contractor as possible. The increased litigious attitudes of the owner have propelled contract change management to a position of major concern to those in the construction industry.

Today, the construction contract that reaches completion as originally designed, planned, and bid is a rare contract indeed. The growing environment of change from original project to ultimate completed project has made proper contract management, especially as it relates to changes, as important to the contractor as effective scheduling, supervision, and performance. Good contract management has advanced far beyond the stage where just bidding and getting the job were the most important things. Today, just getting the job without proper management has put many a contractor under.

The contractor who recognizes that contract changes are a real part of today's construction world and who plans, manages, and documents every construction project in anticipation of these possible changes will be in a much stronger position to recover reasonable compensation for additional work performed when the additional work is the result of third-party actions. Proper contract change management does not necessarily result in litigation. Litigation is only one option available as a last resort to the properly prepared contractor, while long costly litigation is inevitable for the poorly prepared contractor.

Litigation

Although litigation should not be an inevitable part of perfecting a contractor's rights resulting from contract changes, the contractor who must become involved in litigation must understand that claims litigation is an extremely time-consuming and expensive process that has both direct and hidden costs associated with it. Even the contractor with a well-documented claim can be tied up for two to five years with substantial resources in the litigation process. The costs associated with claims litigation include attorney fees, outside consultants, field and administrative personnel productivity, and the encumberance of liquid resources such as held retainage. The fact that a contractor has a valid, well-documented claim does not necessitate that he rush immediately into the litigation arena. The decision to move forward with claim litigation should be made only when all other approaches to settlement have been exhausted and a careful analysis of all the associated costs of pursuing the claim are understood. A dollar recovery three to five years from now may be worth much less than the eighty cents the owner is offering today. The decision to litigate or settle should be as factually based as possible; emotion should play little or no part in the decision.

Contract Management

The construction organization that accepts contract changes as a disruptive but real part of the construction world today and incorporates the basic concepts of good contract change management into the management of all construction projects will benefit in numerous ways. The contract will become a tool to be used by the contractor and not a potential prison in which the contractor must operate. Job problems will be identified early, a fact that will lead to an earlier, more efficient resolution of these problems with less worry about disputes and litigation. In addition, there will be a much better basis for understanding the impact of proposed changes and, therefore, a much better basis for determining the real cost of those changes and the impact on the overall job.

Today, the effective management of the construction contract change and the effective management of the total construction project are one and the same.

MANAGING THE CONTRACT CHANGE

The proper management of the contract and potential changes starts with the original bid documents and job budget and involves every phase of the construction process through the last punch list and acceptance. Each phase of the construction project, if properly organized and managed, produces the basic information necessary to understand, document, and ultimately collect for contract changes. The remainder of this discussion concerns itself with the various documents that are important to contract change management as well as the proper determination of the costs associated with contract changes.

THE CONTRACT

The construction contract is the very foundation for the whole construction process. It defines the project arena and determines what the project is, how the project will be built, what the time frame is, and what the project should cost. The contract also sets forth the specific or implied rights of the parties involved. For these reasons, the very foundation of the properly managed construction project is a complete and thorough understanding of the contract. Without this understanding, the contractor does not know what changes are or are not allowed; what costs, if any, associated with changes are allowed; and to what extent costs associated with changes are recoverable by the contractor. A thoroughly understood construction contract is a management tool; a poorly understood construction contract is a management crisis. Following is a discussion of some of the major areas contained in the construction contract that the contractor should thoroughly understand to manage contract changes properly and to recover sufficient remuneration for the work associated with those changes.

In reviewing a construction contract, it is important to consider who drafted the contract. This generally indicates to the contractor the general point of view underlying the form of the contract. If the contract is developed by the American Institute of Architects (AIA), the contractor can expect to find a general position in the contract that provides protection to the architect. If the contract is developed by the Department of Energy (DOE), the contractor can expect to find that the general tone of the document favors the owner where any type of doubt might exist. Knowing who drafted the contract document can aid the contractor in better evaluating the general attitude of the document.

Contract Types

The construction contract comes in many forms. The most specific form of construction contract is the lump-sum contract, which states that for a specific price, within a specific time frame, a specific contract will be completed by the contractor. This contract can be either bid or negotiated; today, it is typically subject to bid. On the other end of the contract spectrum is the simple time-and-materials contract, which states that the contractor will complete a project of general or specific scope for the time and materials consumed in the project. The remainder of this discussion of contract types takes a brief look at some of the more typical contract types.

Lump-Sum Contract. This is the most typical type of contract on major projects, and is especially prevalent in the public sector. This is the most specific and restrictive of the contract types discussed herein. The contract sets forth a project to be completed, the price to complete the project, the time frame allowed to complete the project, and, typically, specific conditions as to how the project will progress (i.e., scheduling and notifications).

Cost-Plus/Time-and-Materials Contracts. This group of contracts addresses the contractor's performance and compensation from an efforts-expended point of view. The project may be as specifically defined as it is in the lump-sum contract, but it can also be more loosely defined. The means of compensation, rather than the total price, is set forth in terms of specific costs incurred by the contractor or the time and materials consumed by the contractor during the course of the project. The project may have a maximum price or a target price. This type of contract may also allow for bonus payments to the contractor who completes early and/or under original budget. The key factor to this type of contract is the level of effort expended by the contractor. Another variation that may be included in this type of contract is the provision for a fee of some sort to be paid to the contractor in addition to the basic costs.

Unit-Price Contract. This type of contract provides for compensation to the contractor based on a specific price for a specific unit of production. The units can be yards of earth moved, yards of concrete poured, or any other specific unit of production. The unit-price contract may also provide for certain maximums to be set.

Other Types of Contracts. There are a variety of other types of contracts, including those involving construction management. In construction management contracts, the contractor is paid a fee for managing the project that consists of certain combinations of any or all of the above contract types.

General Conditions

The general conditions provisions of the construction contract set forth the broad overall rights of the parties to the contract. Following are some of the more important of these conditions and the reasons why they should be of concern to the contractor.

General Rights. General rights include the contractor's rights as to job planning, job-site access, job manning and manpower scheduling, and other conditions that affect how and at what rate the job progresses. The stated or implied assumptions in this area are that, within the terms of the remainder of the contract, the contractor has the right to expect to be able to plan, to man, and to execute the contract in an optimum manner that best employs his resources, while meeting the terms of the contract.

Contract Changes. These conditions set forth the specific procedures by which changes to the basic contract may be made. Specifically, they include how changes to the contract can be implemented, who can authorize changes, what is timely notification, and how the notification of change is to be made.

Default and Termination. This area sets forth the conditions under which the parties can be considered to be in default. It often includes the specific conditions under which the contractor may close the job down and other specific procedures for terminating the contract prior to completion. The contractor must understand the default and termination provisions should it become necessary to terminate the job early for any reason. He should know what remedies are available if he must terminate, or if the owner terminates, the job.

Exculpatory Clauses. The contract, through the exculpatory clauses, generally attempts to limit the extent of liability for certain actions of one or more of the parties to the contract. The exculpatory clauses are most often drafted to favor the owner or architect. The following are some of the typical exculpatory clauses that may operate to the contractor's disadvantage.

- There are no collectible damages for owner-caused delays unless due to negligence or deliberate misconduct. The importance of this provision to the contractor is especially pertinent if the owner closes down the project for any extended period of time and there is any type of cost exculpation. More than likely, the contractor will not be able to recover for the exculpated costs.

- Certain specifications, such as boring samples, are not to be relied upon. Even though the contractor is given certain pertinent information in the original bid documents, this set of provisions in the exculpatory clauses may stop the contractor from recovering costs if for any reason the ultimate conditions are not as set forth in the bid documents. Any provisions that restrict the extent to which the contractor may rely upon information provided in the bid documents must be completely understood by the contractor at the outset of the project.

- Generally, there are provisions that attempt to require the contractor to accept some responsibility for discovering faulty or defective specifications. These same types of provisions require the contractor to be responsible for adherence to all local building code requirements.

- There are provisions that relate to what, if any, liquidating damages the contractor will be allowed if the project is suspended. Although suspension is a rare event, the contractor should completely understand what will be allowed to close down and demobilize the job. For example, the contractor should know if the allowance will cover such items as overhead and some part of lost profit. Specifically, the contractor should understand which costs are reimbursable and which are not.

- There can be disclaimers as to site availability and crew scheduling. Although the general conditions of the contract usually address this situation, there can often be specific disclaimers incorporated into the exculpatory clauses that alter the usual implied or stated conditions under the general conditions.

CONTRACT CHANGES

Once the general overall contract is understood, the contractor should understand what types of changes can be made to the contract and what rights to reimbursement he may have. Following are some of the more typical change provisions that may be found in the construction contract.

Excusable vs. Inexcusable Changes

These types of change provisions define liability as to changes (i.e., who, if anyone, may be held liable for the impact of a change or changes to the project). This type of change provision generally sets forth what changes are contractually allowed for and what changes are not. Delays, such as those that are owner-caused, that may not provide for additional compensation to the contractor are considered excusable. Delays that are not provided for in the contract or precluded by the contract are considered inexcusable delays or changes.

Reimbursable vs. Unreimbursable Changes

Where excusable versus inexcusable provisions are addressed to the person for whom liability for a contract change would be directed, the provisions dealing with reimbursement address what remuneration, if any, there will be. For purposes of this discussion, changes are assumed to be inexcusable. Even if there is an inexcusable change in the project, however, the contractor may not have the right to monetary reimbursement. Although the contractor may be allowed additional time to complete, there may not be any additional funds forthcoming for the cost impact of a contract change.

Additions vs. Deletions Changes

How the project is to be expanded or contracted in scope, how the costs of the expansion or contraction are determined, and how the costs of these changes are determined are issues addressed by the additions versus deletions provisions of the construction contract. The way in which the monetary adjustments to a contract are determined when in-progress changes are made to the basic contract is an extremely important consideration for the contractor. Some of the important areas to be addressed are impact of costs, overhead, profit provisions, indirect costs, and other somewhat less tangible costs.

Acceleration vs. Delay Changes

The time frame within which the original project is to be completed may be expanded or contracted. Often, there can be a combination of both of these conditions in a single project. In the early stages the project may be delayed but the contractor is still required to complete the project within the original time period. The importance of the contract provisions that deal with acceleration, delay, or both can have an impact on such cost items as escalation of costs, cost of overtime, and job flow.

These are just some of the conditions that are often provided for in the specific provisions of the construction contract. Without a complete understanding of the total contract and how each of the various provisions of that contract can affect its performance and flow, the contractor can only react to the changes that come up during the progress of the project rather than manage the total project regardless of when and if any of these contract-changing conditions occur.

One last consideration is the number of changes involved during a project. There is generally considerable attention given to the impact of substantial changes to the project. However, a series of smaller-type changes, which alone seem insignificant, can, if put together, have a material impact on the contract. This is another reason to have a well-kept contract file and complete job diary to prove the impact of any change or changes to the project.

PROJECT DOCUMENTATION

In any discussion relating to construction contract changes, the inevitable question always is, How much documentation is enough? Simplistic and somewhat fictitious answers to this question are, Whatever it takes to collect full value from a cost point of view for any impact of a contract change, and, in the case of the change order, Whatever it takes to get one's price. The problem with this approach is that the documentation of a change becomes a hit-or-miss process. Without an established, organized approach to contract documentation, the typical contractor, when faced with a contract change, will try to develop the necessary support for the change after the fact, often attempting to document the impact and cost of the change from inadequate records. This after-the-fact approach to documentation can be an extremely expensive process with, at best, marginal results relative to the ultimate recovery.

Good contract documentation should be nothing more than good job management. When the subject of a complete job documentation system is raised, the typical response from the contractor who has not experienced the difficulty of trying to recover for that first major change might be to register complaints such as, It takes too much time and paper work, or, It doesn't get the job done. Another response might be, Overhead is high enough without all the extra paper work to properly document the job. Today, with contract changes a fact of life, the contractor who succeeds is one who has a good, uniform system of job documentation. Of course, a good job documentation system is more expensive than a less organized or less formal system. However, the ultimate recovery from a well-documented and factually based change or claim offsets the extra cost of the documentation system. The contractor who has a uniform system of job documentation will soon discover an interesting by-product of the system: better management with smoother flowing jobs.

The following discussion describes some of the general groups of job documents that are critical in the proper management of the construction project and the successful collection of value for additional work.

Contract and Subsequent Changes

These documents contain the general and specific frameworks within which the job will develop. They set forth the rights of each party, the description of the project, and how much and in what way the contractor will be compensated. The contract file should never be a document file that is put together when the original contract is signed and then filed. Rather, it should be an active file that is updated regularly as the job develops and changes are made. The file should contain a summary schedule that sets out the original contract information. As subsequent changes are made to the original contract, the summary schedule should be updated with the details of the change and the impact of the change on the total project. Management should be able to refer to this file and its summary to get an up-to-date picture of the total job.

Job Schedules and Manpower Plans

Most contracts require the contractor to present a job schedule and manpower plan to the owner within a reasonable time of the outset of the job. The job schedule and manpower plan set forth the flow of the job and the application of resources. Original schedules and plans are extremely important in showing the impact of later job changes on the flow of the job and the dedication of resources. Unfortunately, many contractors prepare the original job schedule, file the schedule with the owner, and then ignore the scheduling process from that point forward. Job schedules and manpower plans should, however, be maintained on a current basis during the entire job. If there are changes as the job progresses, the job schedule should be updated to reflect their impact.

Critical Path Method and Bar Chart Method

The particular type of job-scheduling system the contractor uses depends upon a variety of factors, of which the size and complexity of the contractor's jobs is probably the most important. The contractor can also use a combination of scheduling systems on some jobs. On the larger jobs, he may use the critical path method (CPM) for the total job, while using a variation of the bar chart method (BCM) on particular segments. The CPM is an activity-oriented system of reflecting the job, while the BCM is more time oriented. The CPM for the job may be cost loaded, that is, the budgeted or planned costs for the various activities are included in the scheduling process. No matter which scheduling system the contractor uses, it is very important that he use a scheduling system of some sort. The process of developing a job schedule after the fact to support a claim for additional compensation is an expensive process with questionable results. Any additional cost incurred to develop and maintain a good job schedule is more than recovered in better-run jobs and more effective claims presentations.

Job Budgets and Cash Flow

The job budget and cash flow are as important as the contract file and the job schedule to the proper management of the project in terms of changes. When changes are made to the original contract, the contractor often changes the revenue schedule but does not update the job budget. Depending on the size and number of changes that occur, an updating of the budget as well as the cash flow data can leave the contractor with an erroneous picture of where the project really is in terms of cost to complete.

Job Equipment: Plan and Register

The job equipment plan and register set forth the inventory of equipment that is to be employed on the project and state when that equipment will be needed. Accurate and up-to-date registers are necessary to properly plan equipment usage and availability. They are also valuable to document standby time and other ramifications of possible job changes. Another important value to the equipment register for the job is to ensure that all of the equipment used on the job is charged to the job.

Job Correspondence Files

These files represent the important communication link of the project. The correspondence files document such important areas as proper notification, proper response to unusual situations, and other narrative-type events. A very important point in relation to the correspondence files is the determination of who has the authority and responsibility to initiate and respond to certain types of correspondence. A clear line of authority and responsibility regarding who handles the various notifications when changes occur must be established. The person responsible for the notifications should be handling both the incoming and outgoing correspondence that relates to the job, even though a specific document may have been initiated somewhere else within the construction organization. Everything that relates to the project must be located in one place, under one person's aegis. Although the main office may have all the originals of the correspondence, it is important that the job files also contain a complete set of job-related correspondence. Many job problems could be avoided if the job files contain the information that the main office presumes the job personnel to know.

Job Diary

Probably the most important, and typically the worst maintained, of job documents is the job diary. The job diary is the day-to-day updated picture of the job and its progress. The level of recovery from job changes, delays, or other events is usually tied closely to the quality of the job diary. It is the cornerstone of the whole job documentation system. The exact detail involved in the diary depends on the size and complexity of the specific job. The diary should be the who, what, when, where, why, and how of the job as it progresses. A more specific discussion of the content of the job diary is covered in the discussion of the various cost elements involved in the pricing and documentation of the contract change.

Each of the discussed areas has an important application to the world of proper contract change management. In the remaining discussion in this chapter, the cost of contract changes and the impact that various job conditions and other variables can have on the cost of contract changes are addressed. There are many reasons for contract changes. Changes can be planned, such as those resulting from project expansion or contraction. Changes can also be unplanned, such as those due to strike, weather, material delivery, or owner problems. The responsibility for a contract change and the associated liability are legal issues not within the scope of this discussion. The intent here is to address the management of those changes and to measure, to the extent possible, their cost.

THE COST OF CHANGE

The various documents discussed above are the framework within which job changes and their potential effect on the project are defined, controlled, and identified. The actual measurement of the effect of a contract change or changes is job costs. The job costs associated with a change are obvious. How much time was expended and at what cost, how much material was used and at what cost, and how much additional equipment time has been consumed and at what rate are some of the more obvious costs associated with a change. There are a number of other variables that can affect the cost of job changes that are much more subtle in nature, such as efficiency, crew size, fatigue, crowding, and other such less tangible variables. They can have a much greater impact on job costs than the more obvious direct costs. The following discussion is concerned with the costs of contract changes and some of the variables that can affect the ultimate size of those costs.

CONTRACT COSTS

The costs associated with a construction contract can be broken down into three major groups, as can those associated with contract changes. These groups are: direct costs, indirect costs, and home office overhead, each of which is herein examined.

Direct Costs

Direct costs are subsumed by and directly identifiable with the performance of the contract. They include direct labor, direct material and equipment.

Labor Costs. The labor costs associated with a contract change—whether a change order or a contract claim—consist of some obvious factors that are relatively easy to measure and document once the specific change has been identified. Following are the obvious cost factors associated with the labor cost of a contract change.

- *Time.* The time involved to complete a change to a construction contract is the number of labor hours needed to complete that change. The difficult part in determining these labor hours is isolating those labor hours that relate to the specific change. In some cases, isolating labor hours may be a simple task, such as determining the hours necessary to make a simple addition to some phase of the project. On the other hand, the specific labor hours associated with a change that makes a major addition to the project and thereby changes or delays the remainder of the project can be an extremely difficult task to accomplish, especially if the contractor does not understand all the variables that can affect the labor required.

- *Rate.* Determining the cost of a change, like determining the time involved, may or may not be a relatively easy task. The rate could simply be the normal rate paid for the labor involved. There are, however, a great number of possible variables that can affect the cost of labor, just as there are variables affecting the time involved in a change. Some of these rate variables include overtime and rate changes.

  a *Overtime.* Overtime is the rate differential involved when the work force is required to extend the work day or work week to complete a project that has been altered by a change. The key factor here is isolating the overtime that is related to the change from the overtime that is caused by some reason other than the specific change (such as overtime required by the contractor's plan).

  b *Rate changes.* A change in labor rate for labor mandatory to accomplish the work required by a contract change would be the difference between the original labor rate and the escalated labor rate paid at a later date. Much like the overtime differential, the rate escalation due to work that has been pushed into a later period when labor rates are higher must be isolated from labor cost escalations due to work that is deferred into later periods for reasons other than contract changes, such as contractor-caused labor deferrals that are not related to contract changes.

In the above instances, it is important that the contractor be able to isolate and document the specific costs that are caused by the contract change. Here the contractor's most difficult task is to isolate and document the less obvious factors that affect the labor required by a contract change. Following are some of the less obvious variables that can affect labor costs.

- *Fatigue and efficiency.* These are two of the variables that can materially affect labor costs caused by a change to the contract. When the job crew is required to work extended hours because of a change or because of conditions that arise that were not in the original plan, fatigue may be the result. The amount of overtime required and the associated fatigue can seriously affect the efficiency of the work crew. Due to the fatigue factor, the amount of work produced by the job crew in a given time period may be materially reduced. Various industry studies have attempted to quantify the productivity impact of fatigue.

- *Work crew size, distribution, and supervision.* Because of contract changes, the optimum crew size that had originally been planned may not be possible. The size of the crew—whether larger or smaller than the planned crew size—can affect the planned flow of the job. Another factor involved with the size of the crew is the use of supervision. Altering the size of the crew from the planned size can either underutilize or overuse job supervision. The distribution of the work crews is another factor that can effect the flow of the work and the efficiency of the supervision. If, due to a change, the size of one crew on one segment of the project has to be reduced to supplement a crew on another segment of the project, the change may go smoothly and efficiently but the deprived segment of the project may fall behind plan. The loss of production in the unchanged segment of the project would be a cost of the change.

- *Experience and availability.* The addition of new and possibly inexperienced labor to the work crew can affect the performance of the whole crew. A larger crew does not necessarily mean greater performance. New workers can interrupt the work of the normal crew. Furthermore, new workers do not have the advantage of the learning curve that the regular crew does and may not be of the overall experience level that the contractor wants. In certain situations, the labor force of experienced workers available can be quite limited; therefore, the contractor looking to expand the job crew may have to face a decrease in worker standards.

- *Crowding and trade stacking.* Another somewhat intangible set of variables that can affect the performance of the work crew is crowding on the job-site. Two common reasons for this situation are the crew size (discussed previously) and the possible stacking of various trades that are not normally together on the job-site at the same time. Crowding can be the result of the regular crew working in a location for a longer period than had been originally planned. Thus, the progress of the work on a particular job-site could have progressed beyond its anticipated condition. The environment of the job-site could consequently lack adequate ventilation, lighting, facilities, and space for the crew to work with planned efficiency. In addition to the job environment changes, there could be additional problems added by the stacking of two or more construction trades at the site at the same time. The original schedule may not have called for the carpenters, electricians, and sheet-metal trades working in a particular location at the same time, but the delay caused by a change could easily have caused this to happen. The resulting inefficiency is a cost of the job change, and the contractor must be aware of these intangible variables in order to be able to collect.

Labor Cost and Job Diary. Earlier in this discussion, the importance of the job diary was stressed. There is probably no other area of the job where this importance is more evident. Because of the numerous intangible factors that can affect the cost of labor on a project, the complete job diary is of the upmost importance. The daily job diary should document every factor that relates to the labor force, and the labor section of the diary should not simply document the number of men and hours involved.

The job diary should contain information on the crew that is supposed to be on the job and what they are supposed to be doing. In addition, the diary should contain information as to what the crew did, in fact, do for the day. The diary should also contain a brief description of the job conditions for the day. This description should be as specific in detail about normal or excellent conditions as it is about poor conditions. Furthermore, the diary should detail those things that the labor force is not doing and mention whether or not this can be attributed to planned slack time or to slack time that is caused by some other reason. The objective is to create a picture of the job from a production point of view under a variety of conditions. The following is an example of the use of this type of information in a claim situation.

Assume that a pile-driver contractor has a claim for delay caused by the owner. The contractor wants to prove the extent of the delay damage caused to the job. The job diary shows the number of piles driven each day of the job, including the period of the owner disruption, and details what the job conditions were each day—both good and bad. With this type of information, the contractor can show what a normal crew under normal conditions can produce in a day and, therefore, what the extent of the owner disruption was. The well-kept and properly documented job diary would be the key to the claim. Another important feature of a good job diary under most situations is to regularly take photos of the various segments of the job. These photos would be taken under all conditions and would become a permanent part of the diary; they would show a variety of conditions, both good and bad, and would add support to the rest of the detail in the diary. The job diary should also correspond to the job plan and job schedule. This allows the contractor to show the work progress under various conditions as it relates to the plan and the float time within the plan.

Material Costs. The cost of direct materials that relate to a contract change have the same characteristics as labor costs. There are the obvious costs and the less obvious costs. The obvious costs of additional material required by a contract change, no matter what the reason for that change, consist of basically two variables: volume and cost.

- *Volume.* The amount of additional material required to complete a contract change can usually be determined with relative ease, by multiplying the specific units of material by the numbers of units used on the change. The key factor here is to isolate the additional materials used due to the change from other additional materials that may be required to complete the job resulting from some other source, such as contractor error in estimating.

- *Cost.* This usually is just the unit price of the material multiplied by the number of units used to make the change. There are, however, other variables that can affect the unit cost. The additional materials may not be acquired at the same price as the original materials due to price escalation, odd-lot purchases, necessary substitution, and the need to pay more due to short-order times.

There are other material related costs of which the contractor must be aware. He must also be prepared to document these costs so that they can be recovered. It may be necessary for the contractor to store material because of a change in the job. The materials stored could be the original materials for the job that were delayed in their usage or the additional materials that are required for the change. The storage of materials can generate additional costs such as storage charges if off-site storage facilities are required, additional security, additional insurance if the amount of materials stored is substantial, and additional holding costs if the contractor is required to obtain the materials far in advance of their usage. The contractor may also incur additional transportation costs if the additional materials are acquired from another source.

Once again, the job diary and its relationship to the job schedule and plan are most important to the contractor. The diary, schedule, and plan document both the planned use of the materials and the actual use of the materials, as well as the exact reason for any discrepancies between them. Without a complete job diary, the contractor may lack the knowledge and documentation to recover the additional costs to which he is entitled.

Equipment Costs. Like the cost of labor and materials, the cost of equipment resulting from a change has a number of variables involved. The obvious cost is the additional equipment time and the rate used for the specific equipment. Again, the contractor must be able to isolate the cost of additional equipment related to the change from the cost of additional equipment required for other reasons. Some of the other costs of equipment that may result from a contract change are rental costs, which result from the contractor not having the additional equipment available; additional setup and breakdown costs, which result from the contractor having to bring equipment back to a job-site; and downtime, the time that the equipment is sitting idle because of the change.

The job schedule, plan, and diary are again important to provide the contractor with the necessary documentation, both historical and current, to recover for the additional equipment costs. In addition to these documents, the contractor should also have a detailed equipment log for the job. This log should detail all of the equipment that is dedicated to the job, as well as the time when the equipment is planned to be used on the job. The log should also include every piece of equipment—even the old pickup truck used around the job-site. Many contracts provide for specific dollar amounts per day for equipment. It may be preferable, however, to include every bit of equipment and therefore recover costs directly rather than to recover them through some percentage amount for indirect costs. The equipment log is as important to the equipment cost recovery as the job diary is to the whole job.

Indirect Costs

Costs such as supervision, utilities, supplies, and general security are examples of indirect costs; they are generally provided for in the construction contract as a percentage of the direct costs. The contractor must isolate out of the indirect cost pool as many costs as possible that can be defined as direct costs. Knowing the definitions in the contract as to what are allowable direct costs can increase the contractor's overall recovery, especially where indirect costs are a percentage of direct costs.

Home Office Overhead

Probably the most written-about and litigated phase of costing a change to a construction contract is home office overhead. Home office overhead is the day-to-day general cost of operating the construction entity, covering general and administrative costs. Unlike indirect costs, which are the specific job overhead costs, home office overhead is the general overhead of the construction company. The company has a pool of overhead dollars needed to operate, and this pool of costs must be recovered from the contract work that the company performs in any given period. When there is a change to a contract—whether the change be a change order or a claim—that change should contribute to its fair share of the overhead pool. Disputes may arise, especially in the area of claims, over what is the fair share of the pool to be borne by the change. The general feeling is that the change should cover a share of the costs of the pool as the size of the change relates to the total work performed by the company for a given period. It is not uncommon for some of the approaches to calculating home office overhead to produce, under certain circumstances, a dollar amount that is larger than the remainder of the claim. The most well known of these approaches is the Eichleay Formula, which develops a figure for home office overhead on a per-day basis. From its introduction some years ago, it has gone from acceptance to disrepute and back again. Currently, the Eichleay Formula is not a generally accepted approach. The important factor in successfully recovering a reasonable amount for home office overhead is to use a reasonable approach for the specific circumstances involved.

This discussion has addressed contract changes and the construction project in general as the construction process relates to documentation and cost identification. Some important points made thus far have attempted to communicate that there must be a well-established contract management system in place, manned by personnel who understand the contract and all the various ramifications related to the contract and the project. Following are some of the key ingredients to that system:

- *Established procedures and procedure manual.* There should be a system of established procedures, which should be documented in a project procedure manual. The procedures should outline how the project is to be managed and how various situations are to be handled.

- *Authority and responsibility.* There should be a clear definition of the lines of authority and responsibility. It should be clear who is able to initiate or approve changes, who is responsible for appropriate notifications, and who is responsible for dealings with the various third parties involved with the contract. Clear lines of authority and responsibility are especially important when out-of-the-ordinary situations come up.

- *Documentation and communication.* To the extent possible, there should be standard forms in place to cover the various types of documentation related to the job. The content of the various job files (e.g., correspondence, job diary, notification) should be standardized to the extent possible. There should also be room for expanded discussion where necessary to handle the unexpected. Normal job communications should be standardized as much as possible. Such items as notifications and identification of problems should also be standardized to reduce the chance of error in these critical areas.

- *Problem resolution.* There should be a clear definition of how normal problems are to be resolved when they arise. This would be part of the overall procedure and documentation approach to the management of the contract.

THE CLAIM

Once the contractor is faced with a claim situation, has all the necessary documentation together, and has made a decision to go forward with the claim, there are certain things he should be aware of in relation to this process. A claim of any magnitude will probably require legal assistance and other outside support. This type of support is not normally available within the construction organization. When the contractor has to go outside the organization for specialized help with a construction claim, there are a few things that are important for him to remember. When selecting a law firm to assist in the perfection of the claim, it is vital that the law firm and, more importantly, the specific attorney be experienced in the construction claims area. The company's regular law firm may be excellent, but the construction claims area is a very specialized body of knowledge that, if it is to be handled in the best and least expensive manner, should be approached by people with specific expertise in this area. Other consultants working with the contractor and the law firm should also have experience in the construction claims area. The fact that a particular type of outside firm has experience in the construction industry does not mean that that firm also has the necessary experience in the claims area. The last thing of importance for the contractor to remember is that the better the contractor is prepared before bringing in the outside team, the better the end product and the lower the ultimate cost to the contractor of perfecting the claim. The better prepared the contractor is, the better the whole process is.

Claim Organization

The proper organization of the construction claim is as important as the underlying documents and the content of the claim. The construction claim is a sales document subject to close scrutiny. Therefore, to be most effective, the claim must be a clear, complete, and well-organized document. The claim may contain assumptions and interpretations of facts; however, any false or misleading statements would seriously jeopardize the claim. One false piece of information in the claim, no matter how insignificant, can cause the whole claim to be suspect. Also, before the claim is prepared, the strengths and weaknesses of the contractor's position should be known and understood so that he can help organize the best possible claim under the circumstances. The contract claim should be divided into four parts:

1 *Introduction.* The opening part of the claim gives a general overview of the whole situation. The intent is to review the environment and to set the stage for the factual portions of the claim. If the contractor really feels there has to be some subjective opinion in the claim, this is the place in which to include it. The remainder of the claim is to be kept factual in nature.

2 *Factual information.* This section of the claim contains all of the factual information that relates to the claim. This includes the changes that took place, the conditions surrounding those changes, and any other factual information that will enlighten the reader about the job and the claim.

3 *Problem and liability.* This section of the claim gives a description of the problem(s) that gave rise to the claim and states who was responsible for those problems. This sets the stage for the following section, where the damages are calculated.

4 *Damages.* The last section of the claim sets forth the damages that were sustained by the contractor because of the problems that arose; this section is supported by the factual information set forth in the information section of the claim.

There is a concept in contract claims presentation that is called total cost claims. The total cost claim basically puts forth the theory that if the contractor bid the job at a certain price and the job cost an amount in excess of the bid price, then the contractor is entitled to the difference. The assumption here is that anything that happened that caused the job to run over was not the contractor's fault. Today, the total cost claim is an indication that the contractor does not have the necessary information and documentation to present a detailed and well-supported claim; this type of claim is, therefore, considered a waste of time.

COLLECTING ON THE CLAIM

Once the claim is prepared and properly documented, there are usually three avenues down which the contractor may proceed: (1) negotiation, (2) arbitration, or (3) litigation. Before the contractor gets involved in the actual perfecting of the contract claim, he should understand the ramifications of these various options.

Negotiation

A negotiated settlement of a claim is the simplest of the three approaches. The cost of any recovery is generally small in relation to the claim size. The usual drawback of the negotiated settlement is that the discount from the original amount of the claim may be substantial.

Arbitration

Many of the construction contracts contain clauses that require arbitration of claims. This approach may or may not produce results that are better than those resulting from the negotiation process. Arbitration is not as expensive as litigation, but it is still costly. Depending on the size and complexity of the claim, arbitration could take as long as two or three years to resolve.

Litigation

The last avenue open to the contractor in perfecting a contract claim is litigation. This is by far the most expensive and time-consuming of the claims approaches. The out-of-pocket expenses can at times be matched by the less obvious cost of time. The litigation process can take up to five years to complete and during this time, there may be legal expenses, consultant expenses, and the loss of hundreds of man-hours of time on the part of the contractor and the contractor's personnel.

Analysis

Once the contract claim is complete and all of the facts are known, the contractor should study very carefully the various options available for perfecting the claim. This analysis should include, to the extent possible, all of the costs of each alternative. Only when the contractor has a completed analysis can a rational decision be reached as to which alternative to follow.

CONCLUSION

Today, the area of contract change management in construction is as important to the contractor as any other phase of the construction process. The knowledgeable, well-managed construction organization operates in an environment that makes the management of contract changes an integral part of the overall contract management process. This is not surprising, since it is not uncommon for the contract claim and the recovery from that claim to make the difference between a profit or a substantial loss on a construction project.

CHAPTER 14

Surety Bonds/ Company Agent

Vincent J. Borelli

INTRODUCTION

A surety is a type of guarantor: one who assumes the duties of another in the event of a default. The concept is ancient in its origins. In its early stages an individual acted as surety for another. Corporate sureties ultimately replaced the individual guarantor and have become the primary source of surety guarantees. Most corporate sureties are insurance companies, probably because these companies are large, financial institutions that have the necessary capital to make large commitments.

The author wishes to acknowledge the contributions of Robert Burns to this chapter.

FUNCTIONS AND OBLIGATIONS OF INVOLVED PARTIES

In the construction business, surety is a credit function. The surety bond is an instrument that guarantees the fulfillment of a particular contract. The bond is a three-party instrument. The three parties are the surety (guarantor), the owner (obligee), and the contractor (principal).

The surety guarantees to the owner that the contractor will fulfill the contract. The guarantee is facilitated through the execution and issuance of a surety bond. The owner is the entity or individual to whom the bond is given. The contractor is the person or entity on whose behalf the bond is given.

In addition to guaranteeing the fulfillment of the contract, the surety bond is often an instrument of prequalification. By the issuance of the bond, the surety states that in its opinion the principal is qualified technically and financially to undertake the contract in question.

CONTRACT SURETY

Contract bonds constitute one of a number of classes of surety obligations. The term contract bond ordinarily refers to bonds involving construction-type contracts. Generally, all public contracts require a surety bond. The surety on public work obligations usually issues two types of bonds: the performance bond and the payment bond. Private work, where bonded, is generally done so at the insistence of the lending institution supplying the construction funds. The surety bonds issued in connection with private work could fit any of the following patterns:

- Performance and payment;
- Performance, payment, and lien;
- Lien only; or
- Payment only.

TYPES OF BOND OBLIGATIONS

Letters of Intent

On private work, owners often request a prequalifying letter from the contractor's surety. Generally, the surety will condition the letter with a phrase such as, "it is prepared to consider" or "it is willing to consider" the issuance of a contract bond. This type of wording is utilized to protect both the contractor and the surety from unconditionally supporting a project. The letter of intent should reflect the fact that the surety has a good relationship with the contractor and is willing to underwrite the proposed project.

Bid Bond

Bid bonds are generally required on public work from each contractor who submits a bid for a particular contract. The function of the bid bond is to demonstrate the contractor's good faith and ability to enter into a contract within a certain period of time and to furnish the required performance and payment bonds. These bonds are usually required in order to prequalify the contractor. State and private work generally require a bid security ranging from 5 to 20 percent, while federal work guarantees are typically set at 20 percent of the estimated contract amount.

A bid bond functions in the following manner. A contractor pays a premium, usually a one-time premium, to a surety. The surety then issues a bid bond in the amount required by the owner. If the contractor has the lowest acceptable bid, and thus wins the contract, a performance bond will generally be provided and the protection under the bid bond will lapse. If, however, the contractor with the lowest acceptable bid either fails to sign a contract or fails to provide a performance bond, then the surety is obligated to pay the owner the difference between the contractor's bid and the bid of the next lowest acceptable bidder.

Completion Bond

A completion bond provides assurance to the customer and the lender that a project will be completed by the contractor and that funds will be available for the completion.

Performance Bond

The performance bond states that the contractor will perform a particular contract in accordance with the contract plans and specifications. The bond assures the owner that the contractor will fulfill his contract with respect to both scope of work and price. Should the contractor be unable to complete his obligation, it becomes the responsibility of the surety to take the necessary steps to ensure that the remainder of this obligation is completed satisfactorily.

A performance bond operates as follows:

1 The contractor, subject to dollar and type of contract restrictions imposed by his surety, pays a premium for a bond.

2 This performance bond effectively states that if the contractor should fail to complete a contract in accordance with its terms, the surety will make sure that the job gets done.

This surety relationship is the most important of all the surety relationships in the construction industry and is responsible for establishing the single biggest potential risk that the surety encounters. The cost of the performance bond is, in almost every case, part of the bid cost on a project.

Payment Bond

The payment bond guarantees that those supplying labor and material to the project will be paid. In certain circumstances, this general provision is subject to limitations imposed by state statutes.

In reality, a payment bond protects not only the suppliers of labor and materials, by ensuring that they will be paid; but protects the owner as well. The protection afforded the owner comes about indirectly, since this bond assures him that a job-site will not be shut down by subcontractors who have not been paid. Further, the payment bond probably makes for an earlier "cleaning up" of liens that subcontractors may have placed on title to the property.

Maintenance Bond

Most construction projects provide for a one-year guarantee against any defects in either material or workmanship commencing at the time of substantial completion. This is considered to be a normal condition and is often included as a part of the performance bond guarantee.

Mechanics Lien Bond

A lien law is intended to provide a source of recovery for unpaid laborers and materialmen on private work. The compensation process is initiated by filing a lien against a particular project. The basic justification for the lien law process centers around the notion that a landowner who has had improvements made to his or her property should not have benefit of those improvements without paying for them. A mechanics lien law affords security to the claimants, since the owner must discharge all liens before he or she can obtain a clear title to the property. The only exception to this statutory process is property owned by the federal or state government or its political subdivisions. Many states allow the filing of a lien bond only on private projects. The bond guarantees that the project will be free of liens and essentially affords the same type of protection as is provided by the statutory lien law provision.

Dual Obligee

Since surety bonds may be used in support of private construction projects, it is often necessary to include any interim financer as an additional obligee. There is usually no problem in extending the bond to cover the additional entity as long as the lender is willing to assume the contractual obligations of the owner. The principal obligation involved is the release of funds to the contractor for payment of work performed in the event of a default by the owner.

PREQUALIFICATION PROCESS

There are four major areas that the surety carefully reviews and analyzes prior to extending credit. These are: (1) character, (2) capacity, (3) capital, and (4) considerations and conditions.

Character

Character reflects the qualities of honesty and integrity, and is fundamental to any and all credit relationships. It is developed only over time and by way of an ongoing relationship between the surety and the contractor. Character is clearly the most personal of all surety relationships and is thus a very subjective measure. A contractor with seemingly adequate capacity and capital may be rejected for bonding because of unproven character qualities, while another contractor who appears otherwise inadequate might well achieve bonding because of a long-standing, high-quality reputation.

Capacity

Capacity is analyzed by reviewing several factors, such as how long the company has been in business, who owns the business, and who the company managers are. It is an aid to have resumes on all key people in order to review the business's accomplishments. The underwriter surety attempts to evaluate items such as the firm's prior experience, the organization as a whole, and the organization's strengths and weaknesses.

The firm's prior contracting experience should be reviewed, so it is helpful to have a listing of prior jobs completed. The listing should include the following information for each job:

- Contract price
- Profit earned
- Year completed
- Name and address of the owner
- A basic description of the job itself

References are important. A list of architects, engineers, suppliers, and subcontractors is needed. Continuity of management should also be discussed. The surety will want to have an understanding of the established business arrangement to ensure the completion of existing contracts in the event of the inability of a key management person to perform. The following considerations should be taken into account:

- Buy/sell agreements
- Identity of replacement managers
- The adequacy of life insurance payable to the corporation on the life of the owner and key people

Capital

Financial statements are vital to the prequalification process. The contractor should have financial information that is substantive with respect to both quality and scope. The financial reports should be prepared by a certified public accountant (CPA) in accordance with generally accepted accounting principles (GAAP). At a minimum, the report should include an accountant's report and a complete set of general purpose financial statements. Often included in the financials are work-in-process schedules taken from both completed and in-process contracts. Generally, the surety requests an audited report with an unqualified opinion. Using a CPA firm familiar with and experienced in construction accounting is necessary in order to ensure that the financials are of the best possible quality.

A significant criterion utilized in evaluating the capital aspect of a potential surety account is the strength of a firm's financial condition. The surety, who in this context is the sole measurer of strength, must be able to ensure itself that the contractor's assets are adequate to absorb and indemnify the surety against losses. The financial statements reflect a variety of operating parameters. Over a period of time patterns are established that provide an invaluable indication of a firm's ability to operate successfully in a particular construction market. The financial statements provide induced information such as how well the contractor prices his product and whether or not profitable operations can be maintained.

Work-in-Process Schedules

These schedules are usually reviewed on a quarterly basis. The schedule provides an additional mechanism for the surety to evaluate a firm's overall profitability by making available up-to-date cost and profit data on open and closed projects. The schedules normally provide the following information:

- Billings to date
- Costs to date
- Estimated costs to complete
- Total gross profit
- Profit earned to date
- Profit to be earned
- Overbilling—job borrowing

Considerations and Conditions

Underwriting a contractor's anticipated work program is important to the overall prequalification process. Among other things, specific projects are evaluated to determine whether or not they represent reasonable additions to a firm's ongoing work program. Analysis is focused on such things as geographic spread, which represents an expansion relative to a firm's normal geographic territory; type of work; and size of job. The surety also evaluates the significance of any unusual factors that may result in a change in normal operating procedures.

Specifications are also analyzed. An attempt is made to determine exactly what the nature of the obligation to be guaranteed is, the type of construction required, and whether or not this construction is consistent with the firm's prior experience. Other considerations include whether or not the project involves public or private construction; what is the availability of financing; is there a need for additional equipment; whether or not there are any unusual guarantees present; whether or not the project fits into the firm's present work program; whether or not the project will have an adverse impact on personnel; whether or not there is a reasonable spread of the risk; what other projects are being considered; and whether or not the contractor has made every effort to adequately underwrite the project and minimize his risk. All of these questions should be addressed and answered to the surety's satisfaction.

PROBLEMS, CLAIMS, AND DEFAULTS

The surety's obligations relative to situations involving claims are defined by four factors:

1 Statutes

2 The bond

3 The contract

4 The legal precedent

With respect to public work, the statutes define the circumstances under which a default can be declared, who can be covered by the payment bond, and what has to be done in order to protect one's rights. Claims against the surety that are not properly filled or that in some way do not comply with the statutes will not be satisfied by the surety.

Since the bond guarantees that the contractor will perform his obligations in accordance with the contract, the surety, in effect, stands in the shoes of the contractor. Any claims against a surety, therefore, are only enforceable if they can be enforced against the contractor. If a claim is brought by an ineligible claimant or if it is improperly filed, the bond will not respond.

Contractors may be defaulted for a variety of reasons. The obligee may feel that, among other things, the contractor has failed to perform, is insolvent, or is not competent enough to complete the remaining obligation. Conversely, there have been cases where the contractor himself feels that he cannot continue and calls on his surety to assist him.

Regardless of the surrounding circumstances, claim cases typically involve a lengthy and complicated investigative process. There are three basic options afforded the surety in attempting to resolve contract claim disputes:

1 *Financing the contractor* – this satisfies all of the firm's outstanding obligations.

2 *Reletting the project* — where the surety acts as principal and hires another contractor to complete the remaining work.

3 *Buying back the bond* — the common term for letting the owner/obligee complete the project, then reimbursing the owner.

Payment of Bond Claims

Bond claims may arise even though the project has been completed and the job accepted by the owner. Where such a claim is made, it is necessary for the surety to completely understand why such claim for payment exists. The surety's actions are going to be determined by the attitude of the contractor/principal and the merits of the claim and dispute. From the standpoint of cost effectiveness, every effort should be made to settle all claims as quickly as possible.

Generally, surety bonds are obtained in connection with the law or by regulations of many political bodies and entities in our governmental system. In satisfying the bonding requirements, sureties provide a valuable service in the areas of regulation, indemnification, and selection. These services directly benefit both the political body and the public that is affected by the obligation. Surety bonds not only serve in the public interest as a regulatory function but are also indemnification for default.

Surety bonds serve to prequalify contractors with respect to financial credit and performance. This process of underwriting and prequalifying ensures the obligee that the principal is able to perform the contract.

Overall, however, the most important economic service that the surety performs is that of idemnification. The high degree of corporate security that a surety offers has brought about legislation requiring bonds and provides an incentive for bonding private business ventures.

CHAPTER 15

Risk Management

William S. McIntyre
Jack P. Gibson

INTRODUCTION

During the 1970s, changes took place in the economy that exacerbated the usual insurance industry market cycles. After the recession in the mid-1970s, the insurance industry was hit heavily by falling premiums and increasing inflation, which caused carriers to withdraw sharply from the market. Then, in 1975 and 1976, insurance companies had two of their worst years, when an excruciatingly tight marketplace caused premiums for contractors to skyrocket.

Due to more conservative underwriting in 1977 and 1978, the insurance industry had two of its best years. Beginning in 1979, a traditional buyers' market began to develop and has continued into the early 1980s. This buyers' market continued through 1984, when the insurance industry experienced its worst years in history with an estimated pretax operating loss of $3.55 billion. The poor financial results of 1984 resulted in an almost immediate restriction in insurance availability and drastic price increases in 1985. The sellers' market of 1985 is expected to last at least three years.

Because of these dramatic changes, innovations have taken place in the areas of coverage, cost, and funding arrangements; significant changes have also begun to take place in the basic distribution of insurance and the ways in which services are delivered.

This chapter provides the construction organization with the knowledge to deal with the current and upcoming changes in the insurance marketplace and the ability to take advantage of the changes wherever possible.

This chapter also shows how construction contractors should select and administer a risk management program. There is a great deal of difference between risk management and insurance management. While risk management is dynamic, risk oriented, and creative, insurance management is passive, administratively oriented, and reactive to outside forces.

The construction organization that successfully manages its insurance program is the one that takes an active role in making things happen, not the one that watches things happen or that finally wakes up one day and asks, What happened?

FIG. 15-1   Risk Management Process

---

1. **Identify pure risks.**
2. **Evaluate and analyze pure risks.**
3. **Choose the best risk management technique(s):**
   - ■ Loss control   ■ Noninsurance transfer
   - ■ Risk retention   ■ Insurance
4. **Monitor the results and adjust where necessary.**

RISK MANAGEMENT

Risk management is a relatively new management approach to controlling the costs associated with the possibility of fortuitous loss. It evolved from insurance management, which incorporates the purchase of insurance to protect against every possible loss contingency. A key to understanding the logic of this process is the concept of cost of risk, which is: Insurance Premiums + Cost of Loss Control Efforts + Uninsured Losses + Insurance and Risk Management = Administrative Costs. The risk management approach recognizes that insurance is the most expensive technique to employ and seeks to use the insurance mechanism only as a last resort. The risk management process is summarized in Figure 15-1.

The systematic implementation of a risk management program requires the contractor to take the following steps:

1   *Identify pure risk exposures.* This is the most important function in the risk management process. If a risk is not identified, the process of control and/or finance is eliminated by default. "Pure risk" may be defined as that risk to which a contractor is exposed that presents a possibility for loss but not for gain. In some cases, pure risk is readily apparent—such as the potential loss of buildings and their contents. Other areas of pure risk—such as design/build professional liability, contingent third-party liability resulting from automobiles and equipment leased to others, or liability of others assumed in contracts—are much less obvious.

2   *Evaluate and analyze pure risk.* After an inventory of pure risks is prepared, sound risk management dictates that both

maximum possible loss and probable maximum loss be considered to evaluate and analyze the pure risk. The maximum possible loss can be defined as the worst loss that could possibly occur. The probable maximum loss is the actuarial best estimate of maximum loss should the event take place. These values are used in setting insurance protection limits of liability. In addition, the past loss history of the contractor is analyzed to estimate future probable loss levels and to assist in the implementation of cost-effective loss control programs.

3 *Choose the best risk management techniques.* This handles pure risk, and implements and monitors the chosen techniques. As herein discussed, there are numerous methods for controlling or financing pure risk that can be used either singly or in combination.

The use of these methods clearly identifies for the contractor the position that he is willing to accept regarding risk management. If the contractor decides that loss control is the only viable method to adopt, the impact would be to reduce some costs (e.g., insurance), while increasing others (e.g., training programs). Other contractor decisions would result in other patterns of risk-reward matrices.

Risk Management Techniques

The various techniques for handling risk exposures are generally categorized as loss control, risk retention, noninsurance transfer of risk, and risk transfer through insurance.

Loss Control. Loss control is a technique of risk management employed to lower the probability that a loss will occur, or if it does occur, to reduce the severity (amount) of the loss. This technique is concerned with the elimination of the mechanical or physical causes of loss, the training of personnel to perform their jobs safely, the necessary planning in order to cope with losses that do occur, and the use of security to control crime losses.

Human failure is an integral factor in the implementation of a loss control program. Workers' compensation losses, for example, have traditionally been attributed to mechanical or physical hazards, but all contractors have seen how insignificant the affects of the engineering safeguards prescribed by the Occupational Safety and Health Administration (OSHA) have been on employee injuries. Therefore, steps that the contractor takes to reduce potential loss or injury are beneficial and have a great impact on overall loss control. Obviously, too, there is a threat of loss resulting from human failure in the area of employee infidelity (employee's willful damage or wrongdoing).

Loss avoidance, another control device, can be used to eliminate a risk totally by not undertaking the activity evaluated as risky. For example, the exposures of designer or builder professional liability and contingent liability from leasing automobiles or equipment to others can be totally eliminated by not engaging in these activities. Although often difficult to implement, this technique can be useful in justifying the elimination of marginally profitable operations. Further, the technique emphasizes the need to consider cost of risk in management decisions regarding the expansion of services.

Risk Retention. Risk retention, or the planned acceptance of losses through the use of insurance policy deductibles, the deliberate use of noninsurance, or the implementation of a formal self-insurance program represent alternative methods for financing losses. In many cases, the risk retention technique is the most cost effective, as no agent's commission, insurance company profit, premium taxes, and so on, are incurred. However, careful analysis of the potential severity of retained losses and the contractor's ability to self-fund those losses must be undertaken prior to implementation.

Noninsurance Transfer of Risk. Although risk is often transferred to professional risk bearers (insurance companies), other types of risk transfer are also available. Consideration should be given to transfer to others who are willing or able to assume a portion of the pure risk that may exist through business relationships created by contract.

Examples of noninsurance risk transfers are hold-harmless agreements and/or waivers of subrogation. A hold-harmless agreement is a clause within a contract that transfers liability from one entity to another. A waiver of subrogation simply prevents one entity's insurance company from subrogating against the other party to a contract because it negligently caused a loss covered by the first party's insurance. For example, a waiver of subrogation in the contractor's subcontract agreement can prevent the subcontractor's insurer from pursuing a recovery for damage to its insured's equipment, even if the contractor's employee caused the damage.

Noninsurance risk transfer is one of the most important facets of a contractor's risk management program. Contractors must be cognizant of the attempts by owners to pass disproportionate and frequently uninsurable risks to the contractor in contract clauses. A careful review of all contracts in terms of risk management is a necessity, and unequitable transfers should be avoided wherever possible.

Risk Transfer Through Insurance. After implementation of all of the above techniques of loss control or avoidance, risk retention, and noninsurance transfer of risk, a certain level of residual risk remains. This residual risk, if material in amount, must be transferred to an insurance company—provided the premiums are reasonable where related to the risk being transferred. The reason for buying insurance to treat pure risk is to protect assets from losses which, if they occurred, could have a material effect on earnings or seriously impair the financial stability of the contractor. Additionally, insurance may be required by law or within contractual agreements.

To protect the contractor's assets prudently, the insurance program must be tailored to its loss exposures. Although insurance is heavily regulated, there are many ways coverage can be tailored, as the sections of this chapter dealing with insurance coverage, types of insurance, and market outlets for insurance demonstrate. Only with careful attention to coverage details and knowledgeable negotiation can the proper program be assured.

Risk Management Administration

The risk management process must be incorporated into the mainstream of the contractor's operating environment. Care should be taken to dovetail risk management into overall operations in an orderly manner.

Small Firms. Within small firms, the risk management function can normally be administered on a part-time basis by an officer. Frequently, the function can be administered most smoothly under the auspices of the chief financial officer, treasurer, or sometimes, the corporate counsel. The primary problem with the delegation of risk management responsibilities on a part-time basis is the lack of expertise in the complex field of risk management of the individual assigned these responsibilities. Quite often, the risk management function is intimidating to the individual and other job responsibilities take priority. This leaves the risk management function unattended and results in a higher total cost of risk with less than adequate insurance protection. Nevertheless, with the proper use of insurance agents and consultants, as well as proper emphasis from upper management, the function can be administered effectively on a part-time basis.

Medium-Sized or Larger Contractors. The risk management function for every medium-sized or larger contractor is a major risk. The need for a full-time internal risk manager is becoming increasingly accepted. The staffing of this department depends on the level of authority vested in the risk manager and the contractor's scope of operations. Empirical evidence shows that the risk manager normally needs to be supported by a clerical person and, at times, a claims coordinator. Some very large contractors have even larger risk management staffs to administer loss control programs, adjust claims, and review contracts.

Risk management is not a static process. Risk management techniques that worked for a particular company in the past may not be effective in the future. As firms grow, operational emphasis and construction management techniques change, as do legal environments due to new laws and court decisions.

INSURANCE COVERAGE

Ways to Lower Premium Costs

There are two basic approaches to lowering insurance premium cost or the price paid for insurance: (1) negotiated insurance placements and (2) competitive bidding. The negotiated placement process is highly desirable, especially in instances where long-term relationships have been developed and the contractor is knowledgeable enough to establish some appropriate premium levels to be used as negotiating objectives.

The bidding process, in which various agents or insurance companies bid on providing the insurance package, is appropriate in cases where the contractor wishes to start fresh with a new agent and insurance company or wishes to test his current insurance arrangement in the open market.

If the competitive bidding process is to be utilized, the contractor must take great care in having comprehensive insurance specifications developed and properly presented to the various agents and insurance companies. It is strongly recommended that this approach not be utilized more than once every three to five years in order to ensure that the price paid remains competitive while at the same time allowing both the agent and insurance company the opportunity to serve the contractor in an environment where some loyalty can be developed.

While there are many areas where cost reductions can be made, there are some general points, summarized in Figure 15-2, that should be considered by virtually all construction organizations.

FIG. 15-2   Checklist on How to Reduce Premium Costs

Automobile:
- Use property damage liability deductibles.
- Verify experience credits (40 percent or more).
- Be sure that correct territories are used.
- Verify classification of mobile equipment.
- Question higher collision deductibles or self-insurance.
- Inspect comprehensive deductibles.

General Liability:
- Verify experience credits (60 percent or more).
- Examine the deletion of truck driver payrolls.
- Remove intercompany receipts.
- Test payroll limitations.
- Use property damage liability deductibles.

Workers' Compensation:
- Value dividends.
- Delete overtime surcharges.
- Review computation of experience modifier.

General:
- Review proper classifications on a regular basis.
- Decide on use of higher deductibles.
- Implement competitive quotes.
- Use cash flow programs.
- Use loss control techniques.

Experience Credits. All insurance companies have manual, or book, rates. These rates are similar to sticker prices for automobiles. Virtually all companies apply credits to or discounts on these manual rates.

Using a broad generalized industry average, if the contractor is not receiving a 40 percent credit on automobile liability and physical damage premiums and a 60 percent credit on general liability premiums in the current marketplace, additional negotiations should be held. However, comparisons of such credits are complicated by the fact that different insurance companies use different rate schedules. It might be best to compare the net proposed rates against the manual rates published by the Insurance Services Office (ISO), the rating organization of the insurance industry.

Dividends. Virtually all insurance companies now pay dividends on workers' compensation premiums subject to negotiation between the parties. These dividends, usually payable at some point after the expiration of the policy, reflect reduced losses, lower expense factors, and profits. These dividends can be based on flat percentages or tied to loss ratios.

Another workers' compensation plan that allows premium returns based on good losses is the retrospective rating plan. This plan contracts for a specified premium at the beginning of the term with the understanding between the parties that the rating used in computing the original premium will be retroactively adjusted for the actual loss experience of the contractor. There can be additional premiums where losses are bad and lower premiums where claims are less than expected. Many dividend plans appear to be much more attractive than retrospective rating plans. The insured must remember, however, that dividends are not guaranteed and must be declared by the insurance company's board of directors, whereas retrospective rating adjustments must be made unless the insurance company is insolvent. During the last tight market in the 1970s, many companies elected not to pay dividends as originally proposed.

Since many insurance companies have drastically cut their premium rates in negotiable areas and only pay dividends if there are enough profits, it is highly likely that many insurance companies would elect not to pay dividends or would substantially reduce them. Therefore, in comparing the two different plans, one must consider that dividend programs in general are risky.

One suggested approach is to compare a dividend plan with a retrospective plan at both gross and net amounts, realizing that in the event dividends are not declared, at least the retrospective return may be due.

Proper Classifications. Another area of substantial errors and higher costs is the classification of vehicles, payrolls, and receipts under the various policies. If proper communications are not maintained with the agent and the insurance company, substantial errors, usually unintentional, may occur. A review should be made of all classifications that might possibly apply to the contractor's operations to see whether some operations can be reclassified to lower rate levels.

Exposure units reported for year-end audits need to be carefully checked. For instance, truck driver payrolls should be eliminated with respect to general liability exposures and intercompany sales should not be included in the computation of completed operations premiums. Failure to make these eliminations may result in extra premiums paid with no additional coverage.

Cost Benchmarks

Contractors often ask the question, What percentage of payroll or sales should my insurance costs be? Due to various state benefits for workers' compensation and the great disparity in exposures emanating from construction operations, it is impossible to develop cost guidelines that are stated as a percentage of payroll or receipts.

In evaluating a contractor's account, insurance companies develop loss ratios (losses divided by premiums). On the other hand, managers of construction operations often take a different approach by defining "cost-loss ratios" as insurance cost divided by losses. For instance, if a contractor's premium, or cost, for workers' compensation, general liability, and auto exposures are $500,000 and his losses for the policy period are $250,000, a cost ratio of 2.00 is developed ($500,000 − $250,000). Stated another way, from the contractor's point of view, it is costing $2 to insure each $1 of claims.

After taking into consideration discounts, retrospective returns, and investment income developed from cash flow programs, construction management should follow these guidelines:

Losses	Cost Ratio
$ 0 – 50,000	2.00
50,000 – 100,000	1.75
100,000 – 250,000	1.50
250,000 – 500,000	1.40
500,000 – 750,000	1.30
750,000 – 1,000,000	1.25
1,000,000 and over	1.20

At least three years of premium and losses should be used in developing these ratios. A lesser period would provide an inadequate base for the calculation of the ratios. Also, if abnormally low losses are being experienced, the ratios may be skewed.

According to the guidelines listed above, the ratio in the previous example should be 1.50 rather than 2.00. Therefore, as a negotiating objective, the contractor should work toward effecting a savings of $125,000 ($.50 × $250,000) in the cost of insurance.

FIG. 15-3   Insurance Model—Owner, Contractor, and Subcontractor

Owner's property coverage			
Resulting damage to the work*			
Professional liability	Umbrella liability	General/ automobile liability and worker compensation	Builders' risk and equipment floater
Subcontractor's coverages			

*Due to errors or omissions in design and faulty work or materials.

TYPES OF INSURANCE COVERAGE

Modularized Coverage Approach

The first step in managing the coverage portion of an insurance program is to think of insurance coverage as modularized. Insurance is more readily understandable when visualized as consisting of standard units that are used to build total insurance programs, just as bricks are used to construct buildings.

Some of the more important coverage areas are herein addressed. The illustration model in Figure 15-3 shows how the overall insurance model can be broken down into large modules, or areas of coverage such as professional liability and builders' risk. For a complete listing of items that may be covered in specific areas, see Figures 15-4 through 15-10. Each of these components can be broken down into further modules.

The contractor should understand that the insurance industry has basic policy forms in most coverage areas. Therefore, all the contractor needs to do is make additions to or modifications of these standard documents for specific circumstances. Generally speaking, the average construction organization should make a number of changes to the standard forms as discussed later in the chapter and outlined in the coverage checklists in Figures 15-4 through 15-10.

(text continues on page 15-16)

FIG. 15-4   Automobile Coverage/Cost Checklist

Item	Current Program	Recommendation
Comprehensive deductible	. . . .	. . . .
Collision deductible	. . . .	. . . .
Fire, theft, CAC (combined additional coverage)	. . . .	. . . .
Fleet automatic	. . . .	. . . .
Property damage liability deductible	. . . .	. . . .
Punitive damages covered	. . . .	. . . .
Physical damage to mobile radios	. . . .	. . . .
Blanket waiver of subrogation	. . . .	. . . .
Uninsured motorist	. . . .	. . . .
Medical payments	. . . .	. . . .

FIG. 15-5   Builders' Risk Coverage/Cost Checklist

Item	Current Program	Recommendation
Suspension of coinsurance clause	. . . .	. . . .
Replacement cost	. . . .	. . . .
Collapse coverage	. . . .	. . . .
Resulting damage from faulty work	. . . .	. . . .
Resulting damage from design error	. . . .	. . . .
Miscellaneous unnamed locations	. . . .	. . . .
Earthquake and flood	. . . .	. . . .
Debris removal	. . . .	. . . .
Testing coverage	. . . .	. . . .
Early occupancy provision	. . . .	. . . .
Boiler and machinery	. . . .	. . . .
Damage to electrical panels	. . . .	. . . .
Transit	. . . .	. . . .
Debt service cost	. . . .	. . . .
Interest	. . . .	. . . .
Rents	. . . .	. . . .
Business interruption	. . . .	. . . .
Penalties	. . . .	. . . .
Profit	. . . .	. . . .
Force majeure	. . . .	. . . .

FIG. 15-6   Equipment Floater Coverage/Cost Checklist

Item	Current Program	Recommendation
All risks	. . . .	. . . .
Waterborne exclusion deleted	. . . .	. . . .
Repair/maintenance exclusion deleted	. . . .	. . . .
Automobile exclusion deleted	. . . .	. . . .
Loss occasioned by excessive weight	. . . .	. . . .
Damage to booms	. . . .	. . . .
Newly acquired equipment	. . . .	. . . .
Replacement cost	. . . .	. . . .
Coinsurance requirements deleted	. . . .	. . . .
Deductible	. . . .	. . . .

FIG. 15-7   General Liability Coverage/Cost Checklist

Item	Current Program	Recommendation
Contractual liability:		
■ All written contracts	. . . .	. . . .
■ Oral contracts	. . . .	. . . .
■ Third-party beneficiary exclusion deleted	. . . .	. . . .
■ Railroad exclusion deleted	. . . .	. . . .
■ Architect/Engineer (A/E) exclusion removed	. . . .	. . . .
Personal injury:		
■ Contractual exclusion removed	. . . .	. . . .
■ Employee exclusion removed	. . . .	. . . .
Host liquor liability	. . . .	. . . .
Broad form property damage:	. . . .	. . . .
■ Including completed operations	. . . .	. . . .
■ Modified to cover resulting damage (completed operations)	. . . .	. . . .
Incidental medical marketplace	. . . .	. . . .
Employees as insureds	. . . .	. . . .
Explosion, collapse, and underground property damage coverage	. . . .	. . . .
Employee benefit liability	. . . .	. . . .
Automatic additional insured where required by contract	. . . .	. . . .
Professional services exclusionary endorsement removed	. . . .	. . . .
Foreign operations	. . . .	. . . .

FIG. 15-8 Umbrella Coverage/Cost Checklist

Item	Current Program	Recommendation
Aircraft owned	. . . .	. . . .
Aircraft nonowned	. . . .	. . . .
Watercraft owned	. . . .	. . . .
Watercraft nonowned	. . . .	. . . .
First dollar defense	. . . .	. . . .
Defense in addition to policy limits	. . . .	. . . .
Concurrent with primary policies	. . . .	. . . .
Primary policies properly scheduled	. . . .	. . . .
Discrimination covered	. . . .	. . . .
Punitive damages covered	. . . .	. . . .
Care, custody, or control	. . . .	. . . .
Damage to work	. . . .	. . . .
Explosion, collapse, and underground property damage	. . . .	. . . .
Liberalization endorsement	. . . .	. . . .

FIG. 15-9 Professional Liability Coverage/Cost Checklist

Item	Current Program	Recommendation
Proper definition of services covered	. . . .	. . . .
Construction management coverage	. . . .	. . . .
Design/Build coverage	. . . .	. . . .
Contractual liability for joint negligence	. . . .	. . . .
Prior acts coverage	. . . .	. . . .
Punitive damages	. . . .	. . . .
Liberal discovery clause	. . . .	. . . .

FIG. 15-10   Workers' Compensation Coverage/Cost Checklist

Item	Current Program	Recommendation
Other states' endorsement	. . . .	. . . .
United States longshoremen and harborworkers (USL&H) endorsement	. . . .	. . . .
Jones Act (1920) (maritime)	. . . .	. . . .
Voluntary coverage for employees excluded by some state statutes	. . . .	. . . .
Stop-gap employers' liability	. . . .	. . . .
In rem endorsement (maritime)	. . . .	. . . .
Maintenance, cure, transportation, and wages (maritime)	. . . .	. . . .
Foreign operations	. . . .	. . . .

The assumptions used in preparing the pure risk analysis in Figures 15-11 to 15-13 are:

- The contractor has accepted a job with a total contract price of $3,000,000.

- The contractor will acquire $500,000 of materials (principally sheet metal) for installation on-site.

- No more than 25 percent of the material will be uninstalled and on-site at any time.

- One major piece of equipment will be required on-site at any time, the replacement cost of which is $150,000.

- One vehicle with a replacement value of $30,000 will be on-site throughout the job.

- The contractor will have several employees on-site throughout the job.

- The contractor has assumed by contract all the potential liabilities of the owner/supplier when that owner/supplier is on-site.

- The job-site is relatively secure.

In Figure 15-11 the contractor has attempted to identify prospective pure risk situations based on both industry experience and the nature of this contract.

FIG. 15-11   Pure Risk Analysis (Inventory)

JOB NAME: <u>Global American</u>

AREAS OF RISK

1. On-site materials:
  a. Weather damage _____
  b. Theft _____
  c. Spoilage _____

2. On-site equipment:
  a. Weather damage _____
  b. Theft _____
  c. Breakdown _____
  d. Vandalism _____

3. Vehicles:
  a. Collision _____
  b. Theft _____
  c. Vandalism/breakage _____
  d. Third-party liability _____

4. Personnel:
  a. Injury on job _____
  b. Theft by _____

5. Professional liability _____

6. Losses through owner:
  a. Liabilities assumed _____

7. Liability to the public for bodily injury:
  a. Course of construction _____
  b. Completed operation _____

8. Liability for property damage:
  a. To the completed project _____
  b. To other property _____

Figures 15-11 to 15-13 are intended to be illustrative only and therefore lack the detail and specificity that should be present in an analysis prepared by the contractor or the risk management advisor.

In Figure 15-12, the contractor has attempted to measure both the maximum potential loss and the probable maximum loss in each category identified in the original inventory.

The maximum potential losses include:

- Entire value of materials, equipment, and vehicles

- Total value of the job for professional liability

- Total value of the subcontract risk

- Expected maximum losses in other categories

The probable maximum loss is the contractor evaluation of each of the categories and includes these assumptions:

- No more than 10 percent of the materials will be susceptible to weather damage at any time.

- Theft risk of materials is limited by their lack of portability, theft of equipment is virtually impossible, and theft of vehicles is either total or nonexistent.

- Vandalism is not likely because of active security measures.

- Third-party automobile liability estimate is based on an analysis of past jury awards in the state.

- Employee theft is limited by the nature of the product.

- During the course of construction, bodily injury to the public is a limited risk because of active security measures.

- Completed operations exposure for bodily injury to the public is based upon an analysis of the use of the completed project, past jury awards in the state, and expected "social inflation" during the next 10 years.

- Completed operations exposure for property damage to the work is estimated at 10 percent of the total value.

- Completed operations property damage exposure for other property is estimated to be 10 percent of the value of all property.

FIG. 15-12   Pure Risk Analysis (Measurement)

JOB NAME: Global American

Areas of Risk	Maximum Potential Loss	Probable Maximum Loss
1. On-site materials:		
a. Weather damage	$ 125,000	$ 50,000
b. Theft	125,000	10,000
c. Spoilage	10,000	10,000
2. On-site equipment:		
a. Weather damage	150,000	150,000
b. Theft	150,000	150,000
c. Breakdown	10,000	10,000
d. Vandalism	150,000	20,000
3. Vehicles:		
a. Collision	30,000	30,000
b. Theft	30,000	30,000
c. Vandalism/breakage	5,000	5,000
d. Third-party liability	Unlimited	10,000,000
4. Personnel:		
a. Injury on job	Unknown	Unknown
b. Theft by	5,000	5,000
5. Professional liability	3,000,000	Unknown
6. Losses through owner:		
a. Liabilities assumed	Unlimited	25,000,000
7. Bodily injury to the public:		
a. Course of construction	Unlimited	1,000,000
b. Completed operation	Unlimited	50,000,000
8. Completed operations property damage:		
a. To the completed project	3,000,000	300,000
b. To other property	10,000,000	1,000,000

FIG. 15-13 Pure Risk Analysis (Action Sheet)

JOB NAME: Global American

Areas of Risk	Maximum Potential Loss	Probable Maximum Loss	Loss Analysis and Action to Be Taken	Total Risk Retained
1. On-site materials:				
a. Weather damage	$ 125,000	$ 50,000	Nature of product makes likelihood of loss remote.	$10,000
b. Theft	125,000	10,000	Chance of major loss should be protected. Deductible.	2,000
c. Spoilage	10,000	10,000	Materials are kept in enclosure. Some loss is expected. No insurance.	
2. On-site equipment:				
a. Weather damage	150,000	150,000	Standard equipment protection with large deductible	5,000
b. Theft	150,000	150,000	should protect against all but breakdown. Preventive	–
c. Breakdown	10,000	10,000	maintenance is in force against possible breakdown.	5,000
d. Vandalism	150,000	20,000	Percentage of risk small.	5,000
3. Vehicles:				
a. Collision	30,000	30,000	Only authorized drivers may use. Insure for full value less deductible.	1,000
b. Theft	30,000	30,000	Attractiveness of vehicle partly offset by job-site security. Insure fully.	–
c. Vandalism/ breakage	5,000	5,000	On-site breakage highly probable. Insure with small deductible.	200
d. Third-party liability	Unlimited	Unknown	Proper use of vehicles should reduce some risks. Must cover major catastrophes because of nature of industry. Large deductible.	5,000

4. Personnel:				
a. Injury on job	Unknown	Unknown	Job training before employee is allowed on-site. State-mandated Workers' Compensation.	—
b. Theft by	5,000	5,000	Bonding procedures and personnel forms are used as needed. Contractual.	—
5. Professional liability	3,000,000	Unknown	Unidentifiable risk. Company has standard professional liability policy that provides blanket coverage for all contracts.	—
6. Losses through owner:				
a. Liabilities assumed	Unlimited	25,000,000	Coverage of risk is included in company's other risk portfolio items. This represents no additional risk. No action.	Unknown*
7. Bodily injury to the public:				
a. Course of construction	Unlimited	1,000,000	Coverage for both of these risks is provided by general and umbrella liability insurance with a high limit of liability.	Unknown*
b. Completed operation	Unlimited	50,000,000		
8. Completed operations property damage:				
a. To the completed project	3,000,000	300,000	Limited coverage provided by general and umbrella liability insurance.	
b. To other property	10,000,000	1,000,000	Covered by general and umbrella liability insurance.	Unknown*

*Retained risk only to the extent that policy inclusions preclude coverage or liability exceeds policy limits.

Automobile Insurance

The automobile policy can be used to insure both (1) liability to others arising from the use of an automobile; and (2) physical damage to owned automobiles. The liability insurance portion of the policy protects the contractor from liability arising from automobiles owned, hired, or used by the construction firm, its agents, or employees.

Contractors should be cognizant of the difference between automobiles and mobile equipment. For insurance purposes, an item of mobile equipment is defined as a land vehicle including any machinery or apparatus attached thereto, whether or not self-propelled, that is:

- Not subject to motor vehicle registration;
- Maintained for use exclusively on premises owned by or rented to the named insured, including the ways immediately adjoining;
- Designed for use principally off public roads; *or*
- Designed or maintained for the sole purpose of affording mobility, forming an integral part of, or being permanently attached to equipment of the following types:

  a Power cranes, shovels, diggers, and drills

  b Concrete mixers (other than the mix-in-transit type)

  c Graders, scrapers, rollers, and other road construction or repair equipment

  d Air compressors, pumps, and generators, including spraying, welding, and building-cleaning equipment

  e Geophysical exploration and well-servicing equipment

Liability arising from mobile equipment is automatically covered by the general liability policy with no additional premium necessary. Contractors should avoid the costly and common error of scheduling mobile equipment in the automobile policy with an additional premium charge. Physical damage coverage can and should be provided in an equipment floater.

There are several alternatives for the provision of automobile physical damage coverage. Collision and comprehensive coverages are the types most commonly purchased. Collision insurance indemnifies the contractor for damage to an insured vehicle resulting from a collision with another object or overturn of the vehicle. Comprehensive insurance is an all risk coverage that pays to repair damages resulting from perils other than collision or overturn. Physical damage coverage should usually be written on a fleet automatic basis, which alleviates the need to report newly acquired vehicles during the policy year.

In order to control costs, deductibles should be used with each of the physical damage coverages. An analysis of past loss experience versus the premium involved with various deductible levels usually reveals the optimal deductible amount. Where relatively large fleets are involved (e.g., 35 or more vehicles), it is often logical to retain the collision exposure completely rather than to insure. Frequently retained collision losses are significantly lower than premium costs over an extended period of time. It is also usually prudent to self-insure collision on older (e.g., five or more years) private passenger vehicles because the amount of premium added to the deductible begins to approach the value of the automobile. Retention of the comprehensive exposure, however, is usually not advisable because many vehicles are exposed to common disasters such as windstorms, floods, and other unexpected natural events.

Fire, theft, and combined additional coverage (CAC) is an alternative to comprehensive coverage that can lower insurance costs. This is a "named perils" approach, which insures most of the perils covered by comprehensive, with some exceptions such as windshield breakage. The named perils approach is a listing of each specific type of coverage that an insured requests, as opposed to obtaining an all risk type coverage, which tends to cover both the risks the insured has identified and other risks that probably do not apply to the insured. Since the named perils coverage is more specific, it is frequently less costly than comprehensive coverage. Many contractors opt to purchase comprehensive coverage on personal autos while purchasing fire, theft, and CAC on trucks.

Medical payments coverage and uninsured motorists coverage are two forms of automobile coverage that are optional in many states but mandatory in the others. Medical payments coverage provide reimbursement, without regard to fault, for medical expenses of persons injured while occupying an automobile. Uninsured motorists coverage provides insurance to pay for bodily injury to the occupants of the insured vehicle following an injury sustained due to the fault of another party who has no insurance. Although opinions vary, many risk management professionals believe these two coverages are unnecessary for businesses because workers' compensation insurance and group medical insurance covers injuries to the contractor's employees. For this reason, many contractors forgo these coverages to save the related premium.

Builders' Risk

Builders' risk insurance is one of the most important insurance coverages for contractors, protecting against damage of any kind to the construction project and, depending on the form, damage to materials to be installed in the project. Coverage can be purchased on either a named perils or an all risk basis. Named perils insurance covers only losses that come within the purview of the specific perils named in the policy (e.g., fire, lightning, explosion, and vandalism). All risk coverage, on the other hand, insures against damage caused by any perils not specifically excluded.

Common exclusions in all risk policies are:

- Infidelity of contractor's employees
- Loss or damage caused by vermin, inherent vice, latent defect, or wear and tear
- Breakdown of machinery
- Loss or damage resulting from flood
- Hostile or warlike action
- Loss resulting from fault, defect, error, or omissions in design, plan, or specification
- Cost of correcting faulty or defective workmanship or material

Builders generally prefer all risk coverage for at least two reasons. The first and most obvious reason is that all risk coverage is frequently broader than that provided by named perils insurance. The second reason is that in the burden of proof, a loss is covered by or excluded from the policy. With named perils insurance, the burden is on the insured to prove that a loss was caused by one or more of the perils specified in the policy. With all risk coverage, however, the burden is on the insurance company to prove that the loss is excluded by the policy in order to deny coverage.

As with other forms of property insurance, builders' risk loss recoveries can be valued on either an actual cash value (ACV) or replacement cost basis. The definition of ACV is replacement cost less depreciation.

In conjunction with values insured, there are two approaches typically used for rating purposes—completed value forms and reporting forms. With a completed value form, the insurance policy limit is set to equal the projected completed value of the structure, and the rate on which the premium is based is reduced to reflect the fact that actual exposed values will be less during most of the construction. A reporting form requires the insured to make monthly reports showing updated values as the construction progresses. The reported values must be kept up to date or the insured could sustain a financially crippling loss.

Because of competition, the builders' risk market has been relatively soft for the last 10 years and can be expected to remain so, even if the casualty side of the insurance market tightens considerably. However, this is still an excellent time to negotiate for very broad coverages, including some of those that were not readily available four or five years ago.

Particular attention should be given to resulting damage caused by faulty work or bad materials. In addition, resulting damage from design error and omission can sometimes be covered as well. This is particularly important for contractors involved in design/build construction or construction management operations. In some instances, architects or engineers can be covered under the builders risk policy, including the design error or omission exposure. However, most professional liability markets do not allow a credit to the architect or engineer under their professional liability policies that can, in turn, be passed on to the owner or contractor for providing the broader coverage.

Some builders' risk policies contain provisions that void the builders risk coverage in the event that tenants take early occupancy. This limitation can be and should be removed. There is usually no additional cost for its removal.

Some projects involve circumstances where the builders' risk coverage terminates and the owner provides coverage under the permanent property policy. In these instances, the same coverage that the general contractor and subcontractors had under the builders' risk policy should be continued under the permanent policy, preferably through the warranty period.

With the exception of California and certain areas where flooding is common, the cost for earthquake and flood coverage is reasonable and these coverages should be purchased.

In recent years, a number of new types of coverages have been made available by some builders' risk markets. For example, insurance is now available to guarantee that a particular work product will perform as intended, and coverage can be arranged to ensure debt service cost, interest, rents, penalties, profit, efficacy, and force majeure.

Another developing area is consequential loss. The standard approach to builders' risk insurance is to exclude consequential loss, or, as it is usually referred to in the insurance industry, business interruption. In many instances, the contractor is not held responsible for such loss under the contract documents. While the owners' master property insurance program may extend consequential loss coverage to the new project, often this exposure is self-insured. While this is a coverage area that is very difficult to understand and even more difficult to properly adjust in the event of a loss, markets are developing for this important exposure.

Equipment Floater

Insurers provide coverage for physical damage for contractor's equipment through the use of inland marine policies known as equipment floaters. Equipment floaters may be on either an all risk or named perils basis, although all risk coverage is usually preferred for the same reasons discussed in conjunction with builders' risk: All risk coverage is broader and places on the insurer the burden to prove that the loss is excluded from the policy.

Typical exclusions in all risk policies include:

- Loss arising from employee dishonesty

- Loss occasioned by the weight of a load exceeding the lifting capacity of a machine

- Loss or damage to automobiles

- Loss or damage to insured property while waterborne

- Loss arising from maintenance or repair work

Several of these exclusions, including loss while waterborne, loss arising from repairs, and loss occasioned by weight exceeding lifting capacity can usually be removed for an additional premium. In addition, the automobile exclusion should be reviewed and clarified if any of the insured equipment might come under its purview.

A potentially dangerous clause in many of these policies involves coinsurance. Coinsurance requires the contractor to purchase insurance to a specified percentage of value and invokes a penalty at the time of loss if not complied with. This coverage should be purchased on a non-coinsurance or agreed amount basis. An illustration of the possible impact of a failure to meet coinsurance requirements is shown in Figure 15-14.

Some equipment floaters provide automatic coverage on newly acquired equipment for a certain amount for up to 30 days from the date of acquisition before it must be reported. The amount of automatic coverage is generally specified as a percentage of the policy limit. This is a negotiable area, and the amount of coverage provided should be reviewed for propriety.

Finally, deductibles should be used to control the cost of this insurance coverage. In choosing an appropriate deductible, the premiums associated with several deductible levels should be compared, taking into account losses that would have been assumed in past years if each deductible had been in place.

FIG. 15-14   Illustration of Coinsurance Provisions

The Cable Corporation has insured a piece of equipment that has an actual cash value of $200,000 for the policy amount of $100,000. The policy has a coinsurance requirement of 80 percent. The equipment is damaged and the loss incurred is $30,000. Even though the loss is well below the policy amount, the payment by the insurer on this loss will be computed as follows:

$$\frac{\text{Policy amount}}{\text{Coinsurance requirement} \times \text{Actual cash value of equipment}} \times \text{Loss} = \text{Payment by insurer}$$

$$\frac{\$100,000}{80\% \ (\$200,000)} \times \$30,000 = \$18,750$$

The impact of failing to meet the coinsurance requirement is that Cable Corporation bears an uninsured loss of $11,250.

General Liability

With the possible exception of professional liability, the general liability policy is the most complex of the insurance coverages. Basically, the comprehensive general liability (CGL) policy provides coverage for liability arising from bodily injury or damage to property of third parties. The policy provides coverage for these liabilities arising from four basic hazards: (1) premises and operations; (2) independent contractors; (3) products-completed operations; and (4) incidental contracts.

The basic policy coverage is significantly expanded with the addition of the broad form comprehensive general liability endorsement. This is a package endorsement that adds 12 important coverage extensions, at a lower premium than would be incurred if they were purchased separately. The 12 coverages typically provided include:

1 *Broad form contractual liability* – coverage for the liability of others assumed through contract.

2 *Personal injury liability* – coverage for liability arising from such things as libel, slander, defamation of character, and false arrest.

3 *Premises medical payments* – pays for medical payments on behalf of members of the public on a no fault basis where injured on the contractor's premises.

4 *Host liquor liability* – covers liability for injuries inflicted by an individual to whom the contractor has served alcoholic beverages.

5 *Fire legal liability* – covers liability arising from fire damage to the premises of others that are occupied by the insured.

6 *Broad form property damage* – narrows the application of the "care, custody, or control" and "workmanship" exclusions.

7 *Incidental medical malpractice* – clarifies the application of coverage for bodily injury resulting from errors in the administration of first aid.

8 *Nonowned watercraft liability* – provides coverage for liability arising out of the use of watercraft, under 26 feet in length, that are not owned by the contractor.

9 *Limited worldwide liability* – extends the policy territory to provide coverage for liability arising out of incidental activities of the contractor outside of the United States or Canada if the original suit is brought within the United States or Canada.

10 *Additional persons insured* – includes employees as persons insured.

11 *Extended bodily injury* – provides coverage for liability arising from the use of reasonable force for protecting persons or property.

12 *Automatic coverage* – for newly acquired organizations of the contractor.

The discussion of the broad form endorsement in this chapter addresses the coverages provided by the standard endorsement. Some insurance companies have drafted their own forms, and these forms may either be more restrictive or broader in coverage than the standard form. Such endorsements should be carefully reviewed to ensure that the desired coverages are provided.

Although the broad form general liability endorsement contemplates some of the individual changes, there are numerous additional modifications that need to be considered. Some of the indicated modifications may be difficult for the smaller to mid-size contractor to obtain. However, contractors with $40 million per year or more in receipts can expect to obtain virtually all the modifications.

Perhaps the most difficult modification to negotiate is the removal of the architect/engineer exclusion in the contractual liability section. This exclusion states that no coverage will be provided where the contractor holds the architect/engineer harmless for certain professional activities. Because of the high dollar volume of contracts per year, the very large contractors can sometimes negotiate to have this exclusion removed on a specific job basis or even on a blanket basis.

The most confusing aspect of general liability involves broad form property damage (BFPD) coverage. The endorsement provides the contractor limited coverage for damage to property in his care, custody, or control as well as damage to the work during the course of construction. In addition, damage to his work arising from his subcontractors' work and damage to his subcontractors' work arising out of his work may be covered after the job is completed.

However, the standard BFPD endorsement does not provide coverage for damage to the contractor's work arising out of his own work. In other words, if a contractor builds a building using his own work force, and three years after completion the building is destroyed by a fire that was caused by faulty material or work performed by the contractor, the contractor's completed operations insurance would not apply to the damage to the building itself.

Coverage can sometimes be arranged through negotiation so that only the particular part of the work that was faulty or that malfunctioned would be excluded and resulting damage to good work would be covered. Many of the major insurance companies involved in providing construction insurance now provide this extension of BFPD coverage for a reasonable premium. However, the contractor and his insurance agent must often press the point to obtain such coverage.

In the past, many insurance companies attached a professional services exclusion, normally referred to as a G-307, to the general liability policy. While this limitation is meant to apply to architects/engineers, it has often been construed to apply to design/build contractors as well as construction managers. Under no circumstances should a contractor accept a general liability policy or umbrella policy that contains this limitation.

A majority of contractors have been involved in joint ventures in the past. Most are totally unaware of the fact that their general liability policy does not afford coverage for these past joint ventures with respect to completed operations. To afford such coverage, a residual joint venture endorsement should be added to cover the contractor's own liability arising from past joint ventures, even though the liability arose from actions of another contractor who was a member of the venture. This modification, too, can be obtained only through negotiation.

One of the newer coverages available involves supplemental liability insurance for the contractor dealing with wrap-ups and owner-controlled programs, or programs in which the owner, as opposed to the contractor, provides the insurance for the project and requests that the contractors reduce their bid proposals accordingly. In many instances, the owner-furnished program is not as broad as the contractor's usual program. However, the contractor is still responsible for his negligence and possibly the negligence of others as well. The contractor's master general liability policy can sometimes be endorsed so that it provides the difference between the contractor's master program and the owner-furnished program. As a result, with the supplemental liability insurance, the contractor will have a definite and broad coverage arrangement for all situations.

Many contractors are becoming involved in oil and gas ventures and are even more involved in real estate projects. While any venture in question usually carries specific coverage for liability, this coverage may not be broad enough or have adequate limits. The contractor's master general liability policy can and should be endorsed to cover these nonoperating exposures on a contingency basis.

The general liability policy is quite complicated, but, with proper attention to coverage details and careful negotiations with insurers, a truly broad and comprehensive liability program can be arranged.

Finally, it should be noted that the insurance industry is in the final stages of preparing a new standard general liability form. Actually, there will be two new forms to replace the current one. The new forms will probably go into general use in 1986. The primary changes involve the incorporation of all 12 of the coverages now available in the broad form CGL endorsement into the basic policy, the inclusion of more comprehensive aggregate limits, and an optional claims-made coverage trigger. Contractors should follow these developments closely and be prepared for the changes when they occur.

Umbrella Liability

An umbrella liability insurance policy is designed to provide the insured with high limits of coverage for catastrophic losses. In insurance industry parlance, umbrella liability is referred to as an excess coverage. For the most part, this term means that the umbrella is intended to provide extra, or excess, limits of liability in addition to the limits that are provided by certain primary liability insurance policies.

Thus, a unique coverage over any number of separate and distinct underlying coverages—such as automobile liability, general liability, and the employers' liability portion of the workers' compensation policy—is provided. At times, the umbrella may also provide broader coverage than these primary or underlying insurances. Where the umbrella covers a loss that is not insured by the primary policies, a self-insured retention, which is similar to a deductible, will apply, costing typically $10,000–$25,000.

A contractors' limitation endorsement is invariably attached to the umbrella of any construction contractor, or any organization involved in construction activities. The contractors' limitation endorsement consists of exclusionary endorsements that vary from company to company, and that, in some cases, significantly impair coverage. These endorsements should be reviewed carefully. An acceptable endorsement excludes coverage for, for example, liability arising from faulty workmanship, collapse, underground property damage, explosion, and damage to property in the custody of the contractor *only* if these items are excluded in the underlying general liability policy. Therefore, the exposures will be insured if the general liability policy is properly structured.

In fact, one of the most important reasons for negotiating the broadest possible general liability policy is to lay the groundwork to broaden the umbrella policy. If the liberalization endorsement, which states that any coverages carried in the primary general liability policy will follow through to the umbrella, can be attached to the umbrella policy, a very broad coverage is effected.

For design/build contractors, construction managers, and engineering risks, consideration might be given to obtaining professional errors and omissions for passive liability under the umbrella policy. A few markets are providing these broad umbrellas.

Contractors should cover the personal exposures of key executives. Such exposures, both personal nonbusiness as well as personal business, should be covered under either a master business umbrella or separate personal umbrellas.

Wherever possible, the umbrella should include excess coverage over any aviation policies including liability arising from owned or nonowned aircraft. Most umbrella carriers even provide coverage for liability arising from nonowned aircraft chartered with crew, without requiring an underlying policy or charging an additional premium if this is not a frequent practice of the contractor.

A commonly asked question is, What amount of umbrella limits liability should be purchased? A general rule of thumb is to multiply the contractor's annual volume by a factor of one. If this generates more than $75 million in limits, consideration should be given to multiplying the largest single project by a factor of one. Further consideration should be given to exposures based on property and personnel, with greater exposure using a different factor. For instance, a road contractor working outside the city limits might multiply the indicated amount by 50 percent, while the petrochemical contractor working inside a large complex might multiply the indicated amount by 200 percent.

Limits should not be purchased based on the amount that will be paid, but rather on the potential amount of an ad damnum clause, which contains a statement of the plaintiff's money loss or the damages that he claims. If the insurance is purchased to satisfy the potential claims, an impairment of the contractor's financial statement due to an inadequately insured loss will be avoided, regardless of how remote the possibility of payment.

Professional Liability

Contractors may have professional liability exposures arising from one or more of the following types of services or operations:

- Design/build construction projects in which the contractor's in-house architects/engineers conduct the design phase and place their seal on the working drawings and papers;

- Design/build construction projects in which an outside architectural/engineering firm is employed as a subcontractor and places its seal on the working drawings and papers; and

- Construction management projects, which usually involve inspection and supervision of the work and scheduling, and, sometimes, approval of designs, plans, specifications, and construction processes.

A number of potential loss exposures result from these operations, including design error, delay of completion, failure of completed work to perform, and property damage to the work arising from the work. These loss exposures are not covered by standard builders' risk and general/umbrella liability policies. If the design error and omission limitations are removed from builders' risk, general and umbrella liability policies and amendments are made to the BFPD endorsement of the general liability policy to cover damage to the work; professional liability coverage, however, may not be necessary. Most general contractors and construction management firms should consider retaining the balance of the exposures or transferring them within contract documents.

If the contractor needs professional liability coverage, he should consider a number of factors before buying it, since the available professional liability insurance coverage forms are not identical. Generally speaking, these policies cover the contractor's legal liability arising from negligent acts, errors, or omissions in the performance of design/build or construction management professional services.

These policies are set up on a "claims-made" basis. This means that the policy will only cover a claim brought against the contractor during the policy period. With a claims-made form, the time at which the error, omission, or damage that leads to the claim occurs is irrelevant, with certain exceptions discussed in the next paragraph. In most instances, the actual placement of a claim against the insured triggers coverage. This is in contrast to occurrence policies, such as general, automobile, and umbrella liability, which are triggered by property damage or injury occurring during their policy period (rather than when the claim is made) and which can cover a claim that is brought against the insured years after the policy has expired.

There are situations in which the time of the occurrence of the error or omission affects the applicability of coverage in a claims-made policy. Some claims-made policy forms contain a retroactive date and a provision to exclude coverage on claims brought against the insured arising from acts that took place prior to the retroactive date. Needless to say, it is preferable either not to have such a limitation or to negotiate a date during the period when professional services had first begun.

Another provision exclusive to claims-made policies is the discovery period clause, which provides that the policy covers claims brought after the policy expiration date if the insured had reason to believe that the act resulting in the claim would lead to potential liability and the insured reported the potential claim to the insurer prior to the expiration of the policy.

There are a number of exclusions common to all of the various claims-made policy forms. These include:

- Liability of others assumed under contract
- Liability arising from personal injury (which should be insured in the CGL policy)
- Liability resulting from insolvency or bankruptcy
- Liability arising from failure to advise or require the purchase of insurance or bonds
- Exceeding estimates of probable construction costs
- Damage to property owned by the insured

- Damages arising from failure to complete drawings, specifications, or schedules of specifications on time

The following are additional exclusions that can be avoided through negotiation with the insurer or careful selection of forms:

- Punitive or exemplary damages

- Patent infringement

- Services rendered in connection with tunnels, bridges, or dams

- Services rendered in connection with the surveying of subsurface conditions or the testing of ground or soil

In summary, these policies can provide substantial protection against certain professional liability exposures, particularly design error, but must be selected with care after a thorough exposure identification analysis.

Workers' Compensation

Basically, workers' compensation insurance provides no fault coverage for the benefits mandated by state law to injured employees. It also provides employers' liability coverage in case of a suit by an injured employee under common law rather than workers' compensation.

While workers' compensation generates the majority of the insurance premiums for the contractor, it is generally considered a simple coverage area to select and manage. However, there are a number of basic modifications that should be considered.

The standard policy only provides coverage for the benefits specified by the law in the state(s) identified in the declarations of the policy. Employees frequently travel through other states, and if injured, they have the option of filing for benefits in the state of hire, the state of residency, or the state of injury. Of course, the state with the greatest benefits is usually elected.

Coverage from other states is not sufficient for employees claiming benefits in any of the monopolistic fund states—Nevada, North Dakota, Ohio, Washington, West Virginia, and Wyoming. If a contractor ever engages in operations in any of these states, a workers' compensation policy should be purchased from the state. However, the stop-gap endorsement can be added to the contractors' policy to structure it to respond to employers' liability suits brought in these states.

The USL&H (United States Longshoremen's and Harborworkers') Compensation Act provides employees with greater benefits than those of virtually any state. Although it was originally intended to apply only to longshoremen, it has been extended inland through statutory modification and court interpretation. To cover this contingency, all contractors should attach the USL&H endorsement to their policy. There is a minimum premium charge if no true exposure exists.

Buildings and Contents

Coverages for contractors' buildings and contents are frequently omitted from discussions of construction insurance, since their premium cost is lower than workers' compensation and liability insurances. Such losses do occur, however, and the contractor should review these coverages on an annual basis.

The contractor should normally buy all risk coverage written on a replacement cost basis. The insurance should also have an "agreed amount" endorsement to eliminate the possibility of a coinsurance penalty at the time of loss. It is in the contractor's best interest to estimate building values on the high side rather than risk the possibility of an underinsured loss.

Extra expense insurance may be endorsed into the policy. In the event the building becomes untenantable due to fire or other insured perils, this coverage pays the additional expenses associated with continuing operations at another location (e.g., rent, overtime to employees, and telephone charges).

The contractor should consider buying valuable papers insurance. This coverage pays the costs associated with researching and reconstructing valuable papers lost in a fire or other catastrophe.

If the contractor has a computer system, a specialized electronic data-processing (EDP) policy may also be purchased. EDP policies can be structured to cover damage to the computer or to media as well as extra expenses associated with continuing data-processing operations after a loss.

INSURANCE DISTRIBUTION SYSTEM

Independent Agents

Insurance companies market their products four basic ways: (1) through independent agents; (2) through exclusive agents; (3) through commissioned or salaried employees; and (4) through brokers and broker/agents. Independent agents are independent business organizations that represent several insurance companies nonexclusively. Independent agents are usually compensated on a commission basis by the insurance company, although fees and negotiated commissions are becoming more common. All agency expirations are owned by the agency.

Exclusive Agents

Some insurance companies are represented by "exclusive agents." This marketing method entitles the client to almost exclusive representation from the agent or at least "a first look" at the agent's new business. The agent is compensated on a commission basis, sometimes with certain sales incentives, but he is not a salaried employee of the company.

Commissioned or Salaried Employees

Other insurance companies, usually referred to as direct writers, market their insurance services through commissioned or salaried employees who represent the direct writer exclusively.

Brokers

Another way insurance services are marketed is through brokers. Brokers are independent businesses that do not represent an insurance company as an agent, but shop various markets to obtain appropriate coverage for their clients at competitive premiums. Brokers have traditionally been compensated by commissions paid in whole by the insurance company or, in some instances, by contributions made by the client. A practice that is becoming more and more common is for brokers (and agents) to obtain coverage for their clients on a net basis and then receive a flat or negotiated fee for their services.

Broker/Agents

Some state laws do not permit the licensing of brokers. Therefore, in these situations brokerage firms must represent insurance companies as agents. In states with such laws, the difference between an independent agent and a broker becomes blurred; this has led to the broker being referred to by the term broker/agent.

CASH FLOW PROGRAMS

Alternate risk financing arrangements, normally referred to as cash flow programs, are among the hottest products in the current marketplace. The question is not whether or not to have a cash flow program but what kind of cash flow program to have. The types of plans available vary, based on the size of the premiums and the desirability of the contractor's or owner's business. The plans that generate maximum cash flow are self-insurance plans, paid loss retrospective plans (retros), and captive plans. When properly constructed, these plans avoid the geographical and tax disadvantages of self-insurance and formalize the insurance program as compared to self-insurance. Captive insurance companies insure only the contractor's own business and, in some cases, even issue surety bonds.

Self-insurance, a form of cash flow program, is primarily used for workers' compensation premiums. For such a program to be practical, premiums of $250,000 per year are usually required and operations should be primarily in one state. Such plans have caused some tax problems in that the Internal Revenue Service (IRS) has taken the position that reserves are not tax deductible until actually disbursed. In addition, if a contractor operates in several states, the self-insurance qualifications and filings can be complicated.

In the mid-1970s as insurance companies lost business to self-insurance and captive plans, they began offering paid loss retrospective plans. A key feature was that other lines, such as automobile and general liability coverages, could be readily included. Usually, a contractor will pay at least $500,000 in premiums before an insurance company would offer such a program.

Paid loss retrospectives may not continue to offer the same benefits. The IRS has occasionally treated paid loss retros as if they were self-insurance programs, thereby disallowing deductions for unpaid reserves. On July 1, 1982 new rules went into effect governing cash flow programs in California. One of the key features is that insureds must pay interest on deferred premiums based on the prime rate less 1½ percent. The result is that while the program still provides positive cash flow to the contractor, virtually all of its investment income benefits are negated.

Before paid loss retros became readily available, some contractors formed captives. However, the IRS began treating these wholly owned captives as if they were self-insurance programs. As a result, the contractors lost the tax deductions of the premiums until claims were actually paid. Several group captives were formed by construction groups during the early 1980s, one of the larger being the Bermuda-based American Risk Transfer Insurance Company Ltd. (ARTIC). Some of the principal advantages of group captives involve lower overhead ratios, greater purchasing power in the area of reinsurance and services, and greater risk retention capabilities. In addition, if the ownership is properly structured in an offshore company, substantial tax benefits can be obtained. On the other hand, some captives—both individually owned and group owned—

have been poorly planned, improperly implemented, and undercapitalized.

Although cash flow plans virtually disappeared during the last tight insurance market, they are now considered a permanent fixture. When the market tightens in the future, cash flow programs will become highly negotiable and insurance companies will strive to get a bigger piece of the investment pie—either directly by writing for the contractor or indirectly by participating in captive programs.

WRAP-UP PROGRAMS

Owner-controlled programs, referred to as wrap-ups, are programs where the owner decides to furnish the insurance for the project and requests that the contractor reduce his bid proposals accordingly. In some instances, the programs are contractor-controlled, that is, the contractor controls the placement of casualty insurance and, possibly, surety bonds for subcontractors.

The benefit of wrap-ups to the owner and/or contractor is the resulting returns based on good loss experience. Since cash flow programs are often negotiated, broader coverages can in many cases be obtained along with tailored claim services because of increased purchasing power due to overall size.

One of the biggest problems with wrap-up programs is working out the areas of responsibility and administration and tying this back to the contractor's master policies. There has also been the problem of reimbursement of insurance premiums for gray coverage areas. When the owner provides the coverage, he considers the arrangement more from his standpoint than the contractor's. As a result, he may be overly price conscious and place less priority on coverage and service.

As a result, this may leave the contractor exposed, since he still has potential liability but may not have proper coverage. A project should be at least $50 million in construction values or $500,000 in premiums before a wrap-up is practical.

Wrap-ups can be extremely time-consuming and distracting for the contractor. If not properly monitored, the project manager and other personnel at the job-site may be diverted from their main priority of putting up the project.

Most contractors dislike wrap-ups and will work against them if not properly approached. Contractors' insurance agents find them even more objectionable, since they represent work on the agent's part without receipt of adequate compensation.

There have been some very successful owner-controlled programs and thus such programs are here to stay. Contractors and owners should work together carefully to be sure that the programs are well conceived and carefully executed.

INTERNATIONAL EXPOSURES

At one time, international construction insurance applied to domestic contractors who worked overseas. Now, however, international construction insurance has become a two-way street, since many non-United States contractors have entered the United States market. The difference between United States regulations and laws and those of the domicile of the foreign contractor has caused a new set of problems and concerns for foreign contracts. The foreign contractor and his agents must be careful to use United States insurance firms wherever possible, since they will properly represent his interests.

Adjustments to Other Policies for Foreign Travel

If the United States contractor's employees are involved in foreign travel, certain adjustments to the workers' compensation, general liability, and automobile policies may be necessary. For example, the definition of "territory" under the automobile and general liability policies is the United States, its territories, and Canada. If vehicles are driven by employees in other countries, additional specific insurance policies may be necessary. Workers' compensation policies should be extended to cover travel in these foreign countries. In some cases, the United States policies may be extended to cover these foreign exposures, if the foreign component of the risk is minor. However, if extensive operations are conducted abroad, worldwide coverage should be purchased from an insurance carrier with international capabilities.

Political Risk Insurance

Contractors doing foreign work have a growing need for political risk insurance, which has been reinforced by the situations in Afghanistan and Iran during the early 1980s. In response to this need, several insurers have developed some innovative products in the area of political risks.

Political risks can be substantial. In Islamic countries, for example, unconditional letters of credit or standby letters of credit have been required, creating a loan from which a bank in that country can unilaterally draw down funds. Banks in the United States are more apt to give such letters of credit if political risk insurance has been purchased for the capricious or illegal draw downs.

Other types of political risk coverage involve mobilization loans in case contracts are repudiated or other factors, such as changes in governments, cause the project to be cancelled. Loss of profits can also apply to mobilization loans and other political risk exposures. Export credit insurance, which covers the nonpayment by default of the foreign party or through the use of improper documents, is also available. There is a growing demand for advance profits coverage or business interruption coverage on construction projects.

International Insurance Requirements and Constructor's Responsibilities

There has been a substantial problem involving decennial liability, which is a 10-year guarantee against structural failure that began in France and spread to Saudi Arabia and other areas in the Middle East. Now more countries are picking up this requirement. While there appear to be many underwriters interested in providing a program that will suit all parties, few viable programs have been developed to date.

In January 1982, a committee formed by the Federation Internationale des Ingenieurs-Conseils (FIDIC) issued a report reviewing the existing procedures for complying with the insurance requirements of the current FIDIC conditions and indicating action that might be taken to resolve some of the insurance problems connected with large-scale civil engineering projects in the world.

The report points out that there are certain significant differences between the responsibilities and the corresponding insurance requirements imposed on the contractor under FIDIC conditions of contract on the one hand and the insurance coverage currently available on a best efforts basis on the other. The report contains two main suggestions. The first is the appointment of a risk management adviser at the beginning of the project. The second is that the FIDIC conditions should be revised to include an option for the owner to have a wrap-up insurance program. The report also strongly recommends that the adviser be remunerated by a fee to be financed from project loans. While this particular course of action has only been followed on a small number of projects worldwide, its general introduction can provide substantially improved insurance and risk management services to owners of large projects.

CONCERNS AND PROBLEM AREAS

Workers' Compensation Claims

Many people are of the opinion that workers' compensation will be used to partially solve the social security system's financial problems. Social security payments are currently partially reduced by workers' compensation payments. In the future, workers' compensation payments may increase to offset decreasing social security payments. It is expected that these benefit increases, along with medical inflation, will cause workers' compensation premiums to double for the average contractor during the next five years, even though no additional man-hours will be worked. This continued increase in premiums will, in turn, generate innovative funding programs in the workers' compensation area.

Asbestos Claims

Asbestos claims constitute another substantial problem area. Over 5,000 claims have been logged in London. While each claim is not necessarily large, their combined number may overwhelm some underwriters in the long run—both in the United States and in Great Britain. Similar problems may develop with other disease-related exposures.

Aspect of Limited Liability

In the past, one of the benefits of workers' compensation to the employer has been the aspect of limited liability. However, there have been some disturbing court cases that allowed employees to collect workers' compensation and then sue the employer directly or indirectly where unlimited liability could be imposed. There is a strong need for legislation that would make workers' compensation the sole remedy for employees. The savings in litigation could then be translated into higher benefits for the employee and reduced premiums for the employer.

Completed Work vs. Products

Another developing problem has been the definition of completed work versus products. If a plaintiff's attorney is successful in getting the work of the contractor to be deemed a product, then the doctrine of strict liability may be applicable. This doctrine states that an involved party is liable regardless of negligence. On the other hand, some insurance companies have tried to have the work deemed a product or products in an attempt to apply certain exclusions under the general liability policy that relates to manufacturing operations. Most of the problems to date have involved housing projects. However, all contractors should obtain clarification from their insurance company that all completed work will be deemed completed operations rather than products.

Open Rating for Workers' Compensation Insurance

In the past, workers' compensation has been a heavily regulated line of insurance. However, Arkansas, Colorado, Georgia, Illinois, Kentucky, Michigan, Minnesota, New York, Oregon, and Rhode Island have recently passed statutes that allow open rating for workers' compensation insurance. Similar statutes are pending in other states. Under open rating statutes, insurance companies can deviate a certain percentage from standard rates and charge virtually whatever they choose to charge on an open market basis. While this may translate into some savings for the buying public in the short run, it could further destabilize the marketplace and create even more violent swings. Since the larger buyers of insurance are already receiving most of the available discounts, deregulation will benefit the smaller premium payer the most. In the long run, there is a good chance that the bigger premium payer will end up subsidizing the smaller premium payer under deregulation through state-administered guaranty funds.

Possible Financial Failure of Insurance Companies

Perhaps the most frightening problem facing the average insurance buyer in the future is the possible financial failures of insurance companies. Due to the extended competitive cycle, inflation, and a faltering economy, some insurance companies will simply not be in a position to pay all their claims and will be forced into bankruptcy. While most states have guaranty funds that will guarantee payment up to the first $50,000, this would be of little consolation to a contractor who has a $500,000 outstanding liability claim or an owner who has a $1 million builders' risk claim. In the future, it will be very important to carefully evaluate the financial standing of insurance carriers. It may be worth the extra cost to pay a small premium for financial stability and a solid track record.

TRENDS AND FUTURE INNOVATIONS IN THE INSURANCE INDUSTRY

The London market has often been a bellwether of the United States marketplace. During the last 15 years, there has been substantial vertical integration in the London market, with large holding companies owning both insurance brokers and insurance companies, and a spate of mergers and acquisitions involving these London organizations and the larger United States brokers.

Based on past events, it is reasonable to assume that substantial mergers and acquisitions will occur in the United States, and it is quite possible that by the end of the 1980s, large holding companies may own both top insurance companies and insurance brokers. In fact, some pairing has already begun: Fred S. James has been acquired by Transamerica; Reliance Insurance Company now has a substantial position in E. H. Crump Companies; and the Continental Group has paired off with Reed Stenhouse Companies.

Both insurance companies and brokers have begun to form substantial risk management departments that provide a vast array of engineering, claims, handling, and other technical services to their clients. What will happen to the local agents who are not in a position to merge with the large insurance companies is an obvious question. Insurance is a very personal business, however, and bigger size does not necessarily mean better quality; many of the smaller agents could do quite well. The better insurance talent may stay in smaller firms where entrepreneurial desires can often be best expressed. What all this means to the construction organization is that the 1980s have become a decade of innovation and change in the insurance business, the consequences of which will benefit the insurance buyer.

CHAPTER **16**

Legal Matters

Edward B. Lozowicki

INTRODUCTION

As stated in the first chapter of this book, the construction industry uses a variety of contract types. These include the fixed-price or lump-sum contract, the unit-price contract, a variety of cost-type contracts including the cost-plus, and the time-and-materials contract. Each of these contract forms is explained in the first chapter.

The purpose of this chapter is to explore some of the more common, and at times more complex, intricacies of the legal document around which all construction contracting revolves. This chapter explores the pitfalls of standard contract forms, the problems of collecting on billings; and the requisite contract terms and collection techniques. The legal problems of dealing with change orders and extras and of recovering for unpaid claims are explored through some of the less commonly used techniques. This chapter does not attempt to cover all of the legal aspects of construction contracting, but rather tries to highlight those areas that are of the greatest importance to the nonlegal construction professional.

CONTRACT FORMS

In the typical lump-sum, firm-price contract, bidders make allowances for a variety of contingencies in the performance. These include items from punchlist and clean-up work to potential productivity or site-access problems. Labor productivity rates may be modified or specific line-items included in the bid, depending on the nature of the risk.

In contrast, the budget established for a cost-plus contract seldom includes labor productivity contingencies, since the direct cost of labor is passed through, subject to a guaranteed maximum price. In addition, during negotiations the parties frequently express their intent to minimize contingencies.

Successful negotiations, however, are no guarantee that contingencies will not occur or that the cost will be reimbursed. Thus the written agreement must address these risks and allocate them to one side or the other.

Problems With Standard Forms

The most popular printed forms used for cost-plus contracts are published by the Associated General Contractors of America (AGCA) and the American Institute of Architects (AIA). Unfortunately, the AGCA form is not supplemented with General Conditions and the AIA General Conditions allocate most of the risk of contingencies to the contractor. In many cases, this is not what the parties intend, particularly on fast-track

projects where the design is often incomplete at the time of contract signing. The following is an illustration of some of the problems the contractor can encounter with these forms.

AGCA Document No. 9: Owner-Contractor Negotiated Agreement (Nov. 1975 Ed.)

- The AGCA form refers to "GENERAL CONDITIONS" in
  several places. However, it appears that AGCA has not as yet
  published a complimentary set of General Conditions (see, e.g.,
  Article 9.1, re: Changes).

- Article 6.1 specifies that the time of completion is subject to
  delays beyond contractor's control. However, these delays are
  not specified. Article 9.3 implies that certain types of delays are
  the responsibility of the contractor; again, no details are given.

- Article 10.2.12 provides for reimbursement of the cost of
  corrective work, "provided the cost does not exceed the GMP
  [Guaranteed Maximum Price]." This leaves unanswered the
  question, "Who pays if the corrective work does exceed the
  GMP?"

- Article 15.1.4 provides for the right to suspend work in the
  event of nonpayment. However, there is no provision for interest
  on late payments or attorneys' fees in the event that legal action
  is required to collect payments.

AIA Document No. A111: Cost-Plus Owner-Contractor Agreement (April 1978 Ed.)

The AIA form is generally less favorable to the contractor than the
AGCA form. For example, it does not define the owner's responsibilities,
whereas AGCA Article 5 does. The AGCA form also includes a number
of reimbursable costs not referred to in the AIA form (see AGCA Articles
10.2.17, 18, and 19). Furthermore, the AIA form contains a number of
problematic provisions:

- Form A111 is designed for use with AIA Document A201,
  General Conditions. The latter is written for a traditional lump-
  sum price contract and allocates much of the risk of
  contingencies to the contractor.

- Losses and expenses attributable to the fault or neglect of the
  contractor are not reimbursable under the AIA form (see

Articles A.1.12 and 9.1.6). The AGCA form, however, permits reimbursement of these items to a limited extent.

- The AIA form excludes reimbursement for corrective work (see Article 9.1.6). In reality, some punchlist work is figured into every job and is to be expected.

- The architect remains the administrator of the contract and processes all progress payment billings (see Article 14). In a fast-track job, where the contractor is working closely with the owner and his engineering department, such provisions may not be necessary or appropriate.

AIA Document No. 210: General Conditions (1976 Ed.)

This document is a comprehensive set of terms and conditions prepared with the typical lump-sum, firm-price project in mind and is widely used in the construction industry. It is drafted from the architect's point of view. Thus, it gives the architect a great deal of power to direct the work, process payments, and resolve disputes. It also allocates much of the risk of contingencies to the contractor. Following are a few examples:

- The architect is the agent of the owner for purposes of supervising the work (see Articles 2.2.2, 3.2.6, and 5.2.1). The architect also controls the flow of progress payments (see Article 2.2.6). On fast-track projects the owner and contractor may want a more direct working relationship.

- The contractor is required to report errors and inconsistencies in the plans and specifications to the architect (see Article 4.2.1). On design-build and fast-track projects, the plans and specifications are by no means complete at the beginning of the project.

- Clean-up costs are charged to the contractor (see Article 4.15.2). Such costs should, in ordinary circumstances, be treated as reimbursable labor.

- The contractor is liable for delays in completion (see Article 8.3.4). On many negotiated projects, the completion date cannot be fixed at the time the contract is signed.

- The contractor has very broad indemnity and damage liabilities under this form (see Articles 4.18.2 and 10.2.5). If not insured for these risks, the contractor has substantial exposure.

Sample Remedial Clauses

The following suggested clauses supplement AGCA Form No. 9 and resolve several contingencies in the contractor's favor. These may be used to negotiate changes from the standard AIA Form 201. If used as a supplement to AIA, a precedence clause should be added to resolve conflicts in favor of the supplement.

The sample remedial clauses contained herein cover the following areas: reimbursable costs; insurance and bonds; guarantees; changes in the work; differing site conditions; defaults or delays; and termination.

REIMBURSABLE COSTS

Compensation of the Contractor's employees under Article 10 shall be at the rates specified in the Contractor's then-current Standard Labor Rates. A copy of the rates currently in effect is attached to the contract and is incorporated by this reference. In the event of any increases in wages, fringe benefits, payroll taxes, or insurance, the rates will be adjusted to reflect any such increase, and a copy furnished to the owner.

Costs of tools and equipment rented by Contractor from others shall be reimbursed in accordance with Article 10. Costs of tools, equipment, and vehicles owned by Contractor, with a unit market value in excess of $100, shall be reimbursed at Contractor's Standard Equipment Rates (a copy of which is attached to the contract and is incorporated by this reference).

INSURANCE AND BONDS

Contractor will provide: worker's compensation insurance in the forms and amounts required by the applicable statutes of the state in which the Work is performed; and automobile and general liability insurance in comprehensive form, with policy limits of $300,000 per person and $500,000 per occurrence. Any additional liability insurance or special endorsements will be provided at Owner's request, and the cost of same will be reimbursed to Contractor in accordance with Article 10.

Owner will furnish builder's risk insurance for the project in "all risk" form, in an amount equal to the Guaranteed Maximum Price, endorsed to protect Owner, Contractor, any subcontractors, and any construction lenders, as their interests may appear. Upon request of Owner, Contractor will provide such insurance, and the cost shall be reimbursed in accordance with Article 10.

Owner will bear the risk of loss or damage to the Work or any part of it upon and after its affixturing to the premises. To the extent any loss or damage to the Work is covered by either Contractor's or Owner's property insurance, Contractor and Owner waive in favor of the other any subrogation rights that they or others claiming through them may have.

The costs of any bonds required of Contractor or his subcontractors, including, without limitation, performance, payment, and mechanics lien release bonds, shall be reimbursed to Contractor.

GUARANTEES

For materials and equipment manufactured by others and installed by Contractor, Contractor's liability for defects in such materials and equipment is limited to the extent that, and upon the same conditions as, the manufacturer is liable to Contractor. In the event that Owner makes claim against Contractor for defects in such materials and equipment, Contractor will, within a reasonable time, assign to Owner its rights, if any, to claim against, and recover damages from, the manufacturer for such defects, and Owner agrees to accept such assignment in full satisfaction of any claim it may have against Contractor.

For other work performed hereunder, Contractor will repair or replace, at his sole option, materials that are not in accordance with the specifications and drawings described herein. Contractor will also correct labor that is not performed in a workman-like manner, provided that Owner gives written notice to Contractor of such items within 12 months from the date of Contractor's substantial completion of the Work. The cost of such corrective work will be reimbursed in accordance with the provisions of Article 10.

CONTRACTOR'S LIABILITY FOR DAMAGES DUE TO NEGLIGENT OR DEFECTIVE WORK IS LIMITED TO THE FOREGOING. CONTRACTOR SHALL NOT IN ANY CASE BE LIABLE OTHERWISE. NOR SHALL CONTRACTOR BE LIABLE FOR INDIRECT OR CONSEQUENTIAL DAMAGES OF ANY NATURE. IN NO EVENT SHALL CONTRACTOR'S LIABILITY FOR DAMAGES HEREUNDER, WHEN ADDED TO THE COST OF THE WORK REIMBURSED TO CONTRACTOR UNDER ARTICLE 10 AND TO OTHER DAMAGES FOR WHICH CONTRACTOR IS LIABLE HEREUNDER, CAUSE THE SUM OF SUCH AMOUNTS TO EXCEED THE GUARANTEED MAXIMUM PRICE.

[This paragraph is ordinarily used entirely in upper case, since the limitation of liability provision is of pervasive importance to the contract.]

CHANGES IN THE WORK

The Work may be changed by field order, change order, extra work order, or an order otherwise authorized by Owner, his agents, and his employees. Such changes may include changes in the scope of work, the method of performance, the scheduling, or other elements necessary to complete the Work. In such event, the Guaranteed Maximum Price shall be adjusted as provided in Articles 9 and 10 herein, and the time of completion shall be equitably adjusted. Contractor will notify Owner of such changes within a reasonable period of time after they have been authorized.

DIFFERING SITE CONDITIONS

If Contractor encounters subsurface or latent physical conditions at the site, differing materially from those indicated in the contract documents, or physical conditions at the site of an unusual nature, differing materially from those ordinarily encountered in the trade, Contractor will promptly notify Owner. If such conditions cause an increase in the cost of, or the time required for, performance of the work, such cost will be reimbursed in accordance with the provisions of Article 10; the Guaranteed Maximum Price may be modified accordingly; and the time for completion will be equitably adjusted.

DEFAULTS OR DELAYS

Contractor will not be deemed in default or be liable for damages for any failure or delay in performance of the Work that arises out of causes beyond his reasonable control. Such causes may include, without limitation, acts of God or a public enemy, acts of the government in either its sovereign or contractual capacity, fires, floods, earthquakes, epidemics, quarantine restrictions, strikes, freight embargoes, material shortages, or unusually severe weather. In the event that the Work is delayed by such causes, the time of completion will be extended in proportion thereto.

For other delays in completion within Contractor's exclusive control and not excused by the above paragraph or for other defaults in performance of this Contract, Contractor's liability for damages will not include indirect or consequential damages. In no event shall Contractor's liability for damages hereunder, when added to the Cost of the Work reimbursed to Contractor under Article 10 and to other damages for which Contractor is liable hereunder, cause the sum of such amounts to exceed the Guaranteed Maximum Price.

TERMINATION

If payment is not made to Contractor as required by Article 15; if Owner assigns, transfers, attempts to assign, or attempts to transfer this Contract in whole or in part; if Contractor has reason to believe that Owner is insolvent or unable to make the payments required herein; if Owner becomes the subject of any insolvency or bankruptcy proceeding; or if Owner otherwise defaults in performing the terms and conditions of this Contract, then Contractor may suspend his performance without further notice; remove from the job-site any and all of his labor forces, materials, facilities, and equipment; and render inoperative any device or apparatus previously installed.

In addition, Contractor may, at his election: recover the amounts due for Work performed in accordance with the Contract price plus any sums due for loss of profits and other consequential damages, or, at his election, terminate the Contract and recover the reasonable value of labor and materials that the Contractor contributed to the Work. In any event, Contractor may pursue any rights available to it under applicable private or public bond, stop-notice, or mechanics lien statutes.

In addition, Contractor may recover the cost of enforcing any of his rights, including interest on the sums due from the date ascertained at the maximum legal rate and reasonable attorneys' fees, expert witness fees, costs of arbitration (including arbitrator fees), and court costs associated with such enforcement or with enforcement of any judgment or arbitration award. As used herein, maximum legal rate means the higher of (a) 10 percent per annum or (b) 5 percent per annum plus the rate prevailing on the 25th day of the month preceding the date on which any sum becomes due, established by the Federal Reserve Bank on advances to member banks under Section 13(a) of the Federal Reserve Act (1913) as now in effect or hereafter amended.

COLLECTING ON BILLINGS

Probably the contractor's most frequently encountered problem, especially during times of recession, is the delinquent contract receivable. The old truism, "when money is tight, everyone is tight" becomes all too real. In difficult economic times, delayed payment becomes the rule, not the exception. This problem is compounded during periods of extremely high interest rates, such as those that occurred during the period 1970–1982. The collection of money not only took longer then, but was also more expensive in real dollar terms.

The following is a discussion of certain aspects of the prelitigation collection process, focusing specifically on the use of contract clauses and other collection techniques to leverage contract payments from customers.

The Importance of Contract Terms

Most construction contracts have numerous provisions regarding the contractor's obligations and liabilities for the work, but few provisions regarding his rights. Of the latter, probably the most important provision is the right to payment. Obviously, payment is the quid pro quo for the contractor's continued performance of work.

From the contractor's viewpoint, carefully drafted payment terms serve these three purposes:

1 They nail down the contractor's rights in the event of nonpayment (e.g., the right to suspend performance).

2 They provide insurance against the cost of collection (e.g., interest and legal expense).

3 They provide leverage.

The leverage springs from the principle that written contracts, which provide market level interest on late payments and recovery of attorneys' fees for collection, are enforceable. Thus, the reluctant or stubborn debtor faces a larger exposure if a billing remains unpaid. In fact, he may wind up paying the bill, interest, and attorney's fees plus his own legal expenses.

The following discussion examines these three topics: suspension of work; interest; and attorneys' fees.

Suspension-of-Work Rights

If the owner fails to make progress payments for an extended or unreasonable time period, flatly refuses to pay, or repudiates the contract, the contractor clearly has the right to any of the following remedies: (1) terminate performance; (2) rescind the contract; or (3) sue for the reasonable value of the work performed (California Civil Code § 1511; *Integrated Inc. v. Alec Fergusson Elec. Contractor*, 250 Cal. App. 2d 287, 58 Cal. Rptr. 503 (1967)).

However, these basic remedies have several drawbacks. First, if the contractor waits for nonpayment to reach the unreasonable level, he will continue to pour additional money into the job, thus increasing both his investment and his exposure. Second, if the contractor rescinds and sues for reasonable value, he cannot recover lost profits. (*Porter v. Arrowhead Reservoir Co.*, 100 Cal. 500, 35 P. 146 (1893)). Finally, if the contractor terminates, the owner will invariably attempt to set up backcharges and other counterclaims in defense, in addition to arguing that the contractor wrongfully repudiated the contract because the withholding of payments was justified.

These problems are often minimized by providing an explicit term in the contract giving the right to suspend performance temporarily for *any* failure to make progress payments; such terms are enforceable (*Big Boy Drilling Corp. v. Etheridge,* 44 Cal. App. 2d 114, 111 P.2d 953 (1941)). These provisions avoid the questions of, Who repudiated first?, and How long do I pour money into this job without getting paid? In addition, the contract can expressly provide for the right to recover lost profits upon termination.

By providing expressly for the right to suspend work, the unpaid contractor can present an immediate threat to the owner. His project may be shut down and his lender may cut off further advances. A looser, but frequently used clause, is the AIA standard form of general conditions, Section 9.7.

Interest on Delinquent Payments

During periods of high interest rates, creditors become even more pain-fully aware of the cost of money—particularly in the construction industry where owners use borrowed funds to finance jobs in progress. For this reason, there is a great temptation for owners to withhold payment of receivables until the last possible moment. For the unscrupulous, there is the opportunity to refuse payment over trivial disputes and invest the withheld payables in money market accounts until the dispute is resolved.

To some extent, the latter is cultivated by law. In the absence of an express contract clause, the law, at least in some states, limits the amount of interest recoverable for overdue contract payments to the legal rate. Thus, the unscrupulous creditor could, in times of high interest rates, invest monies otherwise due and make a profit on any spread between the legal rate and the rate paid in the marketplace. In some states, this legal rate applies only where there is no express contract clause. If the contract provides for a different rate, it can be used to calculate interest from the date of the obligation until a judgment is rendered in a lawsuit (California Civil Code § 3289 (West 1970)). After a judgment, a different rate may be provided at law.

For the foregoing reasons, it is important that the payment clause in a construction contract includes a provision for realistic interest rates. The contract rate, in turn, is limited by local usury laws.

Recent revisions of the state usury laws raise several questions. The first is, Are overdue contract receivables loans or forebearances and thus subject to these laws? Second, When is the applicable discount rate determined? Third, What is the penalty for providing a higher rate of interest in the contract?

Loan or Forbearance?

In *Crestwood Lumber Co. v. Citizens Savings & Loan,* 83 Cal. App. 3d 819, 148 Cal. Rptr. 129 (1978), the court deemed that the withholding of payments due was a forbearance. In this case a lumber company had included a provision in its invoices that provided for a charge of 1.5 percent per month (18 percent per annum) on overdue accounts. The court found this provision to be a forbearance subject to the usury law, since the terms of sale clearly provided for payment on a certain date and interest was provided only in the event of default.

This situation should be distinguished from a bona fide sales on credit. If, for example, the contract calls for a sale of real estate with the price to be paid in periodic installment payments over time, the arrangement is not a loan or forbearance subject to the limits in the usury law.

The *Crestwood* case also stated that interest-on-default clauses were a form of liquidated damages and could be subject to certain language requirements. Although this conclusion is debatable, the safest course is to provide such language in the contract clause.

Determination of Maximum Rate

The second question, the determination of the applicable date for establishing a discount rate, is answered in California by the California Usury Law. This law provides: "Five percent per annum plus the rate prevailing on the 25th day of the month preceding the earlier of (i) the date of execution of the contract to make the loan or forebearance, or (ii) the date of making the loan or forebearance. . . ." Similar statutes are in place in many other states.

In the case of contract payments, the relevant date for determining an interest charge in a state with a statute similar to the one above is the 25th of the month before the invoice becomes due. Rates can be obtained directly—usually by telephone—from the local Federal Reserve Bank. In addition, comparable rates are published daily in the Wall Street Journal.

Effect of Usury

The usury statutes of most states specifically provide a penalty against parties attempting to enforce a usurious interest clause. Not only is this clause unenforceable, but the opposing party can recover a variety of other penalties, such as triple the amount of interest paid or complete forgiveness of the debt. For this reason a few words of warning are in order:

1 Do not use the prime rate to determine interest; it may exceed the allowable amount.

2 Do not use arbitrary numbers such as 1.5 percent per month. This rate currently exceeds the legal maximum in some states.

3 Do not compound interest. The rates in most usury laws are simple annual rates. Compounding may lift an otherwise legal rate over the threshold of usury.

Recovery Of Legal Expense

As a general rule, in the absence of a statute or of a contract clause specifically providing for payment of attorneys' fees, they are not recoverable in a civil action. However, clauses in a construction contract specifically providing for reasonable attorneys' fees are enforceable if properly alleged in the lawsuit. In most cases, if the clause provides for payment of attorneys' fees in favor of one party only, the contract will be read to mean that either party may recover attorneys' fees.

Unless the contract provides for the recovery of attorneys' fees, collection of an account receivable may be substantially reduced by legal expense. Outside counsel frequently charges flat percentages in collection matters. These often vary from 25 percent for a pretrial recovery up to 40 percent for recovery after trial plus any out-of-pocket expenses—such as depositions, filing fees, and so forth. In cases where liability is seriously disputed at trial, the legal expense can go even higher. For these reasons, it is important that a construction contract include a clause for attorneys' fees that provides for their recovery in the event that the owner defaults in payment.

The same rule applies in actions against payment bond sureties on federal construction contracts where suit is brought pursuant to the Miller Act (also known as the Federal Bond Act of 1935) (see *U.S. ex rel. Carter Equipment Co. v. H.R. Morgan, Inc.*, 554 F.2d 164 (1977)).

In the case of suits against state and municipal agencies on public work payment bonds, reasonable attorneys' fees can be recovered even in the absence of a contract clause.

Collection Techniques

Involvement in a collection suit will often highlight the need for some basic business procedures, the most important of which is proper paperwork. Written contracts for each job are a must, since only through them can rights be assured. When setting up a job file, subcontractors should promptly provide for the mailing of any prelien notices required to protect lien rights. Billing dates required under the project should be scheduled in advance, and the billings should be issued on a timely basis; late bills get paid late. Finally, the responsibility for collecting on billings should be assigned to one person.

While the job is in progress it is important to anticipate payment problems before they occur. Apart from the obvious case of the bounced check, there are other warning signs: Are other trades complaining about a lack of payment? Is the owner getting paid by the lender? A phone call can frequently provide enlightening answers to these questions. On government projects, subcontractors can obtain information about the prime contractors payments through a request to the owner pursuant to the Freedom of Information Act (1974).

The assigned credit person should follow up on each invoice at the due date specified in the contract. If the owner provides excuses, letters and phone calls should become a daily ritual. Once a payment is overdue by more than the period predefined in time by management—for example, ten days—it is time to apply leverage. If the contract is drafted properly as was suggested earlier, the contractor can do the following: (1) send an invoice for interest costs; (2) advise the owner that it will be liable for payment of the attorneys' fees; and (3) if necessary, suspend work temporarily. These efforts tend to get attention.

After the job is finished, a final progress payment and release of retention will probably be due. If there appears to be any problem of collectability, a mechanics lien or stop notice should be filed before the applicable deadline runs out; this deadline can be as short as 30 days after notice of completion for subcontractors and material men.

In the case of a true delinquency, there is no reason to wait for legal action. Some credit managers flatly state that an account that becomes overdue by more than 90 days stands a poor chance of collection. As a general rule, the older an account, the more uncollectible it becomes. In other words, if legal action is called for, it should be commenced promptly.

As an alternative, there may be cases in which the owner is willing to provide security for an overdue account. For example, an overdue receivable could be converted into a note secured by a deed of trust on other property, with equity sufficient to cover the amount of the debt. A personal guarantee from the officers or shareholders of a corporate customer is another alternative. In either event, it is important to include provisions for recovery of interest and attorneys' fees in any written instrument drawn up by the parties.

Effective collection practice is summarized by two primary principles:

1 It is vital to move faster than one's competing creditors.

2 It is necessary to get security for the debt.

The collection process can often be frustrating, and it is always expensive. Proper contract provisions reduce the contractor's exposure to excessive collection costs and, in addition, provide additional leverage for collecting money prior to litigation.

PRICE INCREASES FOR CHANGES AND EXTRAS

The most frequent claim raised on the job-site is the quotation for extra work or changes. The following discussion is an examination of the practical methods for recognizing and negotiating such claims, and the contractual requirements for entitlement to compensation and time extensions. Many of the rules and techniques discussed herein apply equally to prime contractors and subcontractors. However, there is a significant difference in the negotiating tactics used by these parties, as is discussed below.

Payment for extra work and changes is assumed to be authorized by an express term in the contract. The materials herein do not treat claims based on implied contract terms such as constructive changes, acceleration, and defective specifications; nor are noncontractual theories such as negligence and breach of the implied covenant of good faith and fair dealing explored here.

Recognizing Changes and Extras

Kinds of Changes. The AIA Standard Form of Prime Contract provides a common definition of changes: "The Owner ... may order changes in the work within the general scope of the contract consisting of additions, deletions or other revisions. . . ." This definition is deceptively simple. Some distinctions should be made.

- *Extra work vs. change in work.* Extra work typically increases the scope of work. For example, an architect may write a letter stating, Extend AC paving 30 feet west to edge of landscaping. By contrast, changes usually alter materials, methods, or designs. For example, the architect may order a different mix for concrete, redesign the mechanical system, or prescribe a different finish.

- *Cardinal changes.* Is the contractor required to perform every change order? The answer is a qualified no. Where changes are of great magnitude the contractor's further performance is excused, as long as he negotiates in good faith for a price increase. The change of great magnitude, or cardinal change, is treated as so major a violation to the contract that the contractor cannot be forced to do the work. Federal Administrative Boards and Courts have held that in the construction context, cardinal changes excuse further performance by the contractor and need not be followed (see, e.g., *Keco Industries, Inc. v. United States*, 176 Ct. Cl. 983, 364 F.2d 838 (1966)).

- *Impact of change and extra work—Delays/acceleration.*
  a. Any significant extra work or change in the work may simply take more job-days to perform. Changes in material may stretch out the lead time needed to procure materials or may require a resubmittal; these situations can affect performance time. The AIA Contract Form recognizes this.
  b. The owner's refusal to grant additional time when warranted may cause an acceleration of schedule by requiring the contractor to perform an increased amount of work in a given time. Such events can have an impact on productivity by requiring overtime, inefficient crew loading, and/or extra supervision. Therefore, time requirements must be considered when pricing extra work or changed work.
  c. Caution: The owner in some states can require a contractor to provide reasonable written notice of delays and the intent to claim a price increase or time extension. The AIA Standard Form requires such notice within 20 days of commencement of delay. Failure to request additional time or compensation for delays when the contractor signs the change order may result in a waiver of any further rights based on delay.

- *Interference disruption.* A great number of changes or extras can significantly alter the sequence in which various trades perform their work. For example, a redesign of the mechanical system could change the completion sequence of several trades. Such changes, in turn, may affect overall completion time and/or productivity. These "impact" claims are difficult to negotiate at the end of the job and should be considered in the very beginning when major changes and extras are priced for quotation.

- *Ripple effect.* The term ripple effect was coined by procurement lawyers to describe the impact of a great number of changes that taken individually are small but, in the aggregate, have a substantial effect on the overall performance plan. For example, a great number of small changes can: (1) slow down the submittal process and therefore delay completion; (2) require increased engineering and supervision by the prime contractor; or (3) interfere with the sequence of performing the work. These changes are particularly difficult to recognize in the front-end of the job.

  To avoid waiver problems, some contractors who suspect a ripple effect add exclusionary language on their quotation form, stating in substance that this quotation does not include any allowance for the possible impact of subsequent changes that together with this change alter performance requirements and affect cost.

Detecting Changes on the Job

- *Drawing revisions and clarifications.* Architects frequently issue revised drawings, or clarifications, after the bid is submitted and work has commenced. These items may, in fact, contain extra work or changes in the bid specifications. The contract should be carefully compared with the bid drawings and specifications to determine if a change has been authorized.

- *Oral instructions.* An architect's letter directing extra work is easy to recognize as a change. However, many changes result from the informal directions given on the job-site by inspectors, owners, and others with apparent authority. Oral orders usually raise contractual issues. A more fundamental problem, however, is training job-site personnel to recognize these directions for what they are—changes in the contract that may entitle the contractor to a price increase, time extension, or both. These orders can be deceptively simple. For example, an owner may say, Can't you change the finish on this pad? A superintendent, wanting to oblige the owner, may change the finish without considering the cost or time consequences of following this direction.

- *Pre-job meetings.* On larger projects, the pre-job meeting between owner and prime contractor, or prime and subcontractors, is often the forum for changes in contract procedures and the work. For example, the owner may adopt a practice of issuing field changes—particularly where design is not fully completed—or the architect may take a second look at the bid package and express a clarification, which frequently requires changes from what was bid. Persons attending these meetings should take extensive notes and recheck the specifications and drawings afterwards to determine if any changes or extras have been authorized.

- *Using checklists and diaries.* In the rush of a busy construction day, superintendents may forget the directions given earlier in the day. All superintendents should be given a checklist and a notebook to be used at the end of each day to check for changes and other job conditions (see Figure 16-1). Similarly, estimators may rush out a quotation for changed work without considering time-related issues such as delay/ overtime, or factors that affect productivity in performing a job. A simple checklist can avoid such problems (see Figure 16-2).

- *The change audit.* On major projects, some contractors have weekly meetings with the estimator and the superintendent to review overall job progress. This is an ideal opportunity to review the status of any changes or extras issued during the week. By using the checklist and diary as a starting point, project managers can encourage field personnel to focus on changes and ensure that appropriate contract procedures have been followed where authorized.

- *The cost method.* Once a contractor experiences cost overruns, he may discover too late that they result from changes or extra work previously performed. Few cost systems are so sophisticated that they can provide simultaneous notice of increased labor/matter costs for any element of work. The delay can result in contractual notice problems as well as negotiating problems if the project is in its final stages.

(text continues on page 16-22)

FIG. 16-1 Daily Log Report

DATE: _____ JOB NAME: _____

LOCATION: _____

1. Weather conditions:

	8 A.M.	1 P.M.
Clear	_____	_____
Cloudy	_____	_____
Rain	_____	_____
Snow	_____	_____
Temperature	_____	_____

2. Accidents. (Attach accident report.)

3. a. Work accomplished by forces (in detail).

_____

_____

☐ Base or ☐ Extra

b. Work planned but not performed by our forces (in detail, and include reasons for delay).

_____

_____

4. a. Subcontractors on job and work accomplished (in detail).

_____

_____

b. Work planned but not performed by subcontractors (in detail, and include reasons for delay).

_____

_____

5. Overtime worked by our forces? ☐ Yes ☐ No

a. Who authorized? _____

b. What work performed? _____

c. Payment approved? ☐ Yes ☐ No

d. Adjustment for lost productivity requested? ☐ Yes ☐ No

6. Materials delivered to job-site for use by our forces:

_____

Acceptable quantity? ☐ Yes ☐ No

7. Materials delivered to job-site for subcontractors. (Will use monthly for partial payments.)

_____

8. Work performed by us for subcontractors (e.g., scaffolding erected, unloading, crane rental, clean-up). Number of men, hours, and equipment involved. (Will be used to determine backcharges.)

_____

_____

9. Extra work of subcontractors authorized by superintendent (in detail, verbally or otherwise).

_____

_____

10. Any **changes** made by work order, verbal instruction, or otherwise? ☐ Yes ☐ No

   a. Changed conditions (e.g., overcrowding by other trades, site conditions different from drawings)

_____

   b. Changed schedule (e.g., "holds" on work, orders to accelerate, break in normal sequence)

_____

   c. Change in scope (e.g., actual layout different from drawings, increase in quantities required)

_____

   d. Change in specifications (e.g., engineer's interpretation different from specs, inspector too picky)

_____

   e. Formal changes (e.g., written change order, word order)

_____

   f. If changes, itemize:

     —Who authorized? _____

     —What work performed? _____

     —Men and hours used: _____

     —Does it cause delay or additional cost? ☐ Yes ☐ No

   g. If changes, have we received a written change order? ☐ Yes ☐ No
If not, have we acknowledged with acknowledgment form? ☐ Yes ☐ No

(continued)

FIG. 16-1   (continued)

---

11. Any conversation meeting with customer or engineer?   ☐ Yes   ☐ No
If so, who, what, when, why? Any cost impact? Review item 9 again.
Confirm important conference in letter.

_____

_____

12. I request that the following letters or telegrams be sent by
_____ to _____ (government, architect,
subcontractors, suppliers):

_____

_____

13. I request that the following material be ordered:

_____

_____

14. The following request for interpretation of specifications or plans was made
to _____:

_____

_____

15. Discrepancies noted (between specifications and plans or between one
subcontractor's work and our work, etc.):

_____

_____

16. Other comments (e.g., visits to job-sites by representatives of government,
architect, etc., and comments made; problems with inspectors or
subcontractors; test failures; large layoffs or hiring; unusual occurrences;
description of job progress):

_____

_____

Report submitted by _____

FIG. 16-2   Checklist for Change Orders

Estimating the costs of change orders must frequently be based on factors that do not pertain to estimating the normal job and are therefore overlooked. The following should be taken into consideration:

1. In a multi-story building it costs more to do work on the higher floors.
2. When the proposed change order affects certain work that is partially complete, it is necessary to work around the change area until the nature and extent of the change have been determined.
3. Is adequate manpower available? Is overtime necessary?
4. Weather may play an important part.
5. Are the plans and specifications up to date? Otherwise, there could be a delay.
6. Does the change require installing or moving scaffolding?
7. Does the change require protection of completed work or finished surfaces?
8. Is the job being done at a time or under circumstances calling for daily cleanup?
9. Is there likely to be interference from owners, visitors, or other contractors?
10. Does the size of the work area create problems? If it is too large, time could be lost in transferring materials. If it is too cramped, efficiency may be lowered.
11. Is the lighting or ventilation poor?
12. Will the work be done in cramped spaces where it will be necessary to work around pipes, wires, and duct work?
13. Is the work to be done under hazardous conditions? If so, it will probably be slower and more costly.
14. Will delays and inefficiencies caused by the changes affect job morale to the point of cutting productivity?
15. If overtime is required, will it be so long and so hard that it will cause fatigue, cut down on productivity, and increase the chance of accident or errors?
16. Will starting and stopping or partial shutdowns cut down job momentum that must be regained? If so, how much time will be lost, and what will it cost?
17. Will the disruption of scheduling cause the subcontractors to interfere with each other's work?
18. If manpower must be reassigned because of change, will it be extensive enough to affect productivity?
19. Is the change extensive enough to require new crews to be organized? If so, how much productivity will be lost bringing the new crews to full efficiency?
20. If new crews are needed or old crews on the change order need more supervision, will the supervision on the balance of the work be so diluted as to be costly?
21. Is there an increased likelihood of costly mistakes under the new pressures created by the change? If so, how should these errors be priced to get their cost into the change estimate?

(continued)

FIG. 16-2   (continued)

22. What problems will the change cause in the costs of bringing together men, materials, and equipment? In short, is the change in scope sufficient to affect the logistics of the job?

23. Will the change cause a delay that runs the job into a period covered by new labor agreements with higher wage rates or fringe benefits or a period in which newly enacted tax laws take effect, increasing the cost of the work?

24. Change orders may cause a delay that increases the cost of off-site storage.

25. Every delay increases the dollar amount of general overhead applicable to the job.

Compliance With Contract Amendment Procedures

Written Order Requirement. Almost all modern firms specifically require a written change order to authorize the performance of the extra work, and preclude additional compensation if the contractor proceeds without it. The courts generally uphold such clauses on the basis that the owner has a legitimate interest in limiting the consideration that he is to pay for work performed under the contract (see *Wunderlich Contracting Co. v. United States,* 351 F.2d 956, 173 Ct. Cl. 180 (1965)).

Exceptions. Where the owner has verbally authorized extra work, which the contractor performs, and then refuses to pay for it, an obvious injustice exists. The courts have recognized this and carved out a number of exceptions to the general rule.

- *Waiver.* Where the contractor repeatedly performs extra work on verbal requests and the owner subsequently issues written orders and pays for them, the owner generally is regarded to have waived the clause requiring these formalities as a condition precedent to payment for the extra work.

- *Estoppel.* If the owner asks the contractor to perform additional work and verbally agrees to pay for it without requiring a written order, and the contractor performs the work in reliance on that promise to pay, then the owner may be estopped from enforcing any contract requirement for written authorization.

- *Rescission.* The parties by their conduct may be deemed to have verbally rescinded the provision that all change orders be in writing. For example, in one case numerous oral changes constituted more than half the value of the original contract. In this case, the Court found that the parties had abandoned the written contract and entered into a new oral agreement (*Daugherty Co. v. Kimberly Clark Corp.,* 14 Cal. App. 3d 151 (1971)).

- *Oral modifications.* An oral authorization to perform extra work, which is then performed, may be construed as a modification to the contract (*Ferelli v. Weaver*, 210 Cal. App. 2d 108 (1962); *MacIssac & Menke Co. v. Cardox Corp.*, 193 Cal. App. 2d 661 (1961)). Some statutes permit a contract in writing to be modified by an oral agreement supported by new consideration *unless the contract otherwise expressly provides*, and provided the statute of frauds is satisfied. Since some standard contract forms *expressly* preclude recovery for changes without written authorization, it is not yet safe to rely on this exception to the general rule.

Public Works Differences. In public works contracts, the courts are usually less willing to find a waiver or estoppel of contract clauses designed to protect the public treasury where there has been no express approval by a public board.

Protective Procedures

- *Written quotation.* Obviously the best method of complying with the contract is to get advance approval of a written quotation. Written approval by the architect/owner normally satisfies contract requirements. However, in the "real world," quotations are often disputed or there is insufficient time for preparation and approval in advance of performing the work.

- *Request price-determined-later (PDL) letter.* If time is the only problem, contractors could request the architect/owner to provide a written letter directing the change with the price to be determined later. The AIA Contract Form expressly provides for such a procedure. While the contractor may be unhappy without a firm price for the extra work, this method at least preserves his right to obtain compensation later.

FIG. 16-3   Form of Change Notification Notice

TO: _____ DATE: _____

Re: Project _____

    Contract/Subcontract No. _____

    Change Estimate No. _____

Gentlemen:

This will confirm our agreement with your authorized representative, _____, to modify our contract to perform the following additions, deletions, or revisions in the work on the terms stated.

Description [*identify documents, if any*]:

_____

_____

_____

Price increase (decrease): _____

Extension of completion date: _____ days _____

In reliance on your instructions to proceed without a written change order, we are performing this additional work. If your understanding is otherwise, please advise us immediately.

Very truly yours,

By: _____
Project Manager

- *Send a notice form.* If the owner/architect refuses to issue a letter recognizing that the work has been changed, the contractor has two choices: (1) he can refuse to perform the extra work; or (2) he can create a record. The former may be undesirable, since it frequently leads to poor job-site relations. The latter alternative presents the risk of noncompliance with contract procedure.

  To minimize the latter, the contractor should send a confirming memo to the owner that specifies the oral agreement and states that the contractor is performing the work in reliance thereon unless the owner advises him otherwise (see Figure 16-3 for a sample form).

NEGOTIATING CHANGES AND EXTRAS
Documentation

- *Complete quotation of costs.* The beginning point for any negotiated change is a full and complete quotation detailing all the costs of performance and any time extensions required. Full disclosure is persuasive and avoids delay by requests for more information. The AIA General Conditions require substantial detail (see Figure 16-4).

- *Separate code numbers and files.* A key to good documentation is to assign a separate number to each change on a job in chronological sequence—for example, Job No. 089, Change No. 10. This facilitates the keeping of separate files and separate cost accounting. For major changes, a file should be set up on each change so that appropriate correspondence, drawings, material receipts, payroll cards, and subcontract invoices can be segregated. This file can be used to answer any questions during the negotiating sessions and, furthermore, can preserve evidences that may be needed in subsequent legal proceedings.

- *Visual aids.* In many cases the use of the drawings, schematic sketches, and photographs are valuable tools in negotiations. This is particularly true when the changes are negotiated "after the fact" with owners' representatives who may not be entirely familiar with the drawings or the site.

Early Offensive

With few exceptions, an early resolution of changes is preferable. If disputed changes are left open until the end of the job, construction funds may be used up, the owner may be unhappy about defects, or there may be other disputes that poison the atmosphere for successful negotiations. Furthermore, early in the job the contractor has leverage, since the owner needs his performance. After submission of a quotation, the change request should be scheduled for follow-up on a weekly basis.

Tactics

- *Beware of releases.* Contractors are sometimes called upon to give partial releases to obtain progress payments; frequently, they need to give full releases to obtain final payment of retention funds.

(text continues on page 16-29)

FIG. 16-4   Change Estimate Form

DATE: _____   JOB NO. _____ CHANGE ESTIMATE NO. _____
TO: _____
ATTN: _____   JOB NAME: _____
SUBJECT: _____

Gentlemen:

We transmit herewith the referenced change estimate. Such change estimate is based upon information received from your authorized representative, and is subject to the following conditions:

☐ No change in contract price.

☐ Addition of $_____ in contract price.

☐ Deduction of $_____ in contract price.

☐ Cost of preparation of change estimate is $_____.

☐ _____ days extension of contract time required; retention shall be extended only on value of change estimate.

☐ Change estimate will increase if work is required at overtime rates.

☐ Change estimate is good until _____; after such date, reevaluation will be required.

☐ Work cannot proceed until _____ copies of approved change estimate are returned.

☐ Work is in process as directed.

☐ Additional information will be sent by _____.

                                ASSOCIATED GENERAL
                                CONTRACTORS

                                By: _____

DATE: _____ JOB. NO. _____ CHANGE ESTIMATE NO. _____

JOB NAME: _____

DESCRIPTION OF CHANGE: _____

_____

_____

_____

SUMMARY

A. Total material $ ____

Sales tax ____

B. Labor (including union fringes and payroll taxes) ____

C. Zone pay and/or subsistence ____

D. Equipment ____

E. Engineering and/or support services ____

F. Subtotal $ ____

G. Overhead (10% of line F) ____

H. Total (10% of lines F + G) $ ____

I. Profit (10% of line H) ____

J. Total (10% of lines H + I) $ ____

K. Subcontracts ____

L. Profit (10%) ____

M. Total (10% of lines K + I) $ ____

N. Other direct (see attached breakdown) ____

O. Total price (debit/credit) (lines J + M + N) $ ____

_____

DATE: _____ JOB NO. _____ CHANGE ESTIMATE NO. _____

A. Materials:

Equipment per attached form (@ blue book rate) $ ____

Material per attached listing @ standard ____

Discount on material/Net ____

Special expediting or freight ____

Handling charges on returned goods ____

Total (to Summary, Item A) $ ____

(continued)

FIG. 16-4   (continued)

B. Labor:

	Carpenter	Laborer	Operating Engineer	Other
Total installation hours (from take-off sheets)	_____	_____	_____	_____
Labor multipliers (factors)				
_____	_____	_____	_____	_____
_____	_____	_____	_____	_____
_____	_____	_____	_____	_____
_____	_____	_____	_____	_____
Total hours	=====	=====	=====	=====
1. Base pay	_____	_____	_____	_____
2. Vacation pay	_____	_____	_____	_____
3. Subtotal	_____	_____	_____	_____
4. Union fringes	_____	_____	_____	_____
5. Total hours	_____	_____	_____	_____
6. Payroll taxes	_____	_____	_____	_____
7. Subtotal	_____	_____	_____	_____
8. Foreman	_____	_____	_____	_____
9. Subtotal	_____	_____	_____	_____
10. Insurances	_____	_____	_____	_____
11. Subtotal	_____	_____	_____	_____
12. Retrogression or escalation	_____	_____	_____	_____
13. Total	_____	_____	_____	_____

Labor Summary:

14. Carpenter $_____

15. Laborer   _____

16. Operating Engineer   _____

17. Other   _____

18. Total Labor $_____
(to Summary, Item B)

C. Zone Pay and/or Other Travel Expenses:

Travel $_____

Subsistence   _____

Premium cost   _____

Parking: _____ days @ _____/day   _____

                                        Total $_____
                                (to Summary, Item C)   ======

D. Equipment, Tools, Facilities (per blue book):

_____ $_____

_____   _____

_____   _____

                                        Total $_____
                                (to Summary, Item D)   ======

E. Engineer and/or Support Services:

_____ $_____

_____   _____

_____   _____

                                        Total $_____
                                (to Summary, Item E)   ======

K. Subcontracts—At Quoted Cost:

_____ $_____

_____   _____

_____   _____

                                        Total $_____
                                (to Summary, Item K)   ======

- *Beware of final payments.* Many contract forms provide that the acceptance of final payment constitutes a waiver of any further claims for compensation.

- *Preserve all remedies.* As a general rule, work performed at the owner's specific request, even without formal change order, may be recoverable in a mechanics lien or payment bond action on a quantum meruit theory. However, such security rights are subject to stringent deadlines. Other remedies, such as arbitration, may be conditioned on submittal within specific time limits. Thus, the contractor, although paid for base work, must preserve the remedies regarding any claims for changes and extras by meeting the appropriate deadline.

HANDLING SUBCONTRACTOR CHANGE ORDERS

Use of Flow-Down Clauses in Subcontracts

The biggest problem facing prime contractors with claims for change orders by subcontractors is the frequent inconsistency between prime contract clauses and subcontract clauses. The subcontract may have more generous deadlines for notice of claim or provide for a different method of resolving a dispute (e.g., arbitration instead of litigation). To avoid getting caught in the middle, the prime contractor should include a special clause in his subcontract, which provides that certain terms of the prime contract flow down and bind the subcontractor. Such a clause could incorporate: (1) time limits for submission of change orders; (2) special methods of dispute resolution; and (3) a requirement for a complete release prior to final payment (see Figure 16-5).

Processing Subcontractor Claims

The principles of recognition, notice, and documentation previously discussed apply to subcontractor change orders. For the subcontractor to obtain the benefit of any negotiations between the prime contractor and the owner, where the change is under the prime contract, he must submit his quotation and provide supporting material in time for the contractor to incorporate it in his own quotation to the owner. Similarly, the prime contractor should not ignore subcontractors but should notify them of any deadlines or meetings so that all relevant subquotes can be presented.

Strategy

- *Prime contractor.* The prime contractor's first inquiry regarding a subcontractor claim for change/extras is to determine whether the change is recoverable under the prime contract so that it can be passed on to the owner. If so, then the prime contractor can obtain from the subcontractor appropriate documentation for processing the claim. If the subcontractor does not have a flow-down clause, the prime contractor should obtain a separate claim liquidation agreement, which binds the subcontractor to any decision reached in the dispute resolution procedure in the prime contract.

FIG. 16-5   Sample of Typical Flow-Down Clauses

CHANGES

Contractor may, at any time, by written order and without notice to the sureties, make changes in the Work, and Subcontractor shall proceed with the Work as directed. If the same causes an increase or decrease in the contract price or in the time required for performance, an equitable adjustment shall be made and this Agreement shall be modified in writing accordingly, all subject, however, to the terms of Section 29 hereof. But nothing herein contained shall excuse the Subcontractor from proceeding with the prosecution of the Work as changed. Before proceeding with any change, deviation, addition, or omission, the Subcontractor must first obtain written authorization from the Contractor.

CLAIMS, DISPUTES, AND ARBITRATION

(a) Regardless of any claims, disputes, or action taken or agreed to be taken under this Section 29, Subcontractor, as provided for in Section 19 hereof, shall at all times proceed diligently with its Work as specified and as directed by Contractor.

(b) Subcontractor agrees to submit all claims against Contractor, including but not restricted to claims for extensions of time, damages, extra work, or materials, to the Contractor within the same time limits and in the same manner as provided in the General Contract for similar claims of the Contractor upon the Owner. Contractor shall submit such claims to Owner and prosecute same as provided for in the General Contract. Subcontractor shall cooperate and assist in the preparation and prosecution of all such claims and shall pay or reimburse Contractor for all expenses and costs including but not restricted to Costs of Litigation incurred by Contractor in connection with the preparation and prosecution of such claims. Contractor shall never be liable to Subcontractor on any such claims in any dollar amount greater than Owner finally pays to Contractor in connection therewith. Accordingly, Subcontractor expressly consents to be bound to Contractor to the extent that Contractor is bound to Owner by all decisions and determinations made in accordance with any procedure for the resolution of claims expressly provided for in the General Contract, whether or not Subcontractor is a party to any such proceedings. The terms of this Section 29(a) shall be binding upon Subcontractor whether or not Subcontractor records or files a mechanic's lien or stop notice or prosecutes suit thereon and/or on any bond posted by Contractor; Subcontractor hereby acknowledges that this Section 29 waives, affects, and impairs rights it would otherwise have in connection with such liens, stop notices, and suits on bonds.

(c) In the event the General Contract requires that any claim by Subcontractor against Contractor that Contractor submits to Owner, as provided in Section 29 (b), be decided by arbitration, or the Contractor is given the right therein to require arbitration and so elects, such procedure shall be diligently commenced and prosecuted as provided for in the General Contract. The commencement of such arbitration proceeding shall bar the right of either party to commence or prosecute litigation in connection with such claim(s), including litigation in connection with mechanic's liens or stop notices recorded or filed by Subcontractor or bonds posted by Contractor, except to confirm the award of the arbitrators. If any person(s) or firm(s) other than Owner, Contractor, and

(continued)

FIG. 16-5 (continued)

Subcontractor are involved in the dispute(s) giving rise to Subcontractor's claims, or which might give rise to a claim by Subcontractor, and Contractor is required to or elects to arbitrate such other party's or parties' claim, then Subcontractor agrees to a joint arbitration with Owner, Contractor, Subcontractor, and such other parties, in one arbitration proceeding, which shall finally determine all of the claims and disputes. Contractor shall never be liable to Subcontractor on any of its claim(s) submitted to arbitration in any dollar amount greater than Owner is held liable to Contractor therein.

(d) If Contractor wholly or partially prevails in any litigations and/or arbitration with Subcontractor arising under this Agreement, the court or arbitrators, as the case may be, shall award to Contractor all Costs of Litigation, in addition to any other relief or recovery to which Contractor may be entitled.

(e) If Contractor believes that the commencement or processing of any arbitration or litigation prior to the completion of the Construction Project as a whole may interfere with or delay such completion, Contractor may order that such arbitration or litigation be postponed until such completion, and Subcontractor agrees to such postponement.

- *Subcontractor.* The subcontractor must determine whether his claim is against the prime contractor or the owner. If it is against the owner, then the subcontractor's timing will depend, to a large extent, on the prime contractor. If the change arises under the subcontract rather than the prime contract, the subcontractor should immediately prepare a thorough and comprehensive quotation of the change with supporting documentation for submittal to the prime contractor.

At the same time, the subcontractor must be careful not to allow any deadlines for filing bond claims, mechanics lien claims, or arbitration to run over. Such remedies are often desirable as a means of putting pressure on the prime contractor for settlement.

The subcontractor should continue to perform in accordance with all the terms of the contract so as not to give the prime contractor any counterclaim for delayed performance or unfulfilled warranty obligations.

SUBCONTRACTOR CLAIMS

As a general rule, construction subcontractors are entitled to assert many of the same claims for additional compensation to which prime contractors are entitled. In some cases, prime contractors have been known to invite subcontractors to participate in major claims against owners on such grounds as delay, acceleration, defective specifications, and so forth. In such cases, the prime contractors and the subcontractors act as a team with the prime contractors serving as quarterback in an overall effort to compensate all parties for the changes in scope or scheduling involved.

The more frequent case involves a subcontractor who, without knowing if relief is available from the owner, directs his claim against the prime contractor. In such cases the subcontractor, under modern case law, may have independent grounds for relief from the prime contractor. For example, the prime contractor himself may have interfered with the subcontractor's performance or failed to coordinate the work of all trades in an efficient manner. In a great many cases, there are disputes about payments.

In a typical case, the subcontractor has a greater variety of remedies available to him than the prime contractor, since the latter is often committed to a particular "Disputes" procedure in his prime contract and cannot pursue lien or bond remedies. This presents a serious danger to the prime contractor's legal posture, since he confronts the problem of defending two or more actions in different forums and, if he is pursuing his own remedy against the owner, he must also contend with the problem of making contentions in one forum that may be used against him in another.

In view of the above realities, this section outlines the subject of subcontractor claims from two different perspectives: (1) where the claim appears to be recoverable from the owner under the prime contract; and, (2) where the subcontractor's claim is the fault of the prime contractor. In each situation, the grounds for making a claim, the remedies available to the subcontractor, the procedure used to pursue the claim, and the tactics to be considered in each case are discussed.

Further, the outline focuses on claims for additional compensation under the contract or for breach of contract. These should be distinguished from collection problems, including the area of mechanics lien and bond remedies, which are an entirely separate subject.

CLAIMS UNDER THE PRIME CONTRACT

General Rule

As a general rule, any claim available to the prime contractor under his contract with the owner can flow down to the subcontractor if the subcontractor can show that his performance was directly affected by the owner's act or omission. For example, if the owner failed to make timely delivery of material for incorporation in the work and thereby caused delay to both the prime and the subcontractors, then the subcontractor could join with the prime contractor in a claim for delay costs.

Specific Claims Available

Claims available to the subcontractor are those based on some provision in the *prime* contract.

CLAIMS UNDER THE SUBCONTRACT

Scope Changes

Like most prime contracts, the typical subcontract contains a change clause that allows for the addition or deletion of work from the subcontract scope or changes in the work, such as specification changes. Claims for price adjustments under these clauses are submitted in normal change order fashion. In some cases, the dispute may involve work alleged to be in *another* subcontractor's scope of work. For example, an electrical subcontractor may contest an order to install ceiling light fixtures because the requirement was found in the acoustical section of the specifications. Resolution of these events turns upon an application of the rules of interpretation to the particular subcontracts involved. In such disputes, the work is usually in the prime contractor's scope, and the question is whether it was in a particular subcontractor's scope.

In order to avoid getting caught in the middle, the prime contractor should incorporate in his subcontracts *all* of the project's specifications and drawings, not just those designated for a particular trade.

Schedule Changes

A general contractor is under an implied obligation not to hinder or delay performance by a subcontractor and may incur liability for the latter's damages if he does not take all reasonable steps to insure that the job-site is ready and that work proceeds without undue delay.

A prime contractor has a further duty to avoid hindering or delaying a subcontractor and, thus, to coordinate the work of all subcontractors so that the work is efficiently done.

A prime contractor may avoid the liability discussed above by having specific provisions in his subcontracts allowing him to delay or alter the sequence of work without liability. Some authorities would sustain these provisions on the rationale that the prime contractor is obliged only to make reasonable efforts to avoid delay and interference.

Prime contractors often attempt to preclude liability for any delay in the construction work through use of "no damage for delay" provisions. Other subcontracts purport to limit the subcontractor's remedy to an extension of time only, rather than delay damages. If, in the case of the "no damage for delay" clause, the delay is caused by the prime contractor, the provision may be void in most states. In several cases involving prime-versus-owner claims for delay damages, the courts have very strictly construed exculpatory clauses to avoid a harsh result. The rationale should apply equally to subcontractor-versus-prime disputes.

Most subcontracts provide a force majeure clause, similar to that in prime contracts, which excuses the prime contractor from delay damages due to unforeseen conditions—such as severe weather, fires, and war. Courts generally enforce such clauses. For example, in one recent case the prime was not liable for damages to the subcontractor caused by delays attributable to subcontractor-surface conditions unknown to the prime at the time of the subcontract. A force majeure delay will probably result in the prime contractor granting a time extension to the subcontractor. In turn, the prime contractor should seek a similar extension from the owner.

Payment Disputes

Most subcontract forms tie the subcontractor's right to payment to the prime's receipt of payment from the owner. Traditionally, prime contractors have relied on these clauses to delay subcontractor payments until the owner has paid. However, recent cases indicate that absent a claim against the subcontractor, the prime may no longer exercise the withholding power without incurring liability for interest.

Backcharge Claims

A fertile source of prime-subcontractor disputes is the backcharge power. This simply refers to the prime's right to charge the contractor's account and withhold payments for items of work that the prime has been required to perform although the subcontractor appears to be obligated. Such disputes most often involve cleanup charges, punchlist items, and warranty repairs. The prime's right to charge for these claims is usually sanctioned by standard form subcontracts.

However, general principles of commercial law (UCC § 2-607) would seem to require that due notice of the alleged violation be given the subcontractor before the prime undertakes correction at the subcontractor's expense. In addition, the prime should be judicious in exercising his powers, since the withholding of monies due often precipitates lien and bond litigation by the subcontractor, which can cause negative reactions from the owner and surety on the project.

AVAILABLE REMEDIES

Public Contracts

Equitable Adjustment of Price and/or Time. In cases where the subcontractor can proceed against the public agency owner through the prime contractor, relief can often be found by claiming a price adjustment under the provisions of the prime contract. At the federal level, increased costs due to verbal extras, defective specifications, or acceleration can be recovered by the subcontractor, by a price adjustment under the various clauses of the prime contract, through an indirect appeal by or in the name of the prime (*Gardner Displays Co. v. U.S.*, 346 F.2d 585 (Ct. Cl. 1965)).

Federal Contracts

On federal construction projects, payment bonds are required for all public works exceeding $2,000 (40 USC § 170a(a)). Such bonds provide protection for subcontractors and suppliers in lieu of the mechanics lien protection available in private construction contracts. Since the bonds are remedial in purpose, the statutes governing them are liberally construed to achieve the purpose of protecting subcontractors and suppliers. The remedy extends only to persons having a direct contractual relationship either with the prime or a first-tier subcontractor. The remedy extends not only to amounts due under the contract price, but also to changes or extra work that had been performed even though there had been no corresponding adjustment of the contract price. In such cases the measure of recovery is the reasonable value of the work performed.

State and Local Contracts

Many states have local laws similar to the Miller Act (1935), which requires every prime contractor awarded a contract in excess of a certain amount by a public entity to post a payment bond. Any person, other than an original contractor, entitled to file a stop notice can make a claim against the payment bond on a public works project. Since stop-notice claimants include all those who can file a lien on private works, and the latter includes anyone furnishing labor or material to a project, the coverage of this type of statute is often broader than the Miller Act itself. The measure of recovery is the reasonable value of the labor and materials supplied.

Private Works Claims

Arbitration. In private works, the use of arbitration as a remedy is widespread. For example, the AIA Standard Form Subcontract includes a broad-form arbitration clause. The pros and cons of arbitration are subjects of frequent debate. As a general rule, arbitration tends to be faster and cheaper than litigation. There are some notable exceptions, however, such as the 1969 case in which a construction arbitration lasted 19 months and cost $400,000 in arbitration expenses, *exclusive* of attorneys' fees. Additionally, an appeal from an adverse arbitration award is extremely difficult to win.

On the balance, it appears that arbitration is a useful remedy for small technical disputes, such as those involving specification interpretations or those that are oriented toward trade practice—such as cleanup and punchlist disputes. However, arbitration is not suited to major delay claims where the stakes are high and effective discovery is a must.

Mechanics Lien, Stop-Notice Litigation, and Payment Bond. Subcontractors and suppliers have a number of security devices available to enforce their claims against the prime contractors. The one most often used is a mechanics lien. This remedy is often available to any person providing labor or materials to a project. Since the relief available is for the reasonable value of the labor and materials supplied, the claim can include any changes or extra incurred under the terms of the subcontract.

Those persons entitled to file mechanics liens are also entitled to file stop notices in some jurisdictions with both the owner and the construction lender financing the project. This usually means that the person receiving the stop notice is obliged to withhold funds or render himself personally liable for the claim if he fails to do so. Since this remedy also includes the reasonable value of the labor and materials furnished, the prime contractor can come under considerable pressure from the lender/owner to settle a subcontractor's claim.

When payment bonds are employed on private contracts they can be used in a procedure that is similar in effect to claims made against performance bonds.

Breach of Contract Litigation. As in any contractual matter, the subcontractor is free to enforce his claim by means of a breach of contract suit against the prime contractor.

Direct Litigation Against the Owner. If the subcontractor's claim is simply for nonpayment, the statutory stop notice and mechanics lien remedy are often the only available remedies. Bringing an action under any of the preceding sections does not prohibit the subcontractor from starting actions based on negligence of the owner or design professionals.

CHAPTER 17

Litigation Support

Roger B. Shlonsky

LITIGIOUS NATURE OF THE INDUSTRY

There are times when it seems that involvement in the construction industry ultimately leads to involvement in litigation. Perhaps it is the magnitude of the costs involved in construction projects or perhaps it is simply the nature of the elements involved in construction that provoke this situation. A surprising number of publications and newsletters are devoted to the subject of construction litigation. In fact, litigation in the industry has become so common that a specialized form of legal practice, involving a large number of attorneys across the United States, has evolved.

In any construction project, there are both contractors and owners. By definition, the owners are the parties or entities for whom a project is undertaken. As contractors themselves are often specialized, so too are the different types of owners. This chapter contains descriptions and considerations that are generally common to both owners and contractors.

Materiality

While construction projects may be large or small relative to each other, most projects are significant in terms of their total cost to the parties involved. The larger the construction project with its various components and costs, the more likely there will be disputes between the contractor and the owner that will result in litigation.

One way to quantify the construction contract disputes that occur is to determine the number handled by the American Arbitration Association (AAA). In 1983, the AAA reported 2,675 construction arbitration claims and counterclaims totalling $466,332,189. Of the construction claims reported for 1983, approximately half involved disputes between contractors and owners. These represent only a portion of the construction litigation that takes place in any given year, since most disputes are never sent to arbitration.

Types of Disputes

The nature of the construction project itself is a great contributor to the amount of litigation existing in the industry. New construction methods and materials, the need for construction projects to meet new environmental requirements, and innovations for use in an age of high technology increase the risk of prospective litigation. During the 1970s, the increased use of design-build approaches to construction, often referred to as fast-track, produced a large amount of disputes that resulted in litigation. Innovations and trends aside, however, it appears that litigation in the industry still revolves primarily around the same kinds of problems that it has for decades. Delays, performance problems, and changes or errors in specifications are the primary causes of disagreements between contractors and their customers.

This chapter describes the nature of litigation in the construction industry, the use of litigation support, the types of disputes, and the nature of damage considerations that contractors concerned with construction claims should be aware of.

NATURE OF LITIGATION

Construction litigation most often involves the subject of claims. Construction claims concern both those matters that cause increases in costs and the actual costs themselves. Whether involving matters that affect the contractor directly, such as labor, materials, or subcontractor's obligations, or indirectly, in terms of project and home office overhead costs, the subject is still one of costs.

To consider the nature of the litigation is really to consider the inherent problems that occur in the construction process. Generally accepted accounting principles (GAAP), cost accounting standards, and even various governmental contract accounting rules and requirements are all part of the accounting methods used for construction costs. Accounting defines, identifies, and assigns those costs.

Timing of the Litigation

Disputes arise often, and sometimes litigation begins while the contract is still in process. More often, the onset of litigation, which arises out of the problems occurring during the construction process, is delayed until the completion of the project or the termination of a contract. This occurs because both contractor and owner are concerned about efficiently completing the work even though they are aware of matters that will be litigated at a later time.

Construction litigation takes on additional aspects when it arises out of contract problems with governmental entities, particularly where these problems involve the federal government and its agencies. These peculiarities are discussed in detail in later parts of this chapter. In most respects, however, the problems that occur and the kinds of costs involved are no different than those involving construction projects with nongovernmental owners. Litigation involving government construction contracts is often more dramatic, however, because of the size and total costs involved in many of these projects.

Cost Measurement as Cause of Litigation

When problems occur in the performance of construction contracts, the economic cost to both the contractor and the owner becomes the subject of disagreement. Although the specific construction item is the same, the cost impact on contractor and owner is often very different. The contractor is primarily concerned with completing the construction project and with the costs involved in finishing the job. The owner is not only concerned about those costs, but also about the effects of the proper and timely completion in terms of the intended use of the construction product. Additionally, the owner is affected by extra costs related to the financing of the project. These extra costs specific to owners frequently constitute the largest part of claims against contractors.

Differences in Litigation Types

Construction litigation is often centered around actions brought by or against either contractors or architects and engineers. These actions differ principally in the subject matter of the litigation, that is, actions by or against architects and engineers are concerned primarily with design and specifications issues whereas actions by or against contractors are concerned primarily with cost items such as materials, labor, and overhead. The focus of this chapter is directed at contractor litigation issues specifically relating to cost items.

LITIGATION SUPPORT AND CONSTRUCTION CLAIMS

Litigation support is the assistance of persons who are informed or knowledgeable about the matters and related costs surrounding the nature of disagreements and claims. Litigation support in the context of the construction industry entails defining, understanding, and calculating the costs of the subject matter of the litigation.

Groups Providing Litigation Support

Those involved in providing support in construction litigation fall into the two basic groups: (1) employees of the contractor or owner; and (2) independent professionals who provide relevant services. Contractor employees who are important to understanding and dealing with claims include not only the project manager and the members of his staff who were involved in the actual project, but also the contractor's own engineers, purchasing department personnel, and accounting staff. The accounting staff consists of employees who understand the nature of the contractor's operating methods, operating systems, and more importantly, the details of the actual project. Often these individuals are people who have been involved with the particular construction project throughout its duration.

The technical, financial, and accounting complexities involved in construction litigation make the use of independent professionals particularly valuable to the contractor. Interrelated issues, large volumes of documents and items of cost, and complex cost considerations all contribute to what is very often the confusing realm of construction damages. Even where the contractor's staff includes persons capable of dealing with these situations, the shortage of time and the demands of additional work may make it difficult for them to devote the energy needed. Outside professionals are experts who can reconstruct the situations and relate the costs in orderly, easy to understand, and well-documented presentations. They can help to understand, and thus lessen or defeat, the claims and calculations made by owners.

Professionals assisting in construction litigation include engineers, claims consultants, and independent accountants. Consulting engineers assist with much more than engineering problems. They can provide a wealth of assistance and experience with scheduling or critical path management techniques that are often important in dealing with delay claims. Independent accountants and claims consultants are particularly helpful to contractors in defining the various types of costs that have impact on the project and in calculating those costs. In addition to dealing with the costs, such persons may be helpful in finding ways to explain and demonstrate to courts and juries the types of costs involved and the ways in which they occurred.

Accountants are, of course, professionals in understanding the nature of costs and of cost accounting principles, particularly as these principles relate to accounting for costs in the construction industry. In addition to objectivity and innovation, Certified Public Accountants (CPAs) bring their professional disciplines, knowledge, and broad experience with the construction industry to the litigation process. Moreover, CPAs often can lend an important measure of credibility to the explanation of a contractor's position and calculations.

The first steps in dealing with a claim usually involve the definition and identification of the nature of the elements of disagreement in terms of the causes of excess costs or damages. If claims in construction litigation are to be reasonable and to the point, it is necessary to attempt to identify specifically the interrelationship of matters as they pertain to the resulting costs. Once an understanding of the nature of the elements of disagreement has been arrived at, claims may be calculated or critiqued and audited.

In the end, litigation support means assistance—assistance in chronicling, understanding, defining, calculating, and explaining claims and defenses to claims that arise out of litigation evolving from construction projects.

THE LITIGATION ARENA

Construction litigation takes place in a broad albeit structured arena. The how and when of litigation depend very much on the nature of the construction contract, the location in which the project or the contractor and owner are located, and the type of owner involved. Litigation can take place in the federal or state court systems and often occurs before arbitration boards. In order for litigation to take place before arbitrators, a provision providing for arbitration in case of disputes would either have to be included in the construction contract or agreed to at some later date by the parties. Litigation on a federal level occurs when the owner is a United States Government department or agency.

Generally, when disputes arise in which the owner is the United States Government, the litigation depends upon the department or agency that acted as contractor. Usually, litigation involving construction for federal agencies will first be dealt with before the administrative board of the appropriate department or agency. The boards were created to resolve such disputes and operate very much like courts of law.

Construction litigation with federal government owners can be broadly categorized into defense-related and civil contracts. The two largest Boards are the Armed Services Board of Contract Appeals (ASBCA) and the General Services Board of Contract Appeals (GSBCA). The ASBCA handles all Department of Defense contract disputes. The GSBCA handles disputes arising out of contracts for departments and agencies for which it acts as the grantor of contracts. Other federal government departments and agencies with boards of contract appeals exist. Disputes involving these boards are much less common. In many instances, the ASBCA and GSBCA also provide the forums for departments and agencies that do not have their own boards of appeals.

Disputes arising out of construction contracts with federal departments and agencies are taken to the boards of appeals when they cannot be settled in direct negotiations with the owners. Contractors who are unsuccessful or unsatisfied with the results of board decisions can sue the United States Government in a federal court. The court of reference depends upon the size of the claim. Claims in excess of $10,000 must be pursued in the United States Claims Court, while smaller claims are filed in the district courts. Breach of contract claims are handled in the United States Claims Court rather than by the various boards.

Litigation with governmental owners other than the United States Government is handled through the court systems having jurisdiction in the various states. Some states have created state boards of contract appeals that operate in a manner similar to the federal boards of appeals.

Litigation involving nongovernmental owners is dealt with in state and local court systems. Very often, however, such disputes are settled by arbitration. Arbitration is a voluntary submission of a dispute for final determination to persons who are independent of the dispute. Some contracts may identify specific arbitrators or the means of selecting such arbitrators. A more common provision, however, involves a submission of arbitration to the AAA.

The AAA provides a means of administering arbitration cases under various specialized rules. Arbitrators are made available through appointment by the Association's construction industry panel of persons recommended by the National Construction Industry Arbitration Committee. The construction industry arbitration rules provide an orderly structure resulting in proceedings that are very trial-like in nature and scope, but provide a simpler means of the settlement of disputes arising between contractor and owner.

CATEGORIES OF LITIGATION TYPES

The problems that could result in litigation arising from a construction project are as diverse as the various matters, concerns, and participants in the construction industry. To a large degree, however, this wide variety may be summarized into several general types that generally follow the nature of the construction project and the costs that are common to it. Although some matters, like those occurring between contractor and owner, are very much interrelated, many are quite separate. In discussing the basic types or matters of damages that may be suffered, the separate problems that often occur between contractors and labor or contractors and their subcontractors are not included, as they are too idiosyncratic.

From the contractor's perspective, problems that may result in litigation arise from situations caused by the owner's actions. These problems include:

- Wrongful termination of a contract by the owner

- Delays to performance

- Accelerated performance

- Defective plans and specifications

- Changes in nature or cost of the construction project that deviate from the original contract

Wrongful Termination

Wrongful termination of a contractor by the owner involves termination of the contract for unreasonable, unfair, or false reasons alleged of the owner or because of an owner's decision to abandon the project. Wrongful termination can result in the possibility of the contractor recovering an amount that would return him to the financial position he would have been in had the contract not been terminated.

Delay

Delays to the construction process caused by an owner usually result in harm to the contractor by way of disruption in the construction process and a related increase in costs. The two basic factors always involved in delay of performance are time and cost. In addition to the more obvious direct costs incurred, there are almost always significant time and cost increases that result in a ripple effect to the other phases of the construction project. When the owner causes delays, the contractor is usually entitled to an appropriate time extension for the completion of the project as well as the recovery of damages relative to the increased direct and indirect costs. The time extensions become very important when the construction contract provides for liquidated damages.

Accelerated Performance

Accelerated performance of a contract usually occurs when the contractor accelerates his work either by direct order or as a result of pressure from the owner in order to make up for delays that were not the contractor's fault. Increased costs almost always are the result of accelerated performance. The kinds of costs that are usually involved include labor costs, such as overtime and shift premiums; costs of expediting work and obtaining materials; costs of extra equipment that may be necessary in order to effect the acceleration of work; increased bond rates; extra tool expenses; loss of efficiency; and additional supervision and overhead. Claims that result from accelerated performance would include such costs and may also include additional profit.

Often a construction project is well in progress before it is realized that the plans or specifications for materials provided by the owner are defective. For example, the type of steel that may have been specified for the framework of communications towers may turn out to be inappropriate for the particular design characteristics. The results of such problems may add significantly to the contractor's costs. Costs may be incurred as a result of the reordering of steel in order to continue construction, wasted costs incurred to date, work that must be undertaken in order to make corrections, and delay in completion of the project.

Changes

Changes in the nature or extent of a project are often unexpected or unintended, but nevertheless result in significant increased costs to the contractor. An excessive number of change orders or individual change orders of substance may be of such significance as to change the nature, timing, and cost of the project. Even those change orders that individually would have little impact may occur together in large numbers and thus cause significant cost increases or even change the nature of the construction product. Because these changes may occur over the entire period of the construction process, their effect may be initially obscured. Individual change orders that are significant in terms of their requirements and costs are easier to identify and understand.

Interrelationship of Types

From a contractor's point of view, the matters of acceleration of performance, the problems of defective plans and specifications, and the changes in the nature or extent of the project are matters that may also involve delays in performance. Observed in that light, it is easy to see why contractors' delay claims often seem the most common of disputes involved in construction litigation.

Liquidated Damages

The owner's perspective of the types of construction litigation and related damages can be very different from that of the contractor. While the owner is very much concerned with problems of the costs that may be incurred as a result of delays in construction performance and completion of the contract, he is also concerned with the possibility of the contractor's abandonment of the project and defective or tardy performance. Often an owner's claim for damages resulting from delays is based on liquidated damages.

Most construction contracts include general provisions for liquidated damages due to unexcused delays in performance by the contractor. Delays are basically a matter of the job not being finished by the completion date specified in the contract. The subject of liquidated damages is often a matter of much dispute, as to both reasonableness and applicability.

When the contractor fails to complete a project in the time required, the owner may easily incur additional costs for financing, rental costs for having to remain in other quarters, loss of business profits, and even loss of value. In the absence of a liquidated damages clause, it is necessary for the owner to identify and deal with each of the kinds of costs incurred as a result of delays.

When an owner terminates a contractor or the contractor abandons a project, the owner is placed in a position of having not only to protect the property but to find other means of completing it. This usually means either finding another contractor to step into the defaulting contractor's place and attending to the associated additional costs or dealing with the losses that may result if the whole project is abandoned.

The principle involved in the matter of defective performance by a contractor is that the owner is entitled to receive what he has contracted for or its equivalent. Claims for costs most often relate to the costs of correcting or completing the work. However, costs may also involve extended costs of financing during construction and other costs similar to those that an owner may incur as a result of tardy performance by the contractor.

DELAYS: THE COMMON PROBLEM

Perhaps the most common and difficult matter encountered in dealing with damages in construction litigation is that of delays to contract performance. Viewing a construction project in terms of its various performance components, it is easy to see how complex the problems of calculations may become. The ripple effects on the sequence and scheduling of the various parts of the project may be significant and difficult to define.

An example will help clarify the ripple effect of construction delays and the ways in which additional costs occur. The construction sequence of an electric power generating plant may be simplified for this example into: (1) site preparation; (2) plant foundations and external utilities; (3) initial structures work; (4) equipment installation and completion of structures; (5) proof of equipment performance; and (6) site completion. The owner has separately contracted for the purchase and delivery to the site of the power generating equipment and machinery. For a cause of delay, the delivery to the site is 120 days behind schedule.

In this simplified example, the completion of sequence part (3) and the start of parts (4) through (6) will be delayed. Without acceleration in construction, these parts will each be delayed at least 120 days behind the scheduled completion dates. Actual delays in a work schedule are, of course, never this simple. Efficient work and time planning involve the interrelationship of the various components of one or more major parts of a construction task segment with other components of other major parts. Examples of the types of costs that may result from these delays are idle or inefficient labor and equipment costs, costs for the reworking of previously completed parts of the job, escalated labor rates and materials costs, and unabsorbed or extended field and general office overhead.

Owner-caused delays are delays other than those that result from labor disputes, weather, and force majeure. Delays occur for many possible reasons, some of which are the owner's procrastination in processing preliminary designs and plans; difficulties in coming to decisions within a reasonable period of time about various construction questions as they arise; failure to provide an unencumbered work site; difficulties in coordination and progress of work of other contractors under the direction of the owner; errors in the specification or design of the project; and large numbers of change orders or individual change orders of significant impact.

Regardless of their cause, delays always involve two subjects: time and money. All related arguments aside, the contractor who experiences delays for which he may claim compensation has to be able to establish the nature of his damages and the increased costs that he has experienced as a result of such delays. The experience of delays, and even that of accelerated performance to make up for the delays, does not necessarily result in increased costs. Even when such costs occur, the contractor has an obligation to take reasonable steps to hold down costs during periods of delay (i.e., not to incur otherwise unnecessary costs during delay periods or while accelerating performance).

A peculiar problem in dealing with the costs of delays is the subject of home office overhead. Often a matter of dispute in terms of recoverability and calculation, consideration of such costs in the courts is now a subject in transition.

The kinds of costs that are involved in delays are different as seen through the eyes of the contractor and the owner. For his part, the contractor is concerned with the matters of additional construction costs. To the owner, the costs of delays may well involve the consequential matters of impact on business operations, project financing costs, and even the diminished value of the construction project itself.

DAMAGES

Determining damages in construction litigation is largely a matter of identifying the instances of dispute, associating the costs that relate to those instances, and determining the excess costs that were incurred as a result. In many instances this is not very difficult, providing the contractor keeps detailed descriptions of problems and effects as they occur and maintains sufficiently detailed records of costs incurred. The problem of calculating damages becomes particularly difficult, however, where there are insufficient records or information has not been retained.

The damages that may be experienced by a contractor for any of the various reasons that may result in litigation against an owner or the owner's representatives inevitably boil down to questions of materials, labor, the use of equipment, and job overhead costs. Over the years, the cost of home office overhead has also been claimed and accepted. Complicated delay claims may involve layers of these basic costs.

Identification of the excess costs resulting from delays in performance can be difficult. Matters that cause excess costs that are not in themselves natural delays may contribute to delays and related additional costs. With reference to the earlier example of error in the specification of steel, corrective work may require the reordering of materials and significant reworking of parts of the project. The time involved in obtaining new materials and reworking completed or partially completed parts of the project may seriously delay the performance of the other parts of the project. As a result, the additional costs incurred may go well beyond the parts of the project directly involved and include the excess costs that result from the ripple effects of the delay.

Questions regarding lost profits, loss of other business that might have been obtained, and interest expense are often involved in litigation by a contractor against an owner. While not necessarily unattainable, such claimed costs have generally been difficult to substantiate.

Additional profits claimed by a contractor may be obtained where equitable adjustment clauses are contained in the construction contract or where the owner has altered the nature or scope of the contract. Such awards tend to be most common to instances of large volumes of contract change orders requiring a significant amount of unanticipated work. Awards of damages are generally thought, however, to pay contractors for losses incurred rather than to provide an additional basis for determining profit. Because of the amount of speculation involved, the recovery of profits that might have been obtained from other jobs that a contractor could have obtained, if there had not been problems or delays caused by an owner, are particularly difficult to assess and obtain.

Interest costs are often claimed and always seem to be a matter of disagreement. In earlier years, such costs were unacceptable in contractors' claims arising from federal government contracts. More recently, interest costs have been given consideration where the situation has clearly resulted in excess costs, although often only from the date of the contractor's legal complaint to date of award of damages.

Defining owners' damages is, like defining the damages of the contractors, a matter of identifying the costs associated with the nature of the damages. In some cases, the owner may even claim a loss of business profits or rental revenues due to delays in the use of the facility being built by the contractor. Additional financing costs may occur and be assessed, particularly where claims are for things other than loss of rental value. An example of financing costs of this nature would be the increased interest costs on the owner's debt incurred to finance the construction project. Overhead costs may also be incurred by an owner as a result of delays in the completion of a project, as well as additional supervisory costs that may be expended on a project by the owner's employees and consultants.

Abandonment of a project by a contractor may result in the excess costs of employing another contractor to complete the project. The costs to an owner as a result of defective performance are most commonly those of repair or replacement and those associated with the possibility of diminished value as a result of accepting the project in a manner other than that in accordance with the plans and specifications.

The matter of liquidated damages that may be assessed a contractor by the owner is a special problem. A contract clause for liquidated damages usually fixes an amount and the method of its application. As an example, such a clause may specify a particular amount of money for each day that the contract is not completed subsequent to the contractual completion date. For the liquidated damages provision to be enforceable, the amount would have to be reasonable and bear some relation to actual damages. Moreover, the damages that may occur would have to be of a type that would have been difficult to ascertain at the time that the contract was signed. The responsibility of proving that these tests were met would be that of the asserting owner. Often the contract clause is used by an owner as a weapon to force the contractor to accelerate performance when delays occur, even when the delays were not caused by the contractor. This is a common complaint in litigation.

However the subject of damages is approached, there are some overriding requirements: they must be reasonable, carefully defined, and calculated on the basis of the nature of the matters alleged. In addition, damages calculations must be supportable by understandable documentation in terms of the types of costs and amounts involved.

CALCULATING CONTRACTOR'S DAMAGES RESULTING FROM DELAYS

The contractor suing for delay damages has to demonstrate that the owner's improper conduct affected critical events in or cost components of the construction project. Often the critical path method (CPM) or similar scheduling techniques are used to demonstrate these disruptions. These techniques depict the parts of work in a project that are dependent upon other parts of work. Their purpose is to demonstate the impact of the delay on the timing of the various segments of the construction process. These segments, in turn, bear direct relationships to the costs of materials, labor, equipment, and overhead.

There are two basic approaches that contractors use to calculate the extra costs that they have incurred in completing a construction contract that has been delayed by the actions of the owner: the segregated approach and the total cost approach. These two approaches are very different.

The total cost approach has often been used by owners completing a project subsequent to terminating a contractor prior to completion. This approach disregards the individual reasons for delay. Instead, the contractor determines his actual costs of performing the project and claims as his damages the total of such costs plus a reasonable profit and overhead, reduced by whatever has already been paid by the owner. While administratively convenient, the total cost approach is usually accepted with great reluctance, as it does not identify or define actual costs with specific reasons for the incurrence of these costs. Instead, it depends upon the assumption that all of the excess costs incurred are the result of an owner's delays.

The larger the construction project and the greater the number of parts of the work, the more difficult it is to make judgments as to the fairness of excess costs claimed under the total cost approach. The contractor using such an approach has to establish that the nature of the losses makes it impossible or impracticable to determine them individually with any reasonable degree of accuracy and that the actual costs incurred throughout the course of the project were reasonable. Moreover, the contractor also has to be able to show that his original bid or estimate was realistic and that he was not responsible for any costs incurred beyond that estimate. However convenient the total cost approach may be in computing damages to be claimed, its use is often subject to successful challenge.

Under the segregated approach, the contractor identifies each matter of delay and defines the related costs incurred as a result of such delay. This is the preferred approach because it deals with costs in terms of specific attribution and provides proof on that basis.

Simply explained, the segregated claim approach deals in terms of excess field costs and additional home office overhead. Extra field costs are usually labor, equipment, and overhead costs defined in terms of the actual amounts involved. Additional materials costs that may be incurred due to delay situations are often not significant. Where such costs are significant, they are usually easily identified. The other costs involved in delays are often not quite as easy to determine.

Additional labor costs incurred as a result of delays caused by the owner may occur due to one or both of the following reasons:

1 Labor hours in excess of those that were otherwise required due to inefficiencies caused by the delays

2 Escalations in labor rates and related employment costs

The contentions are that labor becomes more costly because the same work requires more time to complete than otherwise would be required or that costs increased for the same work because it did not occur when originally scheduled. Labor inefficiency usually is a matter of acceleration of performance, that is, the requirement of work under less efficient conditions than would have otherwise prevailed. The accelerated work effort usually involves overtime or premium pay. Escalations in labor rates involve increases in rates of pay and other employment costs that would not have otherwise occurred within the originally scheduled performance periods for the various phases of construction.

The costs of idle or inefficiently used equipment are a common part of most delay claims. Idle equipment is identified in terms of the time in which equipment was not used as a result of delays and the ownership costs of the equipment. Use of equipment over longer than normal periods or intensified use as a result of acceleration of performance can also result in excessive ownership costs. Costs of ownership include:

- Depreciation
- Major repairs
- Interest on the contractor's investment
- Storage, incidentals, and other equipment overhead
- Insurance and taxes

The accumulation and calculation of such actual costs for each piece of equipment used by the contractor and the further conversion of such costs into average daily or other period rates is usually a complex record-keeping process. For this reason, additional costs of equipment are often determined by using the equipment ownership rates published by the Associated General Contractors of America (AGC). The AGC rates take into account the types of ownership costs described in this paragraph.

Often contractors rent some or all of the equipment employed on a project from other contractors. Some contractors rent their equipment to others when it is not employed on their own projects. The costs of idle or increased use of equipment rented from others can, of course, be determined from the amounts paid the lessors. The matter of lost rental income to a delayed contractor who otherwise would have been able to lease his equipment to others is a matter of much dispute. A claim for lost rental values requires careful substantiation. Rates for this form of loss claim are often taken from rental rates included in the Associated Equipment Distributors Manual. This manual provides national rental averages.

Field overhead costs are those costs of supervision and field support usually incurred at the project site. These are the general costs specific to the project that relate to or benefit all of the various segments of the work. To at least some extent, such costs continue throughout the work until contract completion. The costs include: supervision, insurance, site administration and accounting, and stores and materials management. Perhaps even more than the other types of costs associated with delay claims, additional field overhead costs are largely a matter of extended periods of time over which such costs must continue. These costs are usually calculated on the basis of excess days and average actual costs per day incurred. This, of course, is somewhat of a simplification as the composition and total of costs incurred may fluctuate significantly during periods of extended work or accelerated performance necessitated by delays.

Home office overhead, another construction overhead cost element in delay claims, is the subject of much discussion, literature, and debate. This matter is discussed in the next section.

Although the segregated approach is the preferred approach in computing damages as a result of delays caused by the owner, it is not without difficulties in defining and calculating costs. The ripple effects of delays in some parts of the work on other parts usually require some means of defining and demonstrating the interrelated delay consequences. Scheduling or CPM techniques are methods often used for this purpose. Even where the effects are identified by such methods, defining and calculating the related costs may still be difficult. Working with the costs usually requires the participation of the project manager, the contractor's accountants and engineers, and even the additional assistance of consulting engineers and independent accountants.

CALCULATING HOME OFFICE OVERHEAD

Contractors' claims for additional costs incurred as a result of delays usually include additional home office overhead costs. These costs are the general and administrative costs more or less common to all of a contractor's construction business activities. The kinds of costs involved include the salaries of officers, headquarters operations, and other construction business costs not directly attributable to specific projects.

For some time now, home office overhead costs have been claimed and accepted with little or no required proof that additional overhead costs were actually incurred because of the specific delays. These costs had begun to be regarded by courts as a fixed or constant expense to be assessed and accepted without question.

The amount of home office overhead costs to be included in delay claims has evolved from rough allocations of expense to mathematical computations performed by rote. The most common of computations is known as the Eichleay Formula. The formula and its variations have been a favorite subject of speeches and articles throughout the realm of construction litigation. However, the courts have begun to reject the blind use of the Eichleay Formula or any other means of assigning additional home office overhead costs to delay claims that disregard the related actual incurrence of additional costs.

Because it remains in common use, the following explanation of the Eichleay Formula and its application are provided. The formula deals with the matters of:

- The proportion of revenue from the subject construction contract to that of all contracts in process during the period;

- The home office overhead for the period applicable to the contract; and

- The daily contract overhead amount during that period. The formula is set up in such terms and works in a sequence of the following four computations:

1 Contract proportion = $\dfrac{\text{Billings for the contract during the period}}{\text{Total billings for all contracts during the period}}$

2 Overhead allowable to contract = Contract proportion of billings × Total home office overhead expenses for the period

3 Daily (or weekly) contract overhead rate during the period = $\dfrac{\text{Overhead allowable to contract}}{\text{Days (or weeks) of contract performance}}$

4 Home office overhead claimed = Daily contract overhead rate during the period × Number of days of delay

As a simple example of the operation of the formula, assume:

1 A contract began on January 1, 1984;

2 It was to be completed by July 31, 1985; and

3 It was not finished until November 30, 1985.

A total of 122 delay days are claimed. Further assume that the contractor's total home office costs for the period January 1, 1984 through November 30, 1985 were $1 million; that its total billings for all jobs during the period were $20 million; and that the billings for the delayed contract totalled $3 million. The amount of additional home office overhead claimed by the contractor would be $26,180, as shown in the following application of the formula.

1 *Contract proportion:* $\frac{\$3,000,000}{\$20,000,000} \div 15\%$

2 *Overhead allowable to contract:* $15\% \times \$1,000,000 = \$150,000$

3 *Daily contract overhead rate during the period:* $\frac{\$150,000}{699 \text{ days}} = \frac{\$214.59 \text{ per}}{\text{day}}$

4 *Home office overhead claimed:* $214.59 per day $\times$ 122 days = $26,180

The Eichleay Formula has been attacked in a recent GSBCA decision. In the case of *The Appeals of Capitol Electric Company* (GSBCA Nos. 5316 and 5317), the GSBCA ruled that the contractor was allowed to recover only underabsorbed home office overhead. In essence, the GSBCA made it clear that it was not operating under a rule that required the contractor to prove an actual loss in order to claim home office overhead costs. As a result, it may not be necessary for a contractor claiming such costs to prove that the extended performance period caused by delays actually caused increased home office overhead costs. Clearly, the inclusion of home office overhead costs in delay claims is a subject in transition.

CAPTURING AND DOCUMENTING THE COSTS

The best basis for calculation of claims for damages in construction litigation is a strong foundation of information well grounded in good financial and operational record-keeping systems. In order to calculate and demonstrate the damages incurred in a reasonable fashion, the contractor must be able to define, accumulate, and calculate the costs in a clear and logical manner. To do so, it is always necessary first to identify these costs in terms of what types of costs they are and the periods in which they were incurred. Inevitably, this process points back to a soundly structured accounting system for the particular construction project, if not for the contractor's business as a whole. The absence of a reasonably structured accounting system could leave a contractor in the position of having to prepare a total cost claim—a form of claim that is always difficult to prove.

Earlier, it was explained that delay claims involve the impact and related costs of delays of the progress of one or more parts of the construction process upon other parts. The parts of the construction process underscore the need for an accounting system. An accounting system that records the different types of costs of terms of the various segments of work in the project provides a good base with which to identify costs related to delays. Such information in an accounting system allows for both the identification of the types of costs for each part of the work and the impact of delays on those costs.

Two instances of litigation serve to illustrate the problems of not having sufficiently structured accounting systems; one deals with a total cost claim and the other with a delay claim identified by segments as to costs, but not supportable by the contractor's accounting system. Each arbitration was real, with each claimant employing the support of its own in-house engineers and accountants, as well as consulting engineers. To a lesser extent, independent accountants were also involved.

The total cost claim was made by an owner who completed a large construction contract after terminating the contractor when the work was more than 90 percent complete. In completing the work, the owner hired labor and equipment from different contractors, purchased materials by himself, and provided the project supervision and overhead. Costs were accounted for as one pool segregated only into labor, materials, equipment, and a few overhead accounts. There were a very large number of individual instances of labor crew work assignments in the completion of the contract, which were evidenced by various repair and construction crew reports that defined task assignments. The labor crew reports required as little as half a day to as much as several days of work. The owner's accounting system made no attempt to associate costs with actual instances or types or work performed.

The owner presumably turned to the total cost claim approach in this instance because of his failure to account for his costs in a fashion that would have allowed the association of those specific costs that he would claim of the contractor. Using the help of his own field engineers and outside accountants, the contractor was able to analyze the large volume of labor crew reports and compute the specific amounts of those costs claimed with which he disagreed. The result was an analysis product that amounted to a segregated cost claim, the components of which could be specifically objected to in terms of both instance and cost.

The employment of well-structured and responsive accounting systems cannot be overemphasized. In the instance of an owner defending against what appeared to be a well-prepared delay claim by a contractor, the structure of the claim partially dissolved when the contractor's accounting system for the project was reviewed for data in support of its calculations. Although the claim had been prepared on a segregated cost basis, the contractor had not kept his accounting records for this multi-segment project on the basis of those various segments. Instead, the accounting records represented a single large pool into which costs by materials, labor, equipment, overhead, and subcontract were recorded. Moreover, these records did not readily support the contractor's calculations and contentions as to excess costs. While the owner did not disagree that there had been delays, the contractor experienced a great deal of difficulty in demonstrating and proving his calculations of damages.

A well-structured and responsive accounting system is only one tool, albeit a very important tool, to litigation support in dealing with calculations of damages. Most claims for damages occur and are pursued well after the completion of a construction project. By the time claims are made and damages calculated, the specifics of the events that are the subjects of the litigation are often no longer clearly remembered. Furthermore, the people involved are often no longer available or employed by the contractor. Consequently, the importance of well-maintained daily project field logs and memoranda that clearly define problems as they occur, the potential impact of these problems, and the discussions between owner and contractor that occurred at the time become very important in defining and calculating damages. Often in the rush to complete a job or to accelerate performance, the effects on other segments of the work (although well understood at the time), are lost to the subsequent observation so important in determining their impact on costs.

The effort to define and document costs usually requires the work of related disciplines. This effort represents an important element of litigation support in construction contract litigation. The understanding of the cost effects of problems that occurred on the job is invaluable in the calculation and proof of damage claims. Often both the assistance of consulting engineers applying complex scheduling techniques to determine the effect of different instances of disruption or delay on the job and the assistance of independent accountants to help compute and substantiate the related costs becomes invaluable.

DOCUMENTATION OF COSTS: A CRITICAL FACTOR IN LITIGATION CLAIMS

However the damages are captured and documented, it becomes critical in support of construction litigation claims that costs be carefully defined and calculated, and reasonably documented by the records applicable to the project. If additional labor costs are incurred, it is important to be able to show where the additional labor hours occurred and what the costs of those hours actually were. This is also true with regard to any of the types of costs involved. Litigation support in construction claims is very much a matter of the proper use of the contractor's accounting and project records and personnel, assisted by other accountants and consultants who may be helpful under the specific circumstances.

Taxation and Tax Planning

Joseph F. Potter

The author wishes to acknowledge the contributions of Murray B. Schwartzberg, Robert E. Decker, John F. Eckert, Michael H. Frankel, Victor Grigoraci, Douglas J. Lundell, Bernard W. Nebenzahl, Edward W. Shuey, Jr., and Anthony J. Sullivan.

INTRODUCTION

Construction contractors are faced with all the tax problems that any general business faces, as well as many additional tax situations unique to the construction industry. Sales and excise tax problems are particularly troublesome areas for the construction contractor. However, the single most important tax issue for the contractor is the method of recognizing revenue for income tax purposes.

In response to these unique problems, income tax accounting for contractors has become specialized and is susceptible to extensive planning. The tremendous risks involved and the choice of several different methods of accounting result in the opportunity for the control of items such as timing the payment of tax.

One of the major objectives of tax planning generally is deferral of the tax payment; for long-term contracts, it is possibly the primary objective. Faced with the high cost of capital, especially in the early years of a construction firm's existence when cash shortages are critical, a firm's deferral of income tax payments can be crucial to its survival. Tax planning is an ongoing process to be considered in the formation of the organization; in the drafting of construction contract terms; as the year progresses and estimated taxes are required; at year-end when the annual summing-up is required and returns are filed; and when the organization is finally wound up.

Long-term contracts present unusual and unique problems, and the opportunity to take advantage of available accounting methods is essential to the successful contracting organization. Determining eligibility for these special methods and determining the application of these methods is difficult. The evolution of these methods over the years continues to present unique planning opportunities, as well as significant accounting challenges, to construction contractors.

While the primary expense is the federal income tax, state and other income taxes can be significant, and may apply, even to an unprofitable operation. Corporation franchise and excise taxes not based on income, as well as sales and excise taxes levied on construction contractors as consumers, may make the difference between a profit or a loss on any given contract within any taxing jurisdiction. This chapter gives the reader an insight into federal income tax issues, an overview of some state and local tax considerations, and a basis for planning an appropriate tax strategy for the construction contractor.

The special long-term contract methods of accounting were not included in any Revenue Act before 1939, and were not included in the Internal Revenue Codes of 1939 and 1954. However, Treasury regulations have long recognized the completed-contract method and the percentage-of-completion method as permissible methods of tax accounting for contractors performing long-term contracts. Final approval has come from the courts, which have regularly permitted these special methods for contractors performing eligible work and making appropriate elections.

Early regulations provided only general guidance as to the detailed rules regarding long-term contract methods. Most of the basic principles involving the application of these methods have arisen from court cases and Treasury pronouncements. During the 1950s and 1960s, the courts were deciding cases involving the taxation of prepaid service income as well as early receipts from the sale of goods. As a result of these cases, prepayments for services and for the sale of goods could no longer be deferred until the service was rendered or until the goods were delivered. For contractors using the completed-contract method, deferral had normally been permitted for progress payments until a contract was completed. The question became, Would these new rules also apply to taxpayers using the percentage-of-completion method?

In early 1971, new regulations on accounting for long-term contracts were proposed, providing that a contractor could use the completed-contract method of accounting for tax purposes *only* if such a method was also used for all financial reports. In the case of contractors using the percentage-of-completion method for financial reporting and the completed-contract method for tax purposes, this conformity requirement presented insurmountable obstacles. Generally accepted accounting principles (GAAP) and the needs of users of financial statements normally encouraged percentage-of-completion accounting, while the benefits available from the use of the completed-contract method for tax purposes were both desirable and necessary from a management viewpoint. Controversy over implementation of Treasury regulations has continued almost unabated from the minor revisions of 1971 and 1972, through the "final" regulations of 1976 up to the 1982 Tax Equity and Fiscal Responsibility Act (TEFRA) revisions of the law and through all subsequent revisions.

LONG-TERM CONTRACT ISSUES

Alternative Methods

Special alternative methods of accounting are available for contractors who have agreements to manufacture, build, install, or construct a particular item or items if that performance is not completed within the same tax year in which it commenced.

Percentage-of-Completion Method. Although this method is probably the most realistic measure of income and is preferred for financial statement purposes, it does little to defer income. It does, however, allow for the recognition of loss contracts as incurred.

Completed-Contract Method. The major advantages in using this method are the following:

- The deferral of income recognition until the contract is completed, which may be controllable to some extent

- The recognition of income when it is finally determined

- The clear deferral of advances and retainages until completion

The major disadvantages include:

- The possible peaking of income

- The possibility of being subject to more stringent regulations issued under the authority of TEFRA, which, if applicable, will defer the deduction of many more construction contract costs until the contract is completed, rather than deducting them at the time they are incurred

- The nonrecognition of losses until the contract is complete

If either the percentage-of-completion method or the completed-contract method is elected for long-term contracts, that method must be applied consistently to all long-term contracts within the same trade or business. The exception is that long-term contracts of substantial duration (12 months or more) may be reported on a consistent long-term contract method, while long-term contracts of less than substantial duration (less than 12 months) may be reported on another acceptable method of accounting.

TAX ACCOUNTING UNDER THE COMPLETED-CONTRACT METHOD

The completed-contract method provides that the income or loss on a given contract is not recognized until the contract is completed and accepted. At that time, the gross contract price, including retainages, is included in income, and the related costs incurred are deducted therefrom.

If there is a dispute over the contract involving a customer claim and the contractor is assured of either a profit or a loss, the gross contract price—reduced by the amount in dispute—must be included in gross income. All allocable costs are to be deducted in that year. Additional costs are deducted as incurred. However, if neither a profit nor a loss is assured, then the contract is held open until the taxable year when the additional work is completed.

If the claims are contractor claims, the contract price and associated costs are included in the taxable year in which the contract is completed. Subsequent revenues and costs are included in income, or deducted as received or incurred. Mixed claims require application of the above rules.

Allocation of Costs

Some expenses may be deducted as current or period costs. However, many costs must be allocated to the contract and can be deducted only at the time the job is completed. Both direct and indirect costs must be treated as costs properly allocable to a long-term contract. Direct costs include material costs and labor costs. Material costs include:

- The cost of those materials that become an integral part of the subject matter of the contract

- Those materials consumed in the ordinary course of building, constructing, or manufacturing the subject matter of the contract

Labor costs include:

- The cost of labor identified or associated with a particular contract
- Those elements of direct labor that include basic compensation, overtime pay, vacation and holiday pay, sick leave pay (other than under a Section 105(d) wage continuation plan), shift differential, payroll taxes, and payments to a supplemental unemployment benefit plan paid or incurred on behalf of employees engaged in direct labor

Indirect costs include all costs "which are incident to and necessary for the performance of particular long-term contracts." These costs, if used for the performance of a particular contract, include:

- Repair expense of equipment or facilities
- Maintenance of equipment or facilities
- Utilities, such as heat, light, and power, that relate to equipment or facilities
- Rental of equipment or facilities
- Indirect labor costs and contract supervisory wages incurred (including the same elements as those for direct labor, listed above)
- Indirect material and supplies
- Tools and equipment not capitalized
- Costs of quality control and inspection
- Taxes (other than state, local, and foreign income taxes) to the extent attributable to labor, materials, supplies, equipment, or facilities
- Depreciation, amortization (and cost recovery allowances) reported for financial statement purposes on equipment or facilities (but not in excess of that allowable under Chapter 1 of the Internal Revenue Code)
- Cost depletion
- Administrative costs (exclusive of any cost of selling or return on capital)
- Compensation paid to officers attributable to services (other than incidental or occasional services) for the job

- Insurance incurred, such as on machinery, equipment, or facilities

INCOME TAX ACCOUNTING METHODS

Adoption of a Method

Generally, all taxpayers are required to determine taxable income under the method of accounting by which they compute their income. "Method" applies not only to the overall method of accounting, but also to the accounting treatment of any item. Examples of overall methods of accounting are (1) the cash receipts and disbursements method; (2) the accrual method; (3) a combination of these methods; or (4) combinations of these methods with various specialized methods available for the treatment of certain items.

   With this host of alternatives available to a contractor, it is important to determine—at the time of filing the initial income tax return of the business entity—which method or methods of income tax accounting are to be adopted. In order to be prepared to use these methods advantageously, to have the necessary financial information, and to meet certain further requirements, a contractor should review its proposed operations and make these decisions as early as possible.

   Cash Receipts and Disbursements Method. Under this method, all items that constitute gross income are to be included in the taxable year in which they are actually or constructively received. Expenses are deducted in the taxable year in which they are paid.

   Income is deemed to be constructively received in the taxable year during which it is credited to the contractor's account, set apart for the contractor, or otherwise made available. If the contractor can draw upon income at any time, or could have drawn upon income during the taxable year (if notice of intention of withdrawal had been given), then that income is taxable under the constructive-receipt doctrine. For example, a taxpayer whose bank account is credited with interest on the last day of the taxable year is deemed to be in constructive receipt of that income in the taxable year in which the interest is credited.

   Accrual Method. Under this method, income is included in the taxable year when (1) all events have occurred that fix the right to receive such income; and (2) the amount of income can be determined with reasonable accuracy. Deductions are allowable under the accrual method in the taxable year during which all events have occurred that establish the fact of the liability giving rise to that particular deduction, and if the amount of the liability can be determined with reasonable accuracy. Generally, this method may be used if it is in accordance with GAAP, is consistently applied by the taxpayer from year to year, and is consistent with income tax regulations.

An accrual method of accounting recognizes income when payment under a contract is received, even if performance for that payment has not been completed. However, if there is an advance payment under a contract for services required to be performed before the end of the next succeeding year, such an advance payment may be deferred when received and included in income as the services are performed (but no later than the next succeeding taxable year).

A contractor using the accrual method of accounting may be entitled to receive only partial payment under a contract until certain specified events occur, such as completion or acceptance of part or all of the subject matter of the contract. The amount retained and not paid to the contractor pending completion or acceptance for work performed is commonly referred to as a retainage. The retainage may be excluded from income until the contractor has a right to receive it in whole or in part.

Combination of Methods. A taxpayer may use a combination of the previously discussed methods, for example, using the accrual method of accounting with respect to purchases and sales of goods, but using the cash method in computing all other items of income and expense. A taxpayer who uses the cash method in computing the gross income of a business is also required to use the cash method in computing the expenses of that business. Similarly, a taxpayer using the accrual method of accounting in computing business expenses is required to use the accrual method in computing items affecting gross income.

A contractor using one method of accounting for a particular trade or business may compute other items of income and deductions not connected with that trade or business under a different method of accounting. For example, a contractor engaged in the installation of equipment may use the cash method for the installation work and, if a manufacturing business is also operated within that same entity, may be required to use the accrual method of accounting for the manufacturing business.

Other Methods. Special methods of accounting are permitted under particular circumstances. Methods such as the crop method of accounting and the installment method of accounting allow certain items to receive special treatment. These methods are very specialized; therefore, it is inappropriate to discuss them within the scope of this chapter.

Inventories and Choice of Method

If a contractor has inventory, the accrual method of accounting must be used with regard to purchases and sales. Included in inventory are all finished or partly finished goods; also included are raw materials and supplies that have been acquired for sale or that will become physically part of the merchandise that is intended for sale. Inventory is generally defined to include items held for sale in the normal course of business or items to be sold in a transformed state.

For example, a contractor engaged in the construction of a building has building materials on hand that might normally be considered as inventory. However, the sale of building materials is not the contractor's business; rather, these building materials reflect a deferred cost that will become part of the cost of the structure being built. Once these building materials have become part of a specific project, they are a deferred cost attributable to that contract and are not included in inventory. Accordingly, the contractor is not required to use the accrual method of accounting with respect to such contracts.

Change in Method of Accounting

Once a contractor has adopted an overall method of accounting or an accounting method with respect to particular items, changing the method generally requires the approval of the Internal Revenue Service (IRS). The taxpayer should change its accounting method only when its future treatment of an item may differ from its current treatment of that item.

It is not considered a change of method if the taxpayer has not previously adopted an accounting treatment with respect to a particular item, provided that the proposed treatment of that item is not in conflict with the taxpayer's previously adopted accounting methods. For example, a contractor using the accrual method is entitled to a bad debt deduction for the first time after being in business for a number of years. The contractor may adopt either the direct write-off bad debt deduction method or the reserve method of accounting for bad debts. In contrast, a taxpayer who has adopted the direct write-off bad debt method in a prior year, but now desires to change to the reserve method of accounting for bad debts, must obtain advance permission from the IRS.

Because of the relatively unlimited authority of the IRS to grant or deny requests to change a method of accounting, a contractor must review the methods of accounting available well before filing his initial return. Generally, taxpayers desiring to change a method of accounting must file a request to change the method no later than the 180th day of the taxable year for which the change is to be made.

Financial Statement Considerations

The Internal Revenue Code provides that taxable income is required to be computed under the method of accounting that a taxpayer uses in keeping his books. The term "books" does not mean financial statements; rather it means the method by which a taxpayer maintains his internal financial records.

　　The presentation of items in the financial statement generally has effect on the tax treatment of an item. Contractors, however, should be aware that, in determining whether a method "clearly reflects income" for tax purposes, financial statement presentation may be referred to by the IRS as important evidence in deciding whether an accounting method represents a clear reflection of income.

Extended Period Long-Term Contracts

In addition to the allocations set forth above, other items must be allocated if the contract is an "extended period contract," which is any contract that the taxpayer estimates will not be completed within two years of the contract commencement date. However, a *construction* contract is not an extended period long-term contract if

- The taxpayer estimates the contract will be completed within three years of the contract commencement date, *or*
- The taxpayer's average annual gross receipts over the three taxable years ending before the taxable year the construction contract is signed do not exceed $25 million. (This includes gross receipts from trades or businesses under common control.)

Additional costs to be allocated to extended period contracts include:

- Bidding expenses on contracts awarded to the taxpayer
- Distribution expenses such as shipping costs
- General and administrative expenses directly attributable or allocable to the performance of a particular contract

• Deductible pension and profit-sharing contributions that
  represent current service costs

Commencement Date. The commencement date is generally the first
date that design or engineering costs allocable to the job are incurred,
materials or equipment are shipped to the job-site, or workers are sent to
the job-site. Bidding expenses and contract negotiating expenses cannot
be included. Detailed records and other evidence must be maintained;
otherwise the contract date might be held to be the commencement date.

Completion. A long-term contract shall not be considered completed
until final completion and acceptance has occurred. Nevertheless, a taxpayer may not delay the completion of a contract for the principal purpose of deferring federal income tax. However, completion of a long-term
contract is determined without regard to whether a dispute exists at the
time the taxpayer tenders the subject matter of the contract to the party
with whom the taxpayer has contracted.

Final completion and acceptance is determined without regard to
any contract term providing for

• Additional compensation contingent upon future successful
  performance after being initially accepted by the owner
• Supervision of installation or assembly that is otherwise to be
  performed by the purchaser

A subcontractor's work is completed when it is judged complete and
accepted by the prime contractor.

Separating and Aggregating Contracts

In order to prevent the inordinate deferral of reporting a completed contract, one contract can be divided into several contracts with each portion
treated separately, thus accelerating the reporting of income. The Commissioner of Internal Revenue has broad authority to make these determinations, which will depend on the facts and circumstances of each case.

Generally, one agreement will not be separated unless separate delivery or separate acceptance of portions of the contract is contemplated.
Therefore, to avoid separation, the contract should be written in such a
way that "delivery" or "acceptance" of the contract is not considered satisfied until the entire contract is completed. Separation of an agreement
can also be avoided by stipulating that there is a business purpose for the
chosen delivery or acceptance date, and by providing for damages if all

aspects of the contract are not completed and accepted. In addition, to help ensure that the agreement is not separated, it should cover only one subject matter, and it should not include an exercisable option, or a "change order" made for the "increase" in the number of units to be supplied under the agreement. These kinds of clauses or additions might result in their classification as a "separate contract."

Usually, individual agreements will not be aggregated unless it is a commercial practice of the industry to customarily treat them as one contract, or unless there is no business purpose for the several agreements.

Tax Accounting Under the Percentage-of-Completion Method

The percentage-of-completion method of accounting for long-term contracts includes the portion of the gross contract price corresponding to the percentage of the total contract that has been completed during the taxable year in gross income for the year. For this purpose, the contract price includes any retainages, holdbacks, or other payments.

A contractor computes the percentage of completion on each long-term contract by comparing the following at the end of the taxable year:

- The contract costs incurred with the estimated total contract costs, or

- The contract work performed with the estimated total contract work to be performed

If a contractor uses the cost-completion method, any cost comparisons (e.g., comparison of total direct and indirect costs, total direct costs, or total direct labor costs) may be used. The only limitations with respect to the method employed to compute the percentage of completion are that the method must clearly reflect income and must be used consistently until the completion of the specific contract. The contractor is free, however, to choose a different method for each long-term contract in process.

FIXED ASSETS

Acquisitions—Tax Considerations

All contractors need equipment, and an equipment management team is crucial to the survival of a construction company. It is essential to have equipment that performs the maximum tasks required on each project.

In most cases, any two jobs a construction company undertakes bear little resemblance to each other. A company can be involved in the construction of dams, water and diversion tunnels, sewer lines, pipelines, aqueducts, canals, railroads, highways, subways, buildings for every industry and profession, housing (whether individual dwellings or an entire city), land development, and aggregate and cement batch plants, among others.

In performing these diverse projects, every conceivable climatic, geological, and locational problem has to be met and solved. A project may be in the open area of a desert; on or under the streets of a busy city; several hundred feet underground in any terrain; or at elevations of several thousand feet where the hazards of height, ice, snow, and the seasonally high waters of rivers have to be met.

In addition to the necessary personnel, these projects often require specialized equipment. Although the required equipment may already exist or need only slight modifications, often an entirely new design concept is needed. Therefore, contractor management must keep pace with improvements in design and performance of construction equipment.

The acquisition of equipment should be based on the economic fundamental of need; that is, will it be used for a current project or to fill a definite void? Too often, companies that are undercapitalized purchase heavy equipment before an actual need for it may exist. Unfortunately, such a purchase generally results in excessive carrying charges that, in turn, result in a diminution of current working capital and a decrease in net worth. This type of damaging situation can seriously hamper the company's position to bid on new work.

Because of the hazardous nature of the construction business, it is practically impossible to predict the profit outcome of a construction company's operations in advance, other than for the current year. Therefore, a company should structure its purchases to secure maximum financial and income tax benefits, particularly if high profits are forecasted for the current year.

Lease vs. Buy Decisions. The following are some of the factors to be considered when acquiring equipment:

- Is it new or used?

- Is it purchased outright or leased?

- Is it a trade-in or a part of a sale of related or other company equipment?

- Does it provide beneficial depreciation deductions?

- Does it qualify for an investment tax credit?

When acquiring new or used equipment, the machinery being considered should be evaluated as to modernization of design and the related productive capacity, cost of potential repairs and downtime, ultimate trade-in or sales value, available methods of depreciation, and investment credit. Vital factors in an outright purchase are to secure an equity asset that can be depreciated, that qualifies for an investment credit, and that will have a trade-in or sales value. Negative factors could be the drain on needed capital and the carrying of long-term equipment purchase notes at high interest rates.

On a straight lease, an asset's equity is not secured and the rental price can be excessive. However, leasing can be a more favorable option when the equipment is needed for only a short time, or the funds to make the purchase are not available.

A common practice used by contractors is to lease equipment on a long-term basis and have the option to acquire it as of a certain date, applying all or a part of the lease payments against the total purchase price. The IRS uses certain procedures in evaluating such an agreement to determine if it is in substance a lease or a purchase. If it is determined to be an acquisition, the lessee must capitalize and depreciate the asset. In this connection, the IRS has adopted Revenue Ruling 55-540, which states:

> In the absence of compelling factors to the contrary, implication of an intent to treat a transaction as a sale-purchase, rather than a lease or rental, would exist if one or more of the following conditions are present:
>
> a) Portions of the periodic payments are made specifically applicable to an equity to be acquired by the lessee.
> b) The lessee will acquire title upon the payments of a stated amount of rentals, which under the contract he is required to make.
> c) The total amount which the lessee is required to pay for a relatively short period of use constitutes an inordinately large proportion of the total sum required to be paid to secure the transfer of title.
> d) The agreed rental payments materially exceed the current fair rental value.
> e) The property may be acquired under a purchase option at a price which is nominal in relation to the value of the property at the time when the option may be exercised as determined at the time of entering into the original agreement, or which is a relatively small amount when compared with the total payments which are required to be made.

f) Some portion of the periodic payments is specifically designated as interest or is otherwise readily recognizable as the equivalent of interest.

After issuing this ruling, the IRS released Revenue Procedure 75-21, which sets forth additional guidelines to determine whether the transaction is a true lease or a purchase. These guidelines focus on whether the lessor has a sufficient economic interest in the property to justify treating the lessor as the owner of the property for tax purposes.

On a trade-in, any gain or loss must be capitalized as a part of the cost of the new equipment; on a sale, however, the gain or loss is a taxable transaction. Other factors to consider are the amounts of investment credit recaptured, newly acquired investment credit, and the amounts of depreciation to be secured.

Thus, this section has presented the variety of options facing a construction contractor who is contemplating purchasing and/or leasing equipment in order to plan for maximum financial and tax benefits. Sound management practices, combined with ingenuity and imagination, enable contractors to achieve an optimal relationship between equipment needs and cash flow, and subsequently succeed in their endeavors.

Depreciation Applied

As a general rule, expenditures that a company intends to be benefits in future years are required to be capitalized for tax purposes and depreciated or amortized over future years. Costs of acquiring equipment and other fixed assets are a typical example of expenditures that will reap tax benefits over several years.

Deductions for depreciation have been allowed since the inception of the federal income tax system in 1913. The depreciation deduction is intended to account for the reduction in service value of the equipment resulting from wear and tear and decay, including any decline in value for obsolescence resulting from technological improvement or economic changes.

For many years, the amount of the depreciation that was allowed was based primarily upon the estimated useful life of the asset to the taxpayer. The taxpayer could elect to base depreciation on the class life Asset Depreciation Range (ADR) system. In 1981, a substantially revised system of depreciation, known as the Accelerated Cost Recovery System (ACRS) was introduced. Under ACRS, as originally enacted, all eligible property was assigned a "statutory useful life" of 3, 5, 10, or 15 years, depending on the type of property. In 1984, the life for most domestic real estate was extended to 18 years. The predetermined recovery periods of ACRS are unrelated to, but are generally shorter than, the old rules regarding useful lives. However, the ADR rules will affect the ACRS classification of some property as indicated below. The annual recovery allowance for property in each of these classes is computed by referring to statutory tables that specify the appropriate recovery rate, based upon accelerated recovery methods that are explained below.

For cost recovery purposes, no distinction is made between new and used property. However, the distinction is retained with respect to the investment credit limitation for used property, as explained later in this chapter under the section entitled "Investment Credit."

ACRS applies only to property placed in service by the taxpayer after December 31, 1980; however, property that was used by an unrelated taxpayer prior to 1981 and was acquired by the contractor after 1980 is generally eligible. Taxpayers must continue to depreciate property placed in service before 1981 under the old useful-life depreciation system. As a consequence, most businesses use both the old depreciation system and ACRS until all their property that was purchased and placed in service before 1981 has been either fully depreciated or disposed of.

There is also the question of whether the various states will accept these provisions. California and New York, for example, have indicated that they will not accept them. Thus, for companies residing in California, New York, and any other state that refuses to accept the ACRS provisions, the taxpayer must use that state's methods of depreciation, in addition to the method that applies to pre- and post-1981 additions.

Unless the taxpayer elects to use a longer period, all tangible property must be placed in one of the four class periods, depending upon the type of property. The following lists the types of property assigned to each class:

3-Year Class:

- Automobiles

- Light-duty trucks (unloaded weight less than 13,000 pounds)

- Tangible personal property used in connection with research and experimentation

- Other tangible personal property that generally has a class life of four years or less, and certain horses

5-Year Class:

- Tangible personal property that is not included in the other categories, which includes most machinery, equipment and furniture

- Single-purpose agricultural and horticultural structures

- Bulk storage facilities for commodities, petroleum, or primary petroleum products

- Certain property (other than building or structural components) used as an integral part of manufacturing or production (including agriculture, mining, communications, or transportation)

10-Year Class:

- Public utility property with an ADR midpoint useful life of 25 years or less

- Real property with an ADR midpoint useful life of 12.5 years or less (intended to cover structures and land improvements in theme and amusement parks)

15-Year Class:

- Low-income housing

- Public utility property that generally has an ADR midpoint useful life of more than 25 years

18-Year Class:

- Real property not specifically covered under other classes and acquired after March 15, 1984; such realty acquired before March 16, 1984 is 15-year property

An asset that is assigned to one of these five recovery periods cannot be depreciated over a shorter period, even if the contractor is able to demonstrate that the asset has a significantly shorter life in its trade or business. For example, the acquisition of a five-year-class machine cannot be depreciated over less than five years, even if the taxpayer can demonstrate that the machine historically does not last for more than three years. However, upon disposal of the asset at the end of three years, a loss can be recognized for the remaining undepreciated cost under certain circumstances.

Depreciation Methods. The annual recovery allowance is computed by using statutory rate tables. Accordingly, the deduction for the ACRS allowance for an asset is arrived at by multiplying the adjusted cost basis of the asset by the appropriate recovery percentage for the tax year. Since the same recovery percentage applies to all recovery property in the same class placed in service in the same tax year, the recovery percentage may be applied to the total basis of any property in that category.

A taxpayer purchases a new pickup truck in 1985 for $8,000. The statutory table shows that the first-year recovery percentage for three-year property is 25 percent. Consequently, the first-year recovery allowance is $2,000. This is true regardless of whether the date of purchase is January 1, 1985 or December 31, 1985.

The second-year allowance is computed by multiplying the unadjusted basis of the property by the second-year recovery percentage shown in the table. Therefore, the unadjusted cost basis ($8,000) would be multiplied by 38 percent, resulting in a second-year recovery allowance of $3,040. Salvage value does not enter into the computation.

Except for realty, the rates are based on approximately 150 percent declining-balance method in the early years of the recovery periods, but switch to the straight-line method in later years. Salvage value is disregarded. Before applying the recovery rates, the cost basis may be required to be reduced if an investment credit is claimed on the property. (For further details see the section later in this chapter entitled "Investment Credit.")

The applicable percentage for class property, except for 18-year real property, is shown in Figure 18-1.

The "half-year convention" applies for the year placed in service; that is, all property is treated as if placed in service at mid-year, regardless of whether actual use of the property commenced at the beginning or at the end of that year. This adjustment is already factored into the tables. For example, a calendar-year taxpayer who acquires a light-duty truck and places it in service on December 31 is entitled to a full deduction for 25 percent of the truck's cost.

For 18-year real property, the half-year convention is not built into the tables. Instead, the recovery allowance starts in the middle of the month the property is placed in service and ends in the month the property is disposed of. The recovery percentages for 18-year real property, other than low-income housing, approximate the use of the 175 percent declining-balance method with an eventual switch to the straight-line method. The recovery percentages for low-income housing provide 200 percent declining-balance recovery with a switch to the straight-line method. The mid-month convention is not used.

FIG. 18-1 Applicable Recovery Percentage for Class Property

If the recovery year is:	The applicable percentage is:			
	3-Year	5-Year	10-Year	15-Year Public Utility
1	25	15	8	5
2	38	22	14	10
3	37	21	12	9
4		21	10	8
5		21	10	7
6			10	7
7			9	6
8			9	6
9			9	6
10			9	6
11				6
12				6
13				6
14				6
15				6

Component depreciation for real property is no longer allowed. Accordingly, the recovery period and method used for such separate items as wiring and plumbing must be the same as those used for the building itself. Figures 18-2 and 18-3 provide the applicable recovery percentage for 18 year real property and low-income housing. Elevators and escalators are treated as 18-year real property even though they are also eligible for the investment tax credit.

Since the recovery percentages depend on the month in which an asset is placed in service, separate items of 18-year real property may be grouped together only if they are placed in service in both the same month and the same year.

Substantial improvements to a building are treated as a separate building. Thus, for example, a new roof may qualify as recovery property, even though the building itself may have to be depreciated under pre-1981 rules.

Low-income housing property includes (1) federally assisted housing projects in which the mortgage is insured under the National Housing Act; (2) housing financed or assisted under similar local law provisions; (3) low-income rental housing for which rehabilitation expenditures qualify for depreciation deductions; (4) low-income rental housing held for occupancy by families or persons who qualify for subsidies under the

FIG. 18-2 ACRS Cost Recovery Tables for 18-Year Real Estate (Except Low-Income Housing)

If the recovery year is:	*The applicable percentage is:											
	1	2	3	4	5	6	7	8	9	10	11	12
1	9	9	8	7	6	5	4	4	3	2	1	0.4
2	9	9	9	9	9	9	9	9	9	10	10	10.0
3	8	8	8	8	8	8	8	8	9	9	9	9.0
4	7	7	7	7	7	8	8	8	8	8	8	8.0
5	7	7	7	7	7	7	7	7	7	7	7	7.0
6	6	6	6	6	6	6	6	6	6	6	6	6.0
7	5	5	5	5	6	6	6	6	6	6	6	6.0
8	5	5	5	5	5	5	5	5	5	5	5	5.0
9	5	5	5	5	5	5	5	5	5	5	5	5.0
10	5	5	5	5	5	5	5	5	5	5	5	5.0
11	5	5	5	5	5	5	5	5	5	5	5	5.0
12	5	5	5	5	5	5	5	5	5	5	5	5.0
13	4	4	4	5	4	4	5	4	4	4	5	5.0
14	4	4	4	4	4	4	4	4	4	4	4	4.0
15	4	4	4	4	4	4	4	4	4	4	4	4.0
16	4	4	4	4	4	4	4	4	4	4	4	4.0
17	4	4	4	4	4	4	4	4	4	4	4	4.0
18	4	3	4	4	4	4	4	4	4	4	4	4.0
19	–	1	1	1	2	2	2	3	3	3	3	3.6

*Use the column for the month in the first year the property is placed in service.

FIG. 18-3 ACRS Cost Recovery Tables for Low-Income Housing

If the recovery year is:	*The applicable percentage is:											
	1	2	3	4	5	6	7	8	9	10	11	12
1	13	12	11	10	9	8	7	6	4	3	2	1
2	12	12	12	12	12	12	12	13	13	13	13	13
3	10	10	10	10	11	11	11	11	11	11	11	11
4	9	9	9	9	9	9	9	9	10	10	10	10
5	8	8	8	8	8	8	8	8	8	8	8	9
6	7	7	7	7	7	7	7	7	7	7	7	7
7	6	6	6	6	6	6	6	6	6	6	6	6
8	5	5	5	5	5	5	5	5	5	5	6	6
9	5	5	5	5	5	5	5	5	5	5	5	5
10	5	5	5	5	5	5	5	5	5	5	5	5
11	4	5	5	5	5	5	5	5	5	5	5	5
12	4	4	4	5	4	5	5	5	5	5	5	5
13	4	4	4	4	4	4	5	4	5	5	5	5
14	4	4	4	4	4	4	4	4	4	5	4	4
15	4	4	4	4	4	4	4	4	4	4	4	4
16	–	–	1	1	2	2	3	3	3	4	4	4

*Use the column for the month in the first year the property is placed in service.

National Housing Act or local law that authorizes similar subsidies; or (5) housing that is insured or directly assisted under Title V of the Housing Act of 1949.

Disposition of Property. Treatment of gain or loss on the disposition of post-1980 recovery property other than 15-year real property is generally the same as under the prior law. Generally, gain or loss is recognized when recovery property is disposed of, unless another provision of the Internal Revenue Code allows nonrecognition (e.g., nontaxable exchanges and transfers to controlled corporations). Gain is treated as ordinary income to the extent of prior recovery allowance deductions. Any gain in excess of such deductions is treated as Section 1231 gain, and may be eligible for preferential treatment as long-term capital gain.

Optional Recovery Periods

In lieu of accelerated cost recovery, a taxpayer may elect to use the straight-line recovery method over any of the following recovery periods for one or more classes of recovery property.

Class	Straight-Line Recovery Periods Available
3-year property	3, 5, or 12 years
5-year property	5, 12, or 25 years
10-year property	10, 25, or 35 years
15-year public utility property	15, 35, or 45 years
18-year real property and low- income housing	18, 35, or 45 years

If an optional straight-line recovery period is elected, the election applies to all personal property in that class placed in service during the taxable year in which the election is made. The election is available on a year-to-year basis for each year's purchase of assets. However, once the election is made, it is binding for the entire recovery period for that particular year's purchases. For 18-year real property, the election is made on a property-by-property basis. Contractors may not use an accelerated recovery rate over one of the foregoing optional recovery periods.

In 1984, a contractor elects to use a 12-year recovery period for all 5-year property placed in service during 1984. The election applies only to property placed in service in that class in 1984. The contractor is free to treat 5-year property placed in service in 1985 differently.

The longer optional recovery period is provided primarily to assist those contractors who do not have sufficient income to use the deductions that are available under the shorter recovery period. For example, many contractors who use the completed-contract method of accounting for tax purposes frequently generate net operating loss carryovers in the early years of business, even though they could show a profit for financial statement purposes by using the percentage-of-completion method. On the other hand, net operating losses may now be carried over for 15 years, alleviating the problem of losses in early years. In addition, when faced with these circumstances, consideration should also be given to entering into third-party leases for the equipment rather than purchasing it outright, thereby passing these tax benefits onto the lessor. The lessee-contractor should be able to obtain an economic benefit through reduced lease payments.

Investment Credit

Rates. A dollar-for-dollar credit is allowed against income taxes based upon a certain percentage for investments in depreciable personal property. The applicable rate of investment tax credit (ITC) is determined by the property's recovery period or, under prior law, its estimated useful life. The ITC rates based upon the property's ACRS recovery period are as follows:

Recovery Period	Credit Rate
3-year property	6%
5-year property	10
10-year property	10
15-year public utility property	10

The rates under prior law were based upon the estimated useful life of the property as follows:

Estimated Useful Life	Credit Rate
Less than 3 years	0%
3–4 years	$3\frac{1}{3}$
5–6 years	$6\frac{2}{3}$
7 or more years	10

It should be noted that even if a taxpayer elects to use a longer recovery period (as previously explained), such as five years for 3-year property, the property will still not be eligible for a higher credit than 6 percent.

Eligible Property. Eligible property is defined as either (1) tangible personal property or (2) other tangible property, not including a building or its components, used as an integral part of (a) manufacturing, (b) extraction, (c) production, or (d) furnishing of transportation, communications, electrical energy, gas, water, or sewerage disposal services. In addition, certain other specific assets qualify, such as elevators and escalators, research facilities and facilities for the bulk storage of fungible commodities, single-purpose agricultural or horticultural structures, qualified rehabilitation expenditures, and storage facilities used in connection with the distribution of petroleum or its primary products.

Property used predominantly outside the United States does not qualify for the credit, except for certain property used in furnishing transportation services and certain energy-related activities.

Limitations on the Amount of Credit. Starting in 1983, the maximum amount of credit that may be claimed against the tax liability in any one year is limited to $25,000 plus 85 percent of the tax liability in excess of $25,000.

> A corporation's tax liability is $30,000, its qualified investment is $600,000, and therefore the 10 percent investment credit equals $60,000. The amount of credit that may be claimed in that year, however, is limited to $29,250 ($25,000 plus 85 percent of $5,000, the tax liability over $25,000).

Any ITC that cannot be used as a result of the above limitation can be carried back three years and, if still unused, can be carried forward 15 years.

For property placed in service after 1982, the basis of the property must be reduced by 50 percent of the ITC claimed. This reduction affects the computation of ACRS deductions and the computation of gain or loss upon disposition of the asset. Alternatively, the taxpayer may make an election to decrease the regular investment credit percentage by two points in lieu of the 50 percent basis reduction.

If the qualified property upon which the credit is claimed is not new property, the aggregate amount of used property that can be claimed for credit in any one year is limited to $150,000 for tax years beginning in 1985. For prior years, the credit is limited to $125,000.

Qualified Progress Expenditures. As a general rule, the investment tax credit is only allowed on property placed in service in a particular year. An exception to this general rule is for qualified progress expenditures.

A taxpayer may elect to claim advanced investment credits for construction progress payments made on property with a normal construction period of two or more years. The applicable percentage of each qualified progress expenditure must be based on a reasonable expectation of what the eligible ITC rate would be. The credit is to be applied to the lower of (1) the actual cash expenditures or (2) the cost of the completed project based on the percentage-of-completion method.

Recapture Rules. If property is disposed of (including casualties or thefts) before the end of its recapture period, the tax for the year of disposal is increased by the amount of the credit that is recaptured. This rule also applies if property ceases to be qualified property before the end of its recapture period, as when property is converted to personal use. The recapture amount is a percentage of the original credit that is claimed, depending on how long the property is held before recapture is required.

The following table illustrates the ITC recapture rules:

If the recovery property is disposed of or ceases to qualify within:	The recapture percentage is:	
	For 5-year, 10-year, and 15-year property	For 3-year property
The first full year after placed in service	100%	100%
The second full year after placed in service	80	66
The third full year after placed in service	60	33
The fourth full year after placed in service	40	0
The fifth full year after placed in service	20	0
Anytime thereafter	0	0

In the case of property that was placed in service prior to the application of the ACRS rules, the recapture amount is the excess of the credit originally allowed over the credit that would have been allowed if it had been computed using the actual period that the property was in service.

Credit for Rehabilitation Expenditures. Buildings and their structural components generally do not qualify for an ITC. However, present law excepts from the general rule expenditures to rehabilitate a commercial or industrial building that is at least 30 years old. The credit is 15 percent if the building is at least 30 years old, and 20 percent if the building is at least 40 years old. In addition, the credit can be as high as 25 percent for certain certified historic structures. Neither the regular ITC nor the energy credit is available for expenditures to which the rehabilitation credit applies.

The basis of the qualified rehabilitation property must also be reduced by the amount of the credit. In addition, to qualify for the new credit for rehabilitation expenditures, the taxpayer must elect to use the straight-line recovery method.

Property qualifies for rehabilitation only if expenditures during the 24-month period ending with the tax year are more than the adjusted basis of the property, or $5,000, whichever is greater. In addition, 75 percent or more of the existing external walls must be retained. Enlargements of a building do not qualify.

INCOME TAXATION OF VARIOUS FORMS OF ORGANIZATION

Sole Proprietorship

Operating as a sole proprietor is the simplest form of doing business. A proprietorship is not a separate taxable entity; rather, the individual owner of the business is taxed directly on the profits from the business. Income and expense are reported on the owner's personal income tax return, usually on Schedule C, and business profit or loss is combined with the owner's other income. Thus in the very early stages of a construction enterprise, the owner can offset losses against his other sources of income. Even if the owner has no outside income in the year of the loss, the loss can be carried back three years to offset income from all nonconstruction sources, and prior taxes can be recouped. In a corporate form, such a start-up loss can be carried forward to offset future profits, but it may be some time until it is fully used.

As a business grows, the disadvantages of sole proprietorship may outweigh its advantages. The construction business can involve substantial risk, and a form of business that insulates the owner from a potentially catastrophic loss is often desirable. The corporate form serves this purpose because it protects the owner from personal liability. There may also come a time in a growing business when the owner wishes to raise capital or to share the responsibilities of the business by sharing the ownership, for example, by taking on a partner or by selling corporate stock. Finally, corporations are separate taxable entities, and corporate tax rates may be lower than individual rates.

Partnership

As defined by the Uniform Partnership Act, a partnership is "an association of two or more persons to carry on as co-owners a business for profit." The Internal Revenue Code does not, however, define a partnership. Instead, it enumerates a number of business entities that may, under appropriate circumstances, be treated as partnerships for tax purposes. The enumeration includes "a syndicate, group, pool, joint venture, or other unincorporated organization, through or by means of which any business, financial operation, or venture is carried on, and which is not, . . . a trust or estate or a corporation. . . ." For tax purposes, the characteristics of the entity control what type of entity exists. If an association has more characteristics of a corporation than a true partnership, it will be taxed as a corporation.

For accounting and tax purposes, the partnership is considered a separate accounting entity, but not a separate tax-paying entity. The partnership files an information return and may select its own methods of accounting. Net income or loss for the year is deemed to be distributed to the partners as of the last day of the partnership year. Each partner reports on his income tax return his distributive share of income, loss, credits, and so forth.

The pass-through of income to the partners is an attractive aspect of partnership taxation. The earnings of a partnership are taxed only once, to the partner. Partnership losses can offset a partner's other income. Income has the same character to the partner as to the partnership. Thus, capital gains or losses generated by the partnership will be taxed as such to the partners. If a partnership earns tax-free income, such as interest on municipal bonds, it is tax-free to the partner.

A partner's contribution to the capital of the partnership is generally a nontaxable event. The basis of the property to the partnership is the same as the partner's basis at the time of the contribution. The partner's basis in his partnership interest equals the sum of any cash contributed, plus the adjusted basis of other contributed property.

Corporation

The corporate form of business is common in the construction industry. The corporate form insulates shareholders from personal liability, a benefit not available to proprietorships and general partnerships. Ownership of a corporation is generally more easily transferable than ownership of other business entities. As a legal entity, the corporation provides the important conveniences of being able to acquire, hold, and convey property; to contract, to sue and be sued; and generally, to act as a single unit under its own name, separate and distinct from its shareholders. While these and other nontax considerations (such as the limited liability afforded by a corporation) are traditional reasons for choosing to incorporate, tax considerations also make the corporate form an attractive way of conducting business. The following are some of the most distinctive tax advantages of corporations:

- *Sheltering Income.* The tax rate on corporate earnings is often lower than the tax rate on individual earnings. As of this writing, corporations pay a graduated-rate tax on the first $100,000 of income, and pay 46 percent on income in excess of $100,000. The maximum rate for individuals is 50 percent. To the extent that earnings are left in the corporation, there is more money available for business uses. Also, corporations are not limited on their deductions of investment interest. Investments may therefore be financed with pre-tax dollars.

- *Timing of Income.* Owner-executives have the ability to control income with respect to the timing and form of payment. The ability to control the payment of salary, bonuses, and dividends gives the owner-executive the opportunity to shift income to those years when it will be taxed in a lower tax bracket.

- *Corporate Retirement Plans.* Retirement benefit plans, including deferred-compensation, pension-, and profit-sharing plans have generally been given favorable tax treatment under the Internal Revenue Code. If a plan meets certain basic qualifications (e.g., if it is judged nondiscriminatory with respect to coverage, vesting, and funding), contributions of the corporate employer will be currently deductible, but will not be taxed to the employee until they are paid out. The qualifying criteria for these plans, however, are not so stringent as to preclude the tailoring of a plan that can be both beneficial to the owner-employee and the nonowner-employee.

- *Creating a New Taxpayer.* A newly organized corporation may choose for a taxable year the fiscal year that aids its shareholders the most. For a seasonal business, this may mean selecting a fiscal period that permits profits to be balanced by losses. A new taxpayer also has the ability to select accounting methods that are different from those that may have been used while the business was operated by the previous taxpayer.

- *Fringe Benefits.* Certain fringe benefits are available only in corporate form. Incentive stock options are an attractive means of passing through the earnings of a corporation to key employees at no cost to the corporation. In a qualified medical expense reimbursement plan, the corporation can reimburse the employee for medical expenses and deduct the reimbursement, and the employee will pay no income tax on the reimbursement. The corporation can carry life insurance on its employees, and the premiums on the first $50,000 of coverage will be nontaxable to the employee.

- *Disposing of the Corporation at Capital Gains Rates.* Shareholders can transfer ownership of the corporation by selling their stock through stock redemptions or by liquidations, and pay tax on any gain at a capital gains rate. Such treatment is not always available to the proprietorship or the partnership.

S Corporations

In recent years, conducting business as an S corporation has achieved great popularity. For nontax purposes, the S corporation is a regular corporation providing limited liability for shareholders. For tax purposes only, an S corporation has the tax characteristics of a partnership.

Shareholders are taxed directly on S corporation earnings, whether distributed or not. Items of corporate income, such as long- and short-term capital gains and losses, are treated separately by the corporation and passed through to the stockholders. Other items passed through to the shareholders include: tax-exempt interest, credits earned by the corporation, such as investment tax credit, and deductions for charitable contributions. S corporations offer opportunities to shift income among family members through the ownership of stock.

Certain requirements must be met for a corporation to elect S corporation status. The election must be unanimous by the shareholders, must be made within prescribed time periods, and generally must meet a specific set of qualifying rules.

Joint Ventures

The joint venture is an entity formed for the purpose of carrying out one transaction or a series of related short-term transactions. For tax purposes, it is treated as a partnership. Each participant in the venture picks up his share of profits and losses as they are earned, regardless of when distributions are made. A partnership return is required, and the venture must exercise its own elections relative to accounting methods and periods.

Notwithstanding the similarities in taxation between a joint venture and a partnership, the legal differences between the entities may make the joint venture more attractive from a contractor's point of view. The degree of agency existing between a joint venturer and the venture is generally less than that between a partner and the partnership. Thus, it may be more difficult for one party to bind the venture to debt. It is easier for a joint venturer to sue a fellow member of the joint venture on the contract between them, than for a partner to sue the partnership. Finally, some states prohibit a corporation from being a partner, whereas there are no restrictions on a corporation becoming a joint venturer.

INCOME TAX ACCOUNTING PERIODS
Business Cycle Considerations

In determining its year-end, a contractor should take into consideration its business cycle. For example, a building contractor that performs most of its work during the warm months would generally not desire to have its year-end in June, July, or August, when its activity is at its highest level. Terminating the year-end at such a time and starting a new set of financial reports could make the cutoff very time-consuming. Allocating items to the appropriate period and other considerations is better avoided, if possible. In addition, accounting personnel, as well as the information system that is being used, are busiest at this time. An appropriate year-end is often considered to be at a point when business activity is at or near its lowest level, since a contractor is then able to minimize additional costs and use its accounting personnel and equipment more efficiently. In addition, at this time the number of items requiring allocation between accounting periods is at a minimum.

Generally, most taxpayers use the same year-end for tax purposes as for financial reporting purposes. The requirements and cutoffs needed to close the books at two different year-ends cause excessive recordkeeping and, if all procedures are not followed at the tax year-end, the taxpayer could be subject to attack by the IRS.

Taxpayers using the completed-contract method of accounting may find it beneficial to adopt a year-end that occurs just prior to the month of the year in which profitable contracts are normally completed and accepted. Such a selection defers recognition of income. This tax deferral may have a limited long-range impact, however, because of the increasingly stringent estimated tax payment requirements, expecially for larger taxpayers.

Choice of Year-End

In filing its initial return, a taxpayer may adopt any taxable year-end that results in a first taxable year that does not exceed 12 months. Generally, the year-end will be the same year-end used in bookkeeping. An individual contractor operating a proprietorship must use the same year-end as his personal year-end, which is usually the calendar year-end.

Year-Ends of Joint Ventures and Groups

A partnership may adopt a year-end that is the same as the taxable year-end of all of its principal partners. A principal partner is a partner having an interest of 5 percent or more in profits or capital. If all of the partnership's principal partners are not on the same taxable year, a calendar year must be adopted. Any other year-end for a partnership must be approved by the IRS upon the filing of an application on or before the last day of the month following the proposed year-end. Approval of such an application requires the establishment of a business purpose to the satisfaction of the IRS.

Any corporation other than an S corporation may adopt a year ending on the last day of any month. An S corporation may adopt a taxable year ending on December 31, or must comply with the S corporation rules for adoption of another year-end. Generally, this is part of the election to S corporation status.

While most taxpayers normally adopt the year-end using the last day of a month, a taxpayer may elect a taxable year that varies from 52 to 53 weeks, that always ends on the same day of the week, and that ends either on the same day in that calendar month, or on the same day nearest to the last day of that calendar month.

Changing the Year-End

If a taxpayer desires to change its year-end, it must either receive permission for such a change from the IRS or meet the rules relating to changes in the year-end that do not require advance permission by the IRS. A corporation that becomes a member of an affiliated group filing a consolidated return is generally required to adopt the same year-end as the parent of the group.

INTERNATIONAL TAX CONSIDERATIONS

A U.S. contractor who is expanding abroad can anticipate less flexibility regarding tax accounting policies from foreign taxing jurisdictions than from the United States. For example, some foreign jurisdictions are not likely to allow the deferral of construction profits through use of the completed-contract method of accounting; they are likely to require contractors to use the percentage-of-completion method, as used for the books. At the same time, some countries allow faster recovery of fixed asset investments than that available under U.S. tax laws.

Foreign countries tend to rely more heavily on indirect taxes such as excise, turnover, and value added taxes than does the United States. Even so, most foreign countries do levy some form of income tax. These income taxes, or similar types of taxes, can be credited against U.S. income taxes. Foreign taxes, including income taxes, may be claimed as a deduction for U.S. income tax purposes, where that is more beneficial than the credit.

Many foreign countries in need of capital and desiring new business offer attractive tax incentives to businesses such as construction contractors. These incentives can take the form of exemptions for a period of years (a "tax holiday"), tax exemptions for certain kinds of activities, remission of import duties on certain materials, and other similar tax benefits.

U.S. taxpayers may also benefit from income tax treaties between the United States and foreign countries by gaining tax concessions. Taxpayers relying on treaties, however, are cautioned that the treaties are constantly being revised and interpreted and can only be expected to conclusively determine items specifically mentioned. Items not mentioned in the treaty must be reported according to the tax laws of the country where the work is located.

If the foreign operation is directed toward a single, one-time contract, short-term tax considerations may predominate. If the decision is made to enter the construction business abroad indefinitely, longer-term considerations will play a more important role (see the section later in this chapter entitled "Form of Organization").

Income Tax Rates Encountered

Foreign income tax rates may be higher or lower than U.S. rates, and the statutory rate may not indicate the effective rate because of the manner in which taxable income is computed pursuant to the tax laws of that country. If the income tax rate in the foreign jurisdiction is lower than the U.S. rate, the use of a subsidiary organized abroad, rather than a branch operation, may be more beneficial in deferring the higher U.S. income taxes. The gain or loss suffered by foreign operations that are, either currently or at a later date, to be used for U.S. income tax purposes are to be computed under U.S. tax laws. This can result in foreign taxes being assessed on operations that are showing losses for U.S. tax purposes, thus deferring, or even eliminating, the credit for foreign income taxes.

Form of Organization

For the contractor expanding outside the United States, the first tax consideration is the form of organization. Whether the foreign operation should be a branch of the U.S. company, a separate U.S. corporation, a separate foreign corporation, or a joint venture with a local partner is of primary importance. Often the local business climate dictates the legal form of the foreign operation (e.g., limited liability, need for local representation). Income tax considerations usually depend upon the nature of the business being conducted. Operation of the foreign business by a branch of the U.S. parent corporation, a U.S. subsidiary corporation, or a joint venture composed of such entities would result in U.S. tax rules being applied. This is so because under the Internal Revenue Code all citizens and resident individuals, as well as all domestically incorporated corporations, are taxed on worldwide income, regardless of where the activities are located. The potential for double taxation is greatly alleviated by U.S. tax law that provides for a credit against U.S. income taxes for foreign income taxes paid subject to certain limitations.

Generally, if the foreign operation is likely to incur start-up losses or high foreign tax rates, it is advisable to have it organized as a U.S. entity. On the other hand, if profitable operations can be expected from the beginning, or if the operations will be subject to foreign rates that are lower than U.S. rates, a foreign incorporated entity is probably more beneficial. Special tax rules apply when U.S. taxpayers transfer assets to a foreign corporation.

A branch of the U.S. corporation, a partnership with a U.S. entity as partner, or a domestic U.S. subsidiary can offset foreign operating losses against current U.S. income. A foreign corporation generally allows the deferral of U.S. taxation on overseas profits. However, if the foreign operation will only earn short-term profits and the anticipated profits are to be repatriated to the United States in the near future, only a short-term deferral is available, and it is advisable to organize the operation as a U.S. entity. Nevertheless, if foreign income is to be reinvested in foreign operations, considerable benefits may derive from using a foreign entity that provides deferral of U.S. taxes. In some situations, the undistributed profits of a foreign corporation may be deemed currently taxable to U.S. shareholders. Therefore, careful planning is required when using a foreign subsidiary.

Winding Up the Foreign Activity

If foreign operations are terminated before profits are realized, a branch office can maximize its current tax benefits by having the operating losses offset domestic taxable profits as the losses are realized and recognized. A domestic corporation included in a consolidated U.S. return provides the same beneficial results. With proper planning, an unprofitable foreign subsidiary can be terminated with an ordinary loss deduction that is available to the parent company for U.S. tax purposes.

After profits are realized, winding up a U.S. branch office should result in no additional tax being levied, inasmuch as all current operating results would already have been recognized for U.S. tax purposes. A domestic corporation included in a consolidated U.S. return may likewise be closed without having to pay additional U.S. tax.

After a profitable history, the winding-up of a foreign corporation is likely to result in the U.S. taxation of the parent company on undistributed earnings of the subsidiary, with a tax credit allowed for the foreign income taxes previously paid on those earnings. Any excess of value over the accumulated earnings is eligible for long-term capital gains reporting.

STATE AND LOCAL TAXES ON INCOME AND OTHER TAXES

General Rule

Despite their generally much lower rates, state and local taxes can be more important to the construction company than federal income tax. Federal income tax applies to net taxable income; losses can be carried back 3 years and forward 15 years. Thus, the construction company pays federal income tax only on taxable income that, through proper planning, can be deferred and even avoided. Avoidance is legal; evasion is not. This is not the case for many state and local taxes. Most states do not allow loss carryovers or, if they do allow them, they severely limit them. Moreover, many state and local taxes apply to other than net taxable income; bases of gross receipts and so forth are not unusual.

Construction firms planning to expand into an unfamiliar state or municipality should investigate all the applicable state and local taxes. The State Development Office, Chamber of Commerce, or similar local group can provide useful information. State and local taxing authorities, like foreign countries, have been known to waive or mitigate taxes in order to attract business. Any major move should be made only after consulting local professional tax specialists. Commercially published state and local tax services are often inadequate in explaining the application of such taxes to construction contractors, and many do not include any sort of planning sections.

Most of the 50 states of the United States have some sort of income tax, as do a number of cities, counties, and other smaller subdivisions. All of them differ, in some way, from the federal law and from each other. Some states, while otherwise insisting on apportionment for determining the taxable income within the state, allow taxpayers engaged in construction to take advantage of special factors or to use separate accounting methods.

"Unitary" Jurisdictions

Corporations and other taxpayers operating a business in two or more states must determine how much income to report to the various states. While separate accounting has the desirable attribute of simplicity, it has gradually been rejected by most states. Rather, a uniform system of apportionment between the various jurisdictions is usually required.

Apportionment is the assignment of a part of the total income of a business to a particular state. The apportionment is normally done on the basis of a three-factor formula, which relies on the property, payroll, and sales. Most states have provided uniform administration policies for apportionment by enacting the Uniform Division of Income for Tax Purposes Act.

When the business being operated is conducted by more than one entity, and any particular entity's results are clearly contributory to or dependent upon the activities of another related entity, tax authorities have decided that additional steps may be necessary to obtain a realistic apportionment. In order to do this, the concept of a "unitary" business has been defined, and where present, will result in a grouping of the activities of the various entities into a single, combined report, from which a part will be apportioned to the taxing jurisdiction. Generally, a unitary business exists when two or more corporations, related in some way, such as a parent and a subsidiary; a single corporation having multiple divisions; or a corporation acting as a partner or joint venturer with another operate a single business. Because of the nature of construction activities, contractors often find themselves classified as a unitary business—even to the extent of having joint *venturers* being considered in part as unitary with the joint ventures.

California is a leading proponent of apportionment and the unitary concept. It has written regulations that are specifically applicable to construction contracts that are classified as long-term contracts. Construction contractors not engaged in long-term jobs are subject to normal allocation, apportionment, and unitary business rules.

The normal business income of a contractor is apportioned to California using the three normal factors: payroll, property, and sales. However, special rules have been developed to determine the amounts of each of the factors for construction firms. These special rules are needed most in determining the sales factor, since much distortion could occur from year to year from the inconsistent reporting of gross contract prices, progress billings, and the like.

California and many other states provide a special rule for applying annual apportionment percentages to contracts reported on the completed-contract method of accounting. Because the interim years show an absence or loss of income from such a contract, apportioning during those years is precluded. Instead, at completion, the total income or loss from the contract is determined, and it is then assigned to the years during which the contract was performed, based on when the costs were incurred, and using a weighted average of the annual apportionment percentages. This determination is made on an individual contract basis, not on the total business income of the taxpayer for any year.

When California considers whether or not a construction corporation is unitary with a joint venture in which it has an interest, the normal ownership requirement of "more than 50 percent" is not used. A corporation in the construction business and reporting to California is deemed to be unitary with its percentage ownership share of a partnership or joint venture performing construction activities regardless of the percentage interest owned.

The combination of unitary taxation using apportionment factors in one state and separate accounting rules in another state can result in income escaping tax or in double taxation. Contractors facing double taxation should consider making a special request to the offending state for relief from such rules, especially if they can use other, reasonable methods of apportionment or separate accounting methods that would lessen the total tax burden.

Gross Receipts, Sales, and Use Taxes

Some states' gross receipts tax can be a significant job cost to the construction company. Because construction contracts are often for very large amounts, which include funds to be paid out by the prime contractor, tax levied on gross receipts becomes disproportionately significant. Such a tax may present accounting problems, questions as to the timing of deductions for federal income tax purposes, and opportunities for some credit against the state's normal net income tax.

Most of the states and many local governments have retail sales and use taxes. Construction contractors usually must pay sales tax on the materials, supplies, and equipment they use, because the contractor is deemed the consumer of these goods. In addition, some states' sales tax authorities take the position that the tax is applicable to "fabrication labor." Use tax is levied when the form of the transaction is for some reason not subject to the sales tax, as when goods are used or consumed in the state, even though they are purchased from an out-of-state vendor. Some states exempt materials used in construction from the sales and use tax.

If equipment is sold by the construction firm, either the sales tax should be collected or a resale certificate should be obtained from the purchaser. Equipment rentals may also be subject to sales tax.

As with most other forms of state and local taxation, wide variations in law and rulings applying the law to construction activity make anything but a general statement impossible. Proper tax planning can take place only after the contractor has researched a given taxing jurisdiction.

Property Taxes

Property taxes on tangible personal property are of major importance to construction firms that own equipment or have large quantities of materials and supplies in the warehouse or at the job-site. Real estate that is owned is also subject to property taxes in most states. Usually, the property tax applies to the value of tangible personal property that is present in a given jurisdiction on a given date. Rates vary, as does the method of determining either fair market value or assessed value. The contractor may make substantial savings by carefully planning for the delivery and use of personal property that is necessary to the construction operation.

In many instances, it is not clear whether the subject matter of the contract constitutes real or personal property. Generally, state sales and use taxes are not levied on the former, but are levied on the latter. Because most states regard construction contractors as ultimate consumers who transform tangible personal property into realty, the authorities seek to impose the sales or use tax on the cost of the materials—tangible personal property—used in constructing the property. The construction firm's estimating team should take the treatment of these taxes into serious consideration when making a bid.

Payroll Taxes

Virtually all employers must become familiar with payroll taxes at the federal level, as well as at the state and local levels where local personal income taxes are levied. Many states and other local jurisdictions have income tax statutes that require employers to withhold state and local income taxes.

In addition to income taxes, the employer may be taxed for state unemployment benefits. Payment of state unemployment tax generally reduces the federal unemployment tax by a full or partial credit. The state and federal wage bases and due dates for unemployment taxes may differ. A taxpayer engaged in construction activities may face a rate that is quite high because of the seasonal nature of construction employment.

Another expense related to payroll is workers' compensation insurance or tax, which compensates an employee disabled by job-related sickness or injury. In most cases, this "tax" is paid by the employer and is administered through a state government agency, although some states allow voluntary coverage through a private program that meets state requirements. In addition to considering the rates and coverages of the state fund versus those of a private program, a construction firm should weigh the protection that may be available under state coverage against possible lawsuits from employees.

Construction employees may often be working for two or more related companies at the same time, perhaps as an executive of the parent company and as a project manager for a subsidiary. In this situation, the common paymaster rules may be used to reduce total payroll tax for the group. These rules apply to Federal Insurance Contributions Act (FICA) and Federal Unemployment Tax Act (FUTA) taxes, and may be available for comparable state taxes.

Payroll taxes must be withheld, accrued, and paid over to tax authorities regularly and promptly. Failure to do so exposes the persons responsible to substantial penalties.

Excise Taxes

Federal excise taxes are imposed on the sale or use of certain items, on certain transactions, and on certain occupations. There are manufacturer's excise taxes on trucks and equipment, and on petroleum products and firearms. If a business involves liquor, gambling, or firearms, it may be subject to an occupational excise tax. The excise tax affecting the largest number of small businesses is probably the highway use tax, levied on certain motor vehicles, such as trucks, that operate on public roads.

Federal excise taxes are due without assessment or notice; the construction company should inquire at an IRS office to determine if there are any excise taxes that it must pay. Certain state excise taxes have the characteristics of a gross receipts tax. These taxes may or may not be applicable to construction contracts. Excise taxes are also levied by cities or other political subdivisions. Again, the construction company should make inquiries in the taxing jurisdiction about these taxes.

TAX PLANNING FOR INDIVIDUALS

Current Compensation

Current compensation is generally the cash remuneration that an employee receives as he is working. Often, it has two components: a basic amount, and a bonus.

Basic compensation is usually a fixed amount paid periodically. Special pay may be added to the regular basic amount for employees who work under particularly demanding or hazardous situations. For example, those who work at high altitudes, underground, in hazardous areas, with hazardous materials, for long hours, in distant locales, and in other special conditions often receive special pay. This special pay usually ends when the reason for the payment ceases.

Bonuses are extra pay. They can be determined in very simple or very complex ways. A bonus may be an arbitrary amount awarded by a supervisor, or a complicated objective award by the Board of Directors. Bonuses are usually paid in cash, but may be paid in property, such as shares representing ownership in the corporate employer.

All pay must be reasonable in amount to be tax deductible by the payor. Generally speaking, if an objective bonus plan is reasonable when it is established, it will remain reasonable in future periods, even though the annual bonus payments might become quite substantial.

Bonus plans vary in format and potential generosity. It is common for a construction company's chief executive officer to receive 25 percent of his total annual current compensation in his bonus. Though not common, it is not a rarity for such a chief executive's annual bonus to equal or exceed his base compensation.

Employees outside the executive officer corps may also share in a bonus program. Usually, such an arrangement is called a performance pay plan. For example, if a supervisor completed a job budgeted for 1,000 man-hours in 900, he might receive a bonus based on the hours he saved. Likewise, the completion of a job ahead of schedule, under a dollar budget, or using any other objective criteria could also be called a performance pay plan.

Fringe Benefits

Fringe benefits are items of current compensatory benefit that are provided to employees, often at no cost or at a reduced cost, and often in a format that eliminates or reduces their taxability to the employee. Some of these tax advantages are specifically granted in the Internal Revenue Code, others have been granted by IRS administrative action, and others are theoretically taxable items that are traditionally overlooked by the IRS.

Insurance. An employer may provide group term life insurance for employees in coverage of up to $50,000 per employee and $2,000 per dependent, with the cost of coverage tax deductible to the employer and tax-free to the employee. Coverage in excess of these amounts is charged to the employee as compensation in an amount determined by a table published in Treasury regulations. For tax years beginning in and after 1984, group term plans must be nondiscriminatory in order for the benefits provided to key employees to be tax-free to them.

Split dollar life insurance is a contractual arrangement structured around a whole life policy. Simply stated, the employer commonly has the right to the cash value of the policy and the employee has the right to the rest of the policy. The policy is usually structured so that the pure insurance element (face amount less cash value), plus additional coverage provided by dividends and the like, keep the employee's interest at an amount that approximates the policy face amount. The employer generally receives no tax deductions at all for his premium payments in a split dollar program. The employee is taxed on compensation income in an amount equal to the cost of a one-year nonrenewable term policy. This is the theory on which the PS-58 table rates are based, but these rates can often be improved upon by insurance company rate schedules. When creatively combined with other insurance policy techniques, such as minimum deposit, a split dollar program can provide a high benefit at a relatively low cost, particularly for younger employees.

It appears that an employer may provide an employee accidental death and travel insurance that will be tax-free to the employee and tax deductible to the employer. This position is based upon former group term life insurance tax regulations that provided that "general" death benefits did not include amounts paid for travel insurance or accident and health insurance, or amounts payable under a double indemnity clause or rider. The current regulations provide that "life insurance is not group-term life insurance for purposes of Section 79 unless . . . it provides a *general* death benefit. . . ." (Emphasis supplied.) Since this is a specialized coverage, it apparently does not affect a group term program, and the IRS has not suggested that it should be taxable on other grounds.

Employers can provide employees with tax-free medical and disability income insurance. However, employee payment of the disability income premiums is relatively popular because the premiums are low and, if the employee pays the premiums, then any policy benefit payments received by the disabled employee are tax-free.

Medical Plans. Amounts received by employees under a nondiscriminatory or insured medical expense reimbursement plan are tax-free to the employee. However, in the past, most medical expense reimbursement plans were discriminatory, because company-funded plans covered only key employees in amounts greater than those offered under the company's basic medical insurance plan. Such payments would now constitute taxable income to the recipients. However, physical examinations and other diagnostic procedures may be provided to selected employees without constituting taxable income to them.

Executive Plans. Personal use of company property, such as a company car, airplane, yacht, or apartment, is taxable to the user as compensation. Nonetheless, the part-time personal use of these assets may be much more economically efficient than full-time ownership.

Travel on behalf of an employer is a business expense, so it is not an accident that so many conventions and other business meetings are conducted in attractive locales. Per diem allowances that are very adequate can also build morale.

Lodging and meal costs are excludable from an employee's income when the meals are provided on the employer's premises and the employee is required to accept the lodging as a condition of employment. If these conditions are met, the meals and lodging may be provided tax-free to the employee and the employee's spouse and dependents. This benefit can be quite significant to construction personnel on location at a prolonged job. This rule is also used to justify a subsidized employees' cafeteria or executives' dining room, particularly where no restaurants are convenient. Another customary practice is the payment of a supper money allowance to employees who have worked overtime, although the nontaxability of this practice is not specifically sanctioned in the Internal Revenue Code or regulations.

Employers may provide educational assistance under two distinct rules, and in both cases the value is not taxable to the employee. In the first type of program, educational costs for courses that will improve one's job skills may be paid by an employer without creating taxable income to the employee. In tax jargon, these employee training expenses are simply described as an "ordinary and necessary" cost of doing business. Clearly, the employer may arbitrarily determine which employees are to be enrolled and which are not.

In the second type of program, the employer may pay or reimburse the employee for virtually any type of educational expense. In order for this program to be tax-free to employees, it may not discriminate in favor of employees or their dependents who are officers, owners, or are highly compensated. Employees covered by a union contract may be excluded, and certain other limitations also apply.

Employers may also provide their employees with a nondiscriminatory dependent care assistance program on a tax-free basis. This type of program can provide benefits for the care of children, or for the care of another individual who qualifies as a dependent of the employee.

Noncurrent Compensation

Noncurrent compensation can be structured along both "qualified" or "nonqualified" formats. In general, where an arrangement is qualified, current funding is permitted and the employer gets a current tax deduction. However, the employee is not taxed until he actually receives the funds. In a nonqualified arrangement, the rules vary: If the arrangement is funded and nonforfeitable, the employee recognizes the income currently and the employer takes the deduction currently. However, if the arrangement is unfunded or forfeitable, both the income recognition by the employee and the deduction by the employer are deferred until either the cash payments are made to the employee, or the employee's right to the payments becomes nonforfeitable.

Both qualified and nonqualified arrangements could accurately be called forms of deferred compensation but, in common tax parlance, the term "deferred compensation" is generally applied to nonqualified arrangements. Qualified plans basically function as retirement savings plans; however, with certain types of qualified plans it is often possible to withdraw funds prior to retirement without penalty.

A qualified plan permits current employer deductions of amounts contributed to an employees' trust, while deferring the employee's recognition of income until the employee receives distributions from the trust. A compensation plan is considered "qualified" when it conforms with the myriad rules that are, in general, designed to reduce discrimination in coverage and benefits, provide a reasonable vesting schedule, and provide for funding the benefit.

In general, there are two types of qualified plans: defined benefit plans and defined contribution plans. In a defined benefit plan, the projected employee benefit is first determined, and the current employer contribution required to fund the benefits is then actuarially determined. In a defined contribution plan, the current contribution is determined in accordance with the plan; the employee's benefit is a share of the accumulated contributions and their investment yield.

Defined contribution pension plans, often called "money purchase pension plans," are so named because the employee's benefit is equal to what the money in his account can buy. Contributions to a defined contribution pension plan are usually stated as a percentage of an employee's basic wage, often excluding overtime and bonuses, although they can be included. Similarly, defined benefit pension plans have benefit limitations.

A stock bonus plan is a defined contribution plan. Contributions are usually made in employer stock, or in cash used to buy employer stock. Ultimate distributions from the plan may be in the form of cash, options to purchase stock, or solely in employer stock. Company contributions do not depend on profits, and can be tax deductible to the company without incurring a cash outlay. Employees can receive distributions of employer stock without incurring any taxable income at the time of distribution.

A "cash or deferred" arrangement, within specified limits, allows an employee to receive taxable cash compensation or defer a part of his compensation and have his employer contribute it to the employee's account within a profit-sharing or stock bonus plan sponsored by the employer. In addition to the regular qualification requirements applicable to profit-sharing and stock bonus plans, cash or deferred arrangements have additional nondiscrimination requirements, and amounts deferred into the plan by an employee must be immediately and fully vested and nonforfeitable. The benefit such an arrangement is that it allows the employee flexibility in choosing to receive current cash income or a tax-deferred benefit.

Estate Planning

Estate planning involves planning for the creation, preservation, and disposition of an individual's estate. While taxes are an important factor, personal objectives should be the focal point of planning. This section will briefly review the rules governing federal gift and estate tax and planning ideas for the transfer of closely held business interests.

Gift and Estate Tax Planning. The gift and estate tax is a tax on the transfer of property. The 1976 Tax Reform Act unified the gift and estate tax to ensure that the total transfer tax is about the same, whether a person gives away assets during his life or transfers them at death. There is a single unified rate schedule for the gift and estate tax. The rates are progressive on the basis of cumulative lifetime and deathtime transfers. The rates range from .8 percent for the first $10,000 in taxable transfers, to 50 percent for taxable transfers in excess of $3 million. (Prior to 1985 the top rate for such taxable transfers was 55 percent.)

If the gift and estate tax is based upon cumulative transfers, why make gifts at all? Gift transfers are advantageous because there is no tax on annual exclusion transfers; they reduce the estate in the highest bracket, any future appreciation on the property given is excluded, and the income earned on the property goes to the donee. Further, any gift tax paid more than three years before the donor's death is excluded from the estate.

Planning for the Closely Held Business. Generally, the purpose of shifting income is to lower the income tax on the family's overall income; to avoid the need for the donor to use after-tax dollars to give financial assistance to family members; and to avoid the income from building up in the donor's estate, while assisting family members to develop their own estates. The purpose of shifting business interests is not only to shift the income, but also to "freeze" or reduce the size of the donor's estate, and thus reduce estate taxes.

The advantages and disadvantages of the various forms of business organizations were covered earlier in this chapter. In initial business planning, it is a good idea to consider setting up a limited partnership to own any real estate or equipment that can be rented or leased to the operating business entity. During the early years, these may act as tax shelters by diverting tax losses to the owners. In later years, interests in the partnership may be transferred outright, or may be put in trust to other family members and income may be shifted to them. In addition, eventually it may be easier to sell the business without these assets. If family members will be receiving stock in a corporation currently or in the future, issuing common and preferred stock at the inception of the business should be considered; the common stock in order to carry future appreciation, and the preferred stock in order to have a fixed value and a dividend. This structure may be used to transfer control of the company to one group of the family and income to another.

The owner has a number of ways to shift or transfer the value of a going concern to family members. Among them are gifts, a sale, a combination of a gift and sale, or the transfer of "know-how" or opportunities. The valuation of any interest transferred is important. Where the gift or sale is of a minority interest in a company that has no dividend history and whose stock lacks marketability, a discount on the value of the stock from 15 percent to 40 percent is not unreasonable. As long as the donee or purchaser has only a minority interest, this approach may be used to shift a significant part of the value of a business.

In planning transfers, consideration should be given to recapitalizing the company to create two classes of stock, preferred and common. They can be structured so that the common stock will have very little value when issued, but future appreciation will accrue to it. Thus, giving or selling the common stock will tend to freeze the owner's current value, and allow the children or other family members to obtain the appreciation. A recapitalization may also allow the transfer of control of the company to one group of family members, and the transfer of stock that pays a dividend to other family members. The common stock can be structured to have voting control, and the preferred stock can have a fixed or market-rate adjusted dividend. The active family members can be given operating control, and the passive family members given income.

Business interests can also be transferred by sale to family members. This will also shift future appreciation in the value of the company to the purchasers. If the sale is structured as an installment sale, gain on the sale may be deferred until principal payments are made. The seller can also consider forgiving or cancelling all or part of the amounts due. Cancellation of the notes or amounts due will be treated as a gift by the seller. The amount forgiven may qualify for the annual gift tax exclusion and not trigger any gift tax. The cancellation will trigger realization of income attributable to the amount of the debt forgiven.

When new equipment or real estate is to be acquired for use in the business, consideration should be given to forming a partnership of family members to purchase it. The partnership can then lease it to the company. This offers a way of shifting income among family members and keeping any appreciation in these assets out of the company. In addition, if some members of the family will be active in the company and some will not, it provides a method of distributing family assets. The passive income assets can be given to nonactive members, and corporate stock given to active members. This is best done with new acquisitions of equipment or real estate. The partnership could be made up of the owner and family members, and could afford not only a way of shifting corporate income, but could also provide a tax shelter to the partners.

Another tactic is to form a company and have other family members own it, or own the majority of it, and to have the new company become the contractor or joint venturer on selected jobs. Ability, know-how, and contacts will help the family company gain and expand business, acquire investments, develop property, and take advantage of other business opportunities. But in any such arrangement, care must be taken to see that these companies are reasonably capitalized for the risks they are taking, and that they pay their knowledgable and experienced employees reasonable compensation.

For most construction companies, the real value of the company is in their key people, generally the owners. The growth of the company depends upon the talents of these key people and their ability to attract new business, bid and negotiate jobs, and supervise work. If something were to happen to a key person, the value of the business could be substantially impaired. Therefore, the contractor must make plans to deal with this possibility. This can be done by planning for successor management and, if this is impossible, then for the disposition of the business if a key person dies. Depending upon the size and number of people in the company, consideration should be given to the purchase of life insurance to fund unusual costs that may be incurred to complete jobs in the event of the death of key personnel.

Buy-Sell Agreements. A buy-sell agreement generally provides for the sale of a business interest at the occurrence of an event—usually death, disability, or retirement. It can also cover any other circumstances, including disagreements among the parties. Such agreements are advantageous because they not only provide a buyer for what might be an unmarketable asset, but they also avoid a forced sale or the unwelcome participation in the business by the decedent's heirs. The agreement provides for succession of the business ownership and liquidity to the seller for retirement or to pay death taxes.

A key element in the agreement is the determination of the value of the business, or the price. This can be very difficult to do for a construction company, particularly if contracts extend over a long period.

There are a number of ways agreements fix value. Among them are book value, agreed price (adjusted periodically by the parties), appraisal, a combination of any of these three, or a formula providing for other factors. If life insurance proceeds are to be in the company, will they be included in the valuation or not? Generally, they should not be included, but an adjustment might be made to include premiums paid on the policy or policies, or their net cash value. Finding the formula for valuation of a construction company depends a great deal upon the type of contracting being done, the length of jobs, degree of risk involved, types of assets in the company, and the parties who will remain to complete the work.

The Internal Revenue Code has two provisions specifically designed to help alleviate the liquidity problem of a closely held business. Section 303 allows the redemption of stock to pay death taxes and estate administration expenses, and Section 6166 allows deferred payment of estate taxes on business assets in the estate.

In March 1983, The Treasury published proposed regulations reflecting the TEFRA changes in accounting for long-term contracts (Prop. Reg. § 1.451-3 (Fed. Reg. Mar. 14, 1983)). The following outline, Figure 18-4, summarizes accounting for long-term contracts, including the proposals. Note that while the proposals reflect the latest Treasury thinking in this area, they could be changed when published in final form.

FIG. 18-4 Proposed Internal Revenue Service Regulations on Completed-Contract Method of Accounting

I. DEFINITIONS

  A. Long-Term vs. Short-Term:

    1. Short-term—Any contract that is started and completed within the same tax year.

    2. Long-term—any building, installation, construction or manufacturing contract that is not completed within the tax year in which it was entered into (Prop. Reg. § 1.451-3(b)(1)).

  B. Substantial vs. Less Than Substantial Duration (Prop. Reg. § 1.451-3(a)(1)):

    1. Less than substantial duration—less than twelve months.

    2. Substantial duration—twelve months or more.

  C. Extended Period Long-Term Contract (Prop. Reg. § 1.451-3(b)(3)):

    1. Any contract that the taxpayer estimates will not be completed within two years of the contract commencement date.

    2. Exception: a *construction* contract will not be an extended period long-term contract if

      a. The taxpayer estimates the contract will be completed within three years of the contract commencement date, *or*

      b. The taxpayer's average annual gross receipts over the three taxable years ending before the taxable year the construction contract is entered into do not exceed $25,000,000 (this includes gross receipts from trades or business under common control).

    3. "Construction contract" means any contract for the building, construction, or erection of, or the installation of any integral component to, or improvements to real property.

II. METHODS OF ACCOUNTING

  A. Cash Basis

  B. Accrual Basis

  C. Long-Term Contract Methods. Special alternative methods of accounting are available for contractors who have agreements to manufacture, build, install, or construct a particular item or items and such performance is not completed within the same tax year in which it commenced (Prop. Reg. § 1.451-3(b)(1)(i)).

  The election to use either of the following methods must be set forth in a statement attached to the tax return (Prop. Reg. § 1.351-3(a)(2)):

  1. Percentage-of-completion method. Although it is probably the most realistic measure of income and is preferred for financial statement purposes, this method does little to defer income. It does, however, allow for the recognition of loss contracts as incurred.

  2. Completed-contract method.

    a. The major advantages in using this method are:

(continued)

FIG. 18-4   (continued)

    i. The deferral of income recognition until the contract is completed; that event may be controllable to some extent.

    ii. Income is recognized when it is finally determined.

    iii. Advances and retainages are clearly deferred until completion.

  b. Disadvantages include:

    i. The possible peaking of income.

    ii. Now being subjected to stricter regulations issued under the authority of TEFRA Section 229, which, if applicable, will require many more construction contract costs to be deferred until the contract is completed rather than allowing them to be deductible currently.

    iii. Loss jobs cannot be recognized until the contract is complete.

3. If either the percentage-of-completion method or the completed-contract method is elected for long-term contracts, it must be applied consistently to all long-term contracts within the same trade or business. However, long-term contracts of a substantial duration may be reported on a consistent long-term contract method, while long-term contracts of less than substantial duration may be reported on another proper method of accounting (Prop. Reg. § 1.451-3(a)(1)).

4. Long-term contract methods cannot be used in the following instances:

  a. Engineering, procurement, construction management services, and actual construction must be separated for tax reporting purposes into components, since only actual construction is eligible for long-term reporting (Ltr. Rul. 8308005).

  b. By taxpayer who does not assume the responsibility of a contractor (Ltr. Rul. 8203010).

III. TAX ACCOUNTING FOR LONG-TERM CONTRACTS. The following is intended to summarize the tax accounting treatment in determining what costs are allocable to a long-term contract in the case of a taxpayer utilizing the completed-contract method (Prop. Reg. § 1.451-3(d)(5)).

A. Costs Required to Be Allocated. The following costs must be treated as costs properly allocable to a long-term contract (basically unchanged in Prop. Reg. § 1.451-3(d)(5)(i) issued March 14, 1983 pursuant to TEFRA § 229):

1. Direct costs (Prop. Reg. § 1.451-3(d)(5)(i)), including:

  a. Direct material costs, which include:

    i. The cost of those materials that become an integral part of the subject matter of the contract.

    ii. Those materials consumed in the ordinary course of building, constructing or manufacturing the subject matter of the contract.

  b. Direct labor costs, which include:

    i. The cost of labor identified or associated with a particular contract.

      ii. Elements of direct labor include basic compensation, overtime pay, vacation and holiday pay, sick leave pay (other than under a Section 105(d) wage continuation plan), shift differential, payroll taxes and payments to a supplemental unemployment benefit plan paid or incurred on behalf of employees engaged in direct labor.

2. Indirect costs (Prop. Reg. § 1.451-3(d)(5)(ii)) include all costs "which are incident to and necessary for the performance of particular long-term contracts." These costs, if used in specifically performing a particular contract, include:

    a. Repair expense of equipment or facilities.

    b. Maintenance of equipment or facilities.

    c. Utilities, such as heat, light, and power, relating to equipment or facilities.

    d. Rent of equipment or facilities.

    e. Indirect labor and contract supervisory wages incurred (including the same elements as with direct labor, listed above).

    f. Indirect material and supplies.

    g. Tools and equipment not capitalized.

    h. Costs of quality control and inspection.

    i. Taxes (other than state, local, and foreign income taxes) to the extent attributable to labor, materials, supplies, equipment, or facilities.

    j. Depreciation, amortization, and cost recovery allowances reported for financial statement purposes on equipment or facilities (but not in excess of that allowable under Chapter 1 of the Internal Revenue Code).

    k. Cost depletion.

    l. Administrative costs (exclusive of any cost of selling or return on capital).

    m. Compensation paid to officers attributable to services (other than incidental or occasional services) for the job.

    n. Insurance incurred, such as on machinery, equipment, or facilities.

3. The above list of "costs required to be allocated" applies to all long-term contracts, whether or not they are extended period long-term contracts.

4. Certain additional costs, as presented in the Recap of Tax Accounting at the end of this Figure, must be allocated to a long-term contract only if it is an extended period long-term contract.

5. New cost allocation rules for extended period long-term contracts apply only to contracts entered into after December 31, 1982 in taxable years begun after December 31, 1982.

B. Allocation Methods:

1. For taxable years beginning before January 1, 1982, see Prop. Reg. § 1.451-3(d)(6) (revised as of April 1, 1982).

2. For taxable years beginning after December 31, 1982, separate accounts must be maintained for each contract and direct costs allocated thereto (Prop. Reg. § 1.451-3(d)(8)).

(continued)

FIG. 18-4 (continued)

Other allocations of administrative, service, or support costs to extended period contracts are to be made by reference to guidelines set forth in Prop. Reg. § 1.451-3(d)(9), as follows:

a. Introduction—in accordance with the benefits.

b. General rule—if direct benefit, directly allocated.

c. Special rules—for service costs.

  i. If direct, allocated directly.

  ii. If mixed, allocated based upon benefits.

  iii. District director discretion is allowed in determining propriety of allocations.

  iv. Types of such costs required or not required to be allocated are listed.

  v. Illustrations relative to the possible mechanics of making the allocations are provided.
  - Security services
  - Legal services
  - Centralized payroll department
  - Centralized data processing
  - Engineering and design services
  - Safety engineering

C. Phase-In. Those taxpayers required to allocate more costs than previously required for contracts entered into after 1982 for tax years beginning after 1982 (and who are not otherwise exempt from the new rules) are permitted to phase in those additional costs allocable to the contracts as follows:

For taxable years beginning in calendar years	The applicable percentage is:
1983	$33^1/_3$
1984	$66^2/_3$
1985 or thereafter	100

D. Commencement Date. Commencement date will generally be the first date that design or engineering costs allocable to the job are incurred, materials or equipment are shipped to the job-site, or workers are sent to the job-site. Bidding expenses and contract negotiating expenses are not allocable to the cost of the job. Detailed records and other evidence must be maintained; otherwise the contract date might be held to be the commencement date.

E. Completion:

1. "A long-term contract shall not be considered 'completed' until final completion and acceptance have occurred. Nevertheless, a taxpayer may not delay the completion of a contract for the principal purpose of deferring Federal income tax." (Prop. Reg. § 1.451-3(b)(2)). However, "[c]ompletion of a long-term contract is determined without regard to whether a dispute exists at the time the taxpayer tenders the subject matter of the contract to the party with whom the taxpayer has contracted." (Prop. Reg. § 1.451-3(b)(2)(vi)).

2. Effective for tax years ending after December 31, 1982, the "final completion and acceptance" shall be determined without regard to any contract term providing for:

   a. Additional compensation contingent upon future successful performance after being initially accepted by the owner (Prop. Reg. § 1.451-3(b)(2)(iii)).

   b. Supervision of installation or assembly where such is otherwise to be performed by the purchaser (Prop. Reg. § 1.451-3(b)(2)(iv)).

3. Subcontractor's work is completed when complete and accepted by the prime contractor (Prop. Reg. § 1.451-3(b)(2)(v)).

IV. SEVERING AND AGGREGATING CONTRACTS.
Proposed to be effective for tax years ending after December 31, 1982, new rules would be established for severing and aggregating long-term contracts (Prop. Reg. § 1.451-3(e)(1)):

A. The rules will depend on the facts and circumstances of each case.

B. Generally, one agreement will not be separated unless it contemplates separate delivery or acceptance of portions of the contract.

   1. The contract should be written in such a way that: "Delivery" and "acceptance" are not considered satisfied until the entire contract is completed.

      a. It insures that there is a business purpose for delivery or acceptance for instance, by providing for damages if all aspects are not completed and accepted as a condition.

      b. The agreement does not cover two or more subject matters.

      c. It avoids where possible having an option to be exercised or a "change order" made for the "increase" in the number of units to be supplied under the agreement, which might result in a classification of the additions as a "separate contract."

C. Generally, several agreements will not be aggregated unless they would be treated as one contract as a customary commercial practice of the industry, or unless they have no business purpose.

   1. In some circumstances, one contract should avoid any connotation that it was entered into with the expectation that additional contracts would follow.

D. The proposed regulations give the Commissioner of Internal Revenue broad discretion in making the ultimate determinations.

V. OBSERVATIONS

A. For those contracts that fall close to the three-year duration requirement for "extended period long-term contract" status:

   1. Care must be exercised when initial costs are incurred.

   2. Records of dates of incurrence of initial costs should be maintained.

   3. "The taxpayer shall maintain contemporaneous written records setting forth the basis for classifying each contract, and such records shall be in sufficient detail to enable the district director to determine whether a taxpayer's estimate of the time required to complete a contract was made on a reasonable basis. . . . The taxpayer's estimate will not be

(continued)

FIG. 18-4 (continued)

considered unreasonable if the contract was not completed within the expected time solely because of unforeseeable factors not within the control of the taxpayer. . . . Unforeseeable factors include, but are not limited to, prolonged third-party litigation that the taxpayer could not reasonably have anticipated, abnormal weather (considering the season and the job-site), prolonged strikes, and prolonged delays in securing required permits or licenses.'' (Prop. Reg. § 1.451-3(b)(3)(iv)(A))

B. Where applicable, costing systems will have to be refined and perhaps redesigned to comply with the new requirements.

C. A significant number of states have not yet adopted the federal TEFRA changes (or some or all of the Economic Recovery Tax Act of 1981 (ERTA) changes). Therefore, there may result a number of differences between federal, financial statement, and state reporting.

RECAP OF TAX ACCOUNTING FOR OTHER LONG-TERM CONTRACT COSTS[a] PRE- AND POST-TEFRA

	Costs Deductible Currently as Incurred		Costs to Be Deferred Until Contract Completion and No TEFRA Exception Applies
	Under Regs. Issued Under T.D. 7397, 1/14/76[b]	Under TEFRA Prop. Regs. 3/14/83	
Marketing, selling and advertising expenses	X	X	—
Bidding expenses on contracts awarded to the taxpayer	X	—	X
Bidding expenses on contracts not awarded to the taxpayer	X	X	—
Distribution expenses, such as shipping costs	X	—	X
Interest	X	X	—
General and administrative expenses[c] directly attributable or allocable to the performance of a particular contract	X	—	X

	Costs Deductible Currently as Incurred		Costs to Be Deferred Until Contract Completion and No TEFRA Exception Applies
	Under Regs. Issued Under T.D. 7397, 1/14/76[b]	Under TEFRA Prop. Regs. 3/14/83	
General and administrative expenses[d] and compensation paid to officers attributable to services not benefiting any particular contract	X	X	—
Research and experimentation expenses[e] either directly attributable to a particular contact existing when incurred or incurred under agreement to perform R&D	X	—	X
Research and experimentation expenses[f] neither directly attributable to a particular contract existing when incurred nor incurred under agreement to perform R&D	X	X	—
Losses under Section 165[g]	X	X	—
Depreciation, amortization and cost recovery allowances on equipment and facilities in use on a contract, reported for federal income tax purposes in excess of allowances reported in financial statements	X	—	X
Depreciation, amortization and cost recovery allowances on idle equipment and facilities	X	X	—

(continued)

FIG. 18-4 (continued)

	Costs Deductible Currently as Incurred		Costs to Be Deferred Until Contract Completion and No TEFRA Exception Applies
	Under Regs. Issued Under T.D. 7397, 1/14/76[b]	Under TEFRA Prop. Regs. 3/14/83	
Percentage depletion in excess of cost depletion	X	—	X
Income taxes attributable to contract income	X	X	—
Pension and profit-sharing contributions deductible[h] to the extent attributable to past service	X	X	—
Pension and profit-sharing contributions deductible[i] representing current service costs	X	—	X
Payments under a wage continuation plan[j]	X	X	—
Other employee benefit costs	X	X	—
Costs attributable to strikes	X	X	—
Rework labor, scrap, and spoilage incurred on a particular contract	X	—	X
Each of the above items for state tax purposes[k]	?	?	?

(a) Other than those costs specified in Reg. §§ 1.453(d)(5)(i) and 1.453(d)(5)(ii).

(b) The inventory treatment of certain of these costs for federal tax purposes may depend upon their financial statement treatment pursuant to Reg. § 1.471-11.

(c) Whether or not performed on the job-site.

(d) Other than those specified in Prop. Reg. § 1.451(d)(6)(ii)(1) or 1.451(d)(6)(ii).

(e) As described in Section 174 and the regulations thereunder.

(f) Id.

(g) Including regulations thereunder.

(h) Under Section 404.

(i) Id.

(j) Described in Section 105(d).

(k) Reference must be made to the applicable state law, regulation, and/or ruling as it might relate and apply to the treatment of each specified item (except in those states that may have totally conformed to all federal provisions).

Appendix:
Accounting for Performance of Construction-Type and Certain Production-Type Contracts

Introduction

1. This statement of position provides guidance on the application of generally accepted accounting principles in accounting for the performance of contracts for which specifications are provided by the customer for the construction of facilities or the production of goods or for the provision of related services. Changes in the business environment have increased significantly the variety and uses of those types of contracts and the types of business enterprises that use them. In the present business environment, diverse types of contracts, ranging from relatively simple to highly complex and from relatively short- to long-term, are widely used in many industries for construction, production, or provision of a broad range of goods and services. However, existing principles related to accounting for contracts were written in terms of long-term construction-type contracts, and they are not stated in sufficient detail for the scope of activities to which they presently are applied. Those activities range far beyond the traditional construction-type activity (the design and physical construction of facilities such as buildings, roads, dams, and bridges) to include, for example, the development and production of military and commercial aircraft, weapons delivery systems, space exploration hardware, and computer software. The accounting standards division believes that guidance is now needed in this area of accounting.

The Basic Accounting Issue

2. The determination of the point or points at which revenue should be recognized as earned and costs should be recognized as

Reprinted, by permission, from the Audit and Accounting Guide, *Construction Contractors* (New York: American Institute of Certified Public Accountants, Inc.) Statement of Position 81-1, pp. 109–157. © 1981 by the American Institute of Certified Public Accountants, Inc.

expenses is a major accounting issue common to all business enterprises engaged in the performance of contracts of the types covered by this statement. Accounting for such contracts is essentially a process of measuring the results of relatively long-term events and allocating those results to relatively short-term accounting periods. This involves considerable use of estimates in determining revenues, costs, and profits and in assigning the amounts to accounting periods. The process is complicated by the need to evaluate continually the uncertainties inherent in the performance of contracts and by the need to rely on estimates of revenues, costs, and the extent of progress toward completion.

Present Accounting Requirements and Practices

3.   The pervasive principle of realization and its exceptions and modifications are central factors underlying accounting for contracts. APB Statement 4 states:

> Revenue is generally recognized when both of the following conditions are met: (1) the earnings process is complete or virtually complete, and (2) an exchange has taken place. [Paragraph 150]

> Revenue is sometimes recognized on bases other than the realization rule. For example, on long-term construction contracts revenue may be recognized as construction progresses. This exception to the realization principle is based on the availability of evidence of the ultimate proceeds and the consensus that a better measure of periodic income results. [Paragraph 152]

> The exception to the usual revenue realization rule for long-term construction-type contracts, for example, is justified in part because strict adherence to realization at the time of sale would produce results that are considered to be unreasonable. The judgment of the profession is that revenue should be recognized in this situation as construction progresses. [Paragraph 174]

4.   Accounting Research Bulletin no. 45 (ARB 45), *Long-Term Construction-Type Contracts*, issued by the AICPA Committee on Accounting Procedure in 1955, describes the two generally accepted methods of accounting for long-term construction-type contracts for financial reporting purposes:

- *The percentage-of-completion method* recognizes income as work on a contract progresses; recognition of revenues and profits generally is related to costs incurred in providing the services required under the contract.

• *The completed-contract method* recognizes income only when the contract is completed, or substantially so, and all costs and related revenues are reported as deferred items in the balance sheet until that time.

The AICPA Industry Audit Guide, *Audits of Government Contractors*, describes units-of-delivery as a modification of the percentage-of-completion method of accounting for contracts.

• *The units-of-delivery method* recognizes as revenue the contract price of units of a basic production product delivered during a period and as the cost of earned revenue the costs allocable to the delivered units; costs allocable to undelivered units are reported in the balance sheet as inventory or work in progress. The method is used in circumstances in which an entity produces units of a basic product under production-type contracts in a continuous or sequential production process to buyers' specifications.

The use of either of the two generally accepted methods of accounting involves, to a greater or lesser extent, three key areas of estimates and uncertainties: (*a*) the extent of progress toward completion, (*b*) contract revenues, and (*c*) contract costs. Although the ultimate amount of contract revenue is often subject to numerous uncertainties, the accounting literature has given little attention to the difficulties of estimating contract revenue.

5.   ARB 45, paragraph 15, describes the circumstances in which each method is preferable as follows:

> The committee believes that in general when estimates of costs to complete and extent of progress toward completion of long-term contracts are reasonably dependable, the percentage-of-completion method is preferable. When lack of dependable estimates or inherent hazards cause forecasts to be doubtful, the completed-contract method is preferable.

Both of the two generally accepted methods are widely used in practice. However, the two methods are frequently applied differently in similar circumstances. The division believes that the two methods should be used in specified circumstances and should not be used as acceptable alternatives for the same circumstances. Accordingly, identifying the circumstances in which either of the methods is preferable and the accounting that should be followed in the application of those methods are among the primary objectives of this statement of position. This statement provides guidance on the application of ARB 45 and does not amend that bulletin.

6. In practice, methods are sometimes found that allocate contract costs and revenues to accounting periods on (a) the basis of cash receipts and payments or (b) the basis of contract billings and costs incurred. Those practices are not generally accepted methods of accounting for financial reporting purposes. However, those methods are appropriate for other purposes, such as the measurement of income for income tax purposes, for which the timing of cash transactions is a controlling factor. Recording the amounts billed or billable on a contract during a period as contract revenue of the period, and the costs incurred on the contract as expenses of the period, is not acceptable for financial reporting purposes because the amounts billed or billable on a contract during a period are determined by contract terms and do not necessarily measure performance on the contract. Only by coincidence might those unacceptable methods produce results that approximate the results of the generally accepted method of accounting for contracts that are appropriate in the circumstances.

Other Pronouncements and Regulations Affecting Contract Accounting

7. Accounting Research Bulletin no. 43, chapter 11, "Government Contracts," prescribes generally accepted principles in three areas of accounting for government contracts. Section A of that chapter deals with accounting problems arising under cost-plus-fixed-fee contracts. Section B deals with certain aspects of the accounting for government contracts and subcontracts that are subject to renegotiation. Section C deals with problems involved in accounting for certain terminated war and defense contracts. Those pronouncements govern accounting for contracts in the areas indicated.

8. The pricing and costing of federal government contracts are governed by cost principles contained in procurement regulations such as the Federal Procurement Regulation (FPR) and the Defense Acquisition Regulation (DAR). Also, most major government contractors are subject to cost accounting standards issued by the Cost Accounting Standards Board (CASB). CASB standards apply

to the cost accounting procedures that government contractors use to allocate costs to contracts; CASB standards are not intended for financial reporting.

9. Accounting for contracts for income tax purposes is prescribed by the Internal Revenue Code and the related rules and regulations. The methods of accounting for contracts under those requirements are not limited to the two generally accepted methods for financial reporting. For numerous historical and practical reasons, tax accounting rules and regulations differ from generally accepted accounting principles. Numerous nonaccounting considerations are appropriate in determining income tax accounting. This statement deals exclusively with the application of generally accepted accounting principles to accounting for contracts in financial reporting. It does not apply to income tax accounting and is not intended to influence income tax accounting.

Need for Guidance

10. Because of the complexities and uncertainties in accounting for contracts, the increased use of diverse types of contracts for the construction of facilities, the production of goods, or the provision of related services, and present conditions and practices in industries in which contracts are performed for those purposes, additional guidance on the application of generally accepted accounting principles is needed. This statement of position provides that guidance. Appendix A contains a schematic chart showing the organization of the statement.

Scope of Statement of Position

11. This statement of position applies to accounting for performance of contracts for which specifications are provided by the customer for the construction of facilities or the production of goods or the provision of related services that are reported in financial statements prepared in conformity with generally accepted accounting principles.[1] Existing authoritative accounting literature

---

[1]This statement is not intended to apply to "service transactions" as defined in the FASB's October 23, 1978 Invitation to Comment, *Accounting for Certain Service Transactions*. However, it applies to separate contracts to provide services essential to the construction or production of tangible property, such as design, engineering, procurement, and construction management (see paragraph 13 for examples).

uses the terms "long-term" and "construction-type" in identifying the types of contracts that are the primary focus of interest. The term "long-term" is not used in this statement of position as an identifying characteristic because other characteristics are considered more relevant for identifying the types of contracts covered. However, accounting for contracts by an entity that primarily has relatively short-term contracts is recommended in paragraph 31 of this statement. The scope of the statement is not limited to construction-type contracts.

Contracts Covered

12.   Contracts covered by this statement of position are binding agreements between buyers and sellers in which the seller agrees, for compensation, to perform a service to the buyer's specifications.[2] Contracts consist of legally enforceable agreements in any form and include amendments, revisions, and extensions of such agreements. Performance will often extend over long periods, and the seller's right to receive payment depends on his performance in accordance with the agreement. The service may consist of designing, engineering, fabricating, constructing, or manufacturing related to the construction or the production of tangible assets. Contracts such as leases and real estate agreements, for which authoritative accounting literature provides special methods of accounting, are not covered by this statement.

13.   Contracts covered by this statement include, but are not limited to, the following:

- Contracts in the construction industry, such as those of general building, heavy earth moving, dredging, demolition, design-build contractors, and specialty contractors (for example, mechanical, electrical, or paving).
- Contracts to design and build ships and transport vessels.
- Contracts to design, develop, manufacture, or modify complex aerospace or electronic equipment to a buyer's specification or to provide services related to the performance of such contracts.
- Contracts for construction consulting service, such as under agency contracts or construction management agreements.

---

[2]Specifications imposed on the buyer by a third party (for example, a government or regulatory agency or a financial institution) or by conditions in the marketplace are deemed to be "buyer's specifications."

• Contracts for services performed by architects, engineers, or architectural or engineering design firms.

14.   Contracts not covered by this statement include, but are not limited to, the following:

• Sales by a manufacturer of goods produced in a standard manufacturing operation, even if produced to buyers' specifications, and sold in the ordinary course of business through the manufacturer's regular marketing channels if such sales are normally recognized as revenue in accordance with the realization principle for sales of products and if their costs are accounted for in accordance with generally accepted principles of inventory costing.

• Sales or supply contracts to provide goods from inventory or from homogeneous continuing production over a period of time.

• Contracts included in a program and accounted for under the program method of accounting. For accounting purposes, a program consists of a specified number of units of a basic product expected to be produced over a long period in a continuing production effort under a series of existing and anticipated contracts.[3]

• Service contracts of health clubs, correspondence schools, and similar consumer-oriented organizations that provide their services to their clients over an extended period.

• Magazine subscriptions.

• Contracts of nonprofit organizations to provide benefits to their members over a period of time in return for membership dues.

15.   Contracts covered by this statement may be classified into four broad types based on methods of pricing: (a) fixed-price or lump-sum contracts, (b) cost-type (including cost-plus) contracts, (c) time-and-material contracts, and (d) unit-price contracts. A fixed-price contract is an agreement to perform all acts under the contract for a stated price. A cost-type contract is an agreement to perform under a contract for a price determined on the basis of a defined relationship to the costs to be incurred, for example, the

---

[3]The division is preparing a separate statement of position on program accounting, which will provide guidance on the circumstances in which existing and anticipated production-type contracts may be combined for the purpose of accumulating and allocating production costs.

costs of all acts required plus a fee, which may be a fixed amount or a fixed percentage of the costs incurred. A time-and-material contract is an agreement to perform all acts required under the contract for a price based on fixed hourly rates for some measure of the labor hours required (for example, direct labor hours) and the cost of materials. A unit-price contract is an agreement to perform all acts required under the contract for a specified price for each unit of output. Each of the various types of contracts may have incentive, penalty, or other provisions that modify their basic pricing terms. The pricing features of the various types are discussed in greater detail in Appendix B.

Definition of a Contractor

16.   The term "contractor" as used in this statement refers to a person or entity that enters into a contract to construct facilities, produce goods, or render services to the specifications of a buyer either as a general or prime contractor, as a subcontractor to a general contractor, or as a construction manager.

Definition of a Profit Center

17.   For the purpose of this statement, a "profit center" is the unit for the accumulation of revenues and costs and the measurement of income. For business enterprises engaged in the performance of contracts, the profit center for accounting purposes is usually a single contract; but under some specified circumstances it may be a combination of two or more contracts, a segment of a contract or of a group of combined contracts. This statement of position provides guidance on the selection of the appropriate profit center. The accounting recommendations, usually stated in terms of a single contract, also apply to alternative profit centers in circumstances in which alternative centers are appropriate.

Application and Effect on Existing Audit Guides and SOPs

18.   This statement of position presents the division's recommendations on accounting for contracts (as specified in paragraphs 11 to 17) in all industries. The recommendations in this statement need not be applied to immaterial items. Two existing AICPA Industry Audit Guides, *Audits of Construction Contractors* and *Audits of Government Contractors*, provide additional guidance on the application of generally accepted accounting principles to the

construction industry and to government contracts, respectively. The recommendations in this statement take precedence in those areas. *Audits of Construction Contractors* is being revised concurrently with this statement to conform to its provisions.

19.   The guidance on contract accounting and financial reporting in *Audits of Government Contractors* is essentially consistent with the recommendations in this statement except that this statement recommends the cumulative catch-up method for accounting for changes in estimates under the percentage-of-completion method of accounting, whereas either the cumulative catch-up method or the reallocation method is acceptable under the guide. Therefore, *Audits of Government Contractors* is amended so that its guidance on accounting for changes in estimates conforms to the recommendations in this statement. Also, since the recommendations in this statement provide more comprehensive and explicit guidance on the application of generally accepted accounting principles to contract accounting than does the guide, *Audits of Government Contractors*, the guide is amended to incorporate this statement as an appendix. The provisions of that guide should be interpreted and applied in the context of the recommendations in this statement.

20.   This statement is not intended to supersede recommendations on accounting in other AICPA industry accounting or audit guides or in other statements of position.

The Division's Conclusions

Determining a Basic Accounting Policy for Contracts

21.   In accounting for contracts, the basic accounting policy decision is the choice between the two generally accepted methods: the percentage-of-completion method including units of delivery and the completed-contract method. The determination of which of the two methods is preferable should be based on a careful evaluation of circumstances because the two methods should not be acceptable alternatives for the same circumstances. The division's recommendations on basic accounting policy are set forth in the sections on "The Percentage-of-Completion Method" and "The Completed-Contract Method," which identify the circumstances

appropriate to the methods, the bases of applying the methods, and the reasons for the recommendations. The recommendations apply to accounting for individual contracts and to accounting for other profit centers in accordance with the recommendations in the section on "Determining the Profit Center." As a result of evaluating individual contracts and profit centers, a contractor should be able to establish a basic policy that should be followed in accounting for most of his contracts. In accordance with the requirements of APB Opinion 22, *Disclosure of Accounting Policies*, a contractor should disclose in the note to the financial statements on accounting policies the method or methods of determining earned revenue and the cost of earned revenue including the policies relating to combining and segmenting, if applicable. Appendix C contains a summary of the disclosure requirements in this statement.

The Percentage-of-Completion Method

22.   This section sets forth the recommended basis for using the percentage-of-completion method and the reasons for the recommendation. Under most contracts for construction of facilities, production of goods, or provision of related services to a buyer's specifications, both the buyer and the seller (contractor) obtain enforceable rights. The legal right of the buyer to require specific performance of the contract means that the contractor has, in effect, agreed to sell his rights to work-in-progress as the work progresses. This view is consistent with the contractor's legal rights; he typically has no ownership claim to the work-in-progress but has lien rights. Furthermore, the contractor has the right to require the buyer, under most financing arrangements, to make progress payments to support his ownership investment and to approve the facilities constructed (or goods produced or services performed) to date if they meet the contract requirements. The buyer's right to take over the work-in-progress at his option (usually with a penalty) provides additional evidence to support that view. Accordingly, the business activity taking place supports the concept that in an economic sense performance is, in effect, a continuous sale (transfer of ownership rights) that occurs as the work progresses. Also under most contracts for the production of goods and the provision of related services that are accounted for on the basis of units delivered, both the contractor and the customer obtain enforceable rights as the goods are produced or the services are performed. As units are delivered, title to and the risk of loss on those units

normally transfer to the customer, whose acceptance of the items indicates that they meet the contractual specifications. For such contracts, delivery and acceptance are objective measurements of the extent to which the contracts have been performed. The percentage-of-completion method recognizes the legal and economic results of contract performance on a timely basis. Financial statements based on the percentage-of-completion method present the economic substance of a company's transactions and events more clearly and more timely than financial statements based on the completed-contract method, and they present more accurately the relationships between gross profit from contracts and related period costs. The percentage-of-completion method informs the users of the general purpose financial statements of the volume of economic activity of a company.

Circumstances Appropriate to the Method

23. The use of the percentage-of-completion method depends on the ability to make reasonably dependable estimates. For the purposes of this statement, "the ability to make reasonably dependable estimates" relates to estimates of the extent of progress toward completion, contract revenues, and contract costs. The division believes that the percentage-of-completion method is preferable as an accounting policy in circumstances in which reasonably dependable estimates can be made and in which all the following conditions exist:

- Contracts executed by the parties normally include provisions that clearly specify the enforceable rights regarding goods or services to be provided and received by the parties, the consideration to be exchanged, and the manner and terms of settlement.
- The buyer can be expected to satisfy his obligations under the contract.
- The contractor can be expected to perform his contractual obligations.

24. For entities engaged on a continuing basis in the production and delivery of goods or services under contractual arrangements and for whom contracting represents a significant part of their operations, the presumption is that they have the ability to make estimates that are sufficiently dependable to justify the use of

the percentage-of-completion method of accounting.[4] Persuasive evidence to the contrary is necessary to overcome that presumption. The ability to produce reasonably dependable estimates is an essential element of the contracting business. For a contract on which a loss is anticipated, generally accepted accounting principles require recognition of the entire anticipated loss as soon as the loss becomes evident. An entity without the ability to update and revise estimates continually with a degree of confidence could not meet that essential requirement of generally accepted accounting principles.

25. Accordingly, the division believes that entities with significant contracting operations generally have the ability to produce reasonably dependable estimates and that for such entities the percentage-of-completion method of accounting is preferable in most circumstances. The method should be applied to individual contracts or profit centers, as appropriate.

a. Normally, a contractor will be able to estimate total contract revenue and total contract cost in single amounts. Those amounts should normally be used as the basis for accounting for contracts under the percentage-of-completion method.

b. For some contracts, on which some level of profit is assured, a contractor may only be able to estimate total contract revenue and total contract cost in ranges of amounts. If, based on the information arising in estimating the ranges of amounts and all other pertinent data, the contractor can determine the amounts in the ranges that are most likely to occur, those amounts should be used in accounting for the contract under the percentage-of-completion method. If the most likely amounts cannot be determined, the lowest probable level of profit in the range should be used in accounting for the contract until the results can be estimated more precisely.

c. However, in some circumstances, estimating the final outcome may be impractical except to assure that no loss will be incurred. In those circumstances, a contractor should use a zero estimate

---

[4]The division recognizes that many contractors have informal estimating procedures that may result in poorly documented estimates and marginal quality field reporting and job costing systems. Those conditions may influence the ability of an entity to produce reasonably dependable estimates. However, procedures and systems should not influence the development of accounting principles and should be dealt with by management as internal control, financial reporting, and auditing concerns.

of profit; equal amounts of revenue and cost should be recognized until results can be estimated more precisely. A contractor should use this basis only if the bases in (a) or (b) are clearly not appropriate. A change from a zero estimate of profit to a more precise estimate should be accounted for as a change in an accounting estimate.

An entity using the percentage-of-completion method as its basic accounting policy should use the completed-contract method for a single contract or a group of contracts for which reasonably dependable estimates cannot be made or for which inherent hazards make estimates doubtful. Such a departure from the basic policy should be disclosed.

Nature of Reasonable Estimates and Inherent Hazards

26.   In practice, contract revenues and costs are estimated in a wide variety of ways ranging from rudimentary procedures to complex methods and systems. Regardless of the techniques used, a contractor's estimating procedures should provide reasonable assurance of a continuing ability to produce reasonably dependable estimates.[5] Ability to estimate covers more than the estimating and documentation of contract revenues and costs; it covers a contractor's entire contract administration and management control system. The ability to produce reasonably dependable estimates depends on all the procedures and personnel that provide financial or production information on the status of contracts. It encompasses systems and personnel not only of the accounting department but of all areas of the company that participate in production control, cost control, administrative control, or accountability for contracts. Previous reliability of a contractor's estimating process is usually an indication of continuing reliability, particularly if the present circumstances are similar to those that prevailed in the past.

27.   Estimating is an integral part of contractors' business activities, and there is a necessity to revise estimates on contracts continually as the work progresses. The fact that circumstances may necessitate frequent revision of estimates does not indicate that the estimates are unreliable for the purpose for which they are

[5]The type of estimating procedures appropriate in a particular set of circumstances depends on a careful evaluation of the costs and benefits of developing the procedures. The ability to produce reasonably dependable estimates that would justify the use of the percentage-of-completion method as recommended in paragraph 25 does not depend on the elaborateness of the estimating procedures used.

used. Although results may differ widely from original estimates because of the nature of the business, the contractor, in the conduct of his business, may still find the estimates reasonably dependable. Despite these widely recognized conditions, a contractor's estimates of total contract revenue and total contract costs should be regarded as reasonably dependable if the minimum total revenue and the maximum total cost can be estimated with a sufficient degree of confidence to justify the contractor's bids on contracts.

28.   ARB 45 discourages the use of the percentage-of-completion method of accounting in circumstances in which inherent hazards make estimates doubtful. "Inherent hazards" relate to contract conditions or external factors that raise questions about contract estimates and about the ability of either the contractor or the customer to perform his obligations under the contract. Inherent hazards that may cause contract estimates to be doubtful usually differ from inherent business risks. Business enterprises engaged in contracting, like all business enterprises, are exposed to numerous business risks that vary from contract to contract. The reliability of the estimating process in contract accounting does not depend on the absence of such risks. Assessing business risks is a function of users of financial statements.

29.   The present business environment and the refinement of the estimating process have produced conditions under which most business entities engaged in contracting can deal adequately with the normal, recurring business risks in estimating the outcome of contracts. The division believes that inherent hazards that make otherwise reasonably dependable contract estimates doubtful involve events and conditions that would not be considered in the ordinary preparation of contract estimates and that would not be expected to recur frequently, given the contractor's normal business environment. Such hazards are unrelated to, or only incidentally related to, the contractor's typical activities. Such hazards may relate, for example, to contracts whose validity is seriously in question (that is, which are less than fully enforceable), to contracts whose completion may be subject to the outcome of pending legislation or pending litigation, or to contracts exposed to the possibility of the condemnation or expropriation of the resulting properties. Reasonably dependable estimates cannot be produced for a contract with unrealistic or ill-defined terms or for a contract be-

tween unreliable parties. However, the conditions stated in paragraph 23 for the use of the percentage-of-completion method of accounting, which apply to most bona fide contracts, make the existence of some uncertainties, including some of the type described in ARB 45, paragraph 15, unlikely for contracts that meet those conditions. Therefore, the division believes that there should be specific, persuasive evidence of such hazards to indicate that use of the percentage-of-completion method on one of the bases in paragraph 25 is not preferable.

The Completed-Contract Method

30.   This section sets forth the recommended basis for using the completed-contract method and the reasons for the recommendation. Under the completed-contract method, income is recognized only when a contract is completed or substantially completed. During the period of performance, billings and costs are accumulated on the balance sheet, but no profit or income is recorded before completion or substantial completion of the work. This method precludes reporting on the performance that is occurring under the enforceable rights of the contract as work progresses. Although the completed-contract method is based on results as finally determined rather than on estimates for unperformed work, which may involve unforeseen costs and possible losses, it does not reflect current performance when the period of a contract extends beyond one accounting period, and it therefore may result in irregular recognition of income. Financial statements based on this method may not show informative relationships between gross profit reported on contracts and related period costs.

Circumstances of Use

31.   The completed-contract method may be used as an entity's basic accounting policy in circumstances in which financial position and results of operations would not vary materially from those resulting from use of the percentage-of-completion method (for example, in circumstances in which an entity has primarily short-term contracts). Although this statement does not formally distinguish on the basis of length between long-term and short-term contracts, the basis for recording income on contracts of short duration poses relatively few problems. In accounting for such contracts, income ordinarily is recognized when performance is substantially completed and accepted. Under those circumstances,

revenues and costs in the aggregate for all contracts would be expected to result in a matching of gross profit with period overhead or fixed costs similar to that achieved by use of the percentage-of-completion method. For example, the completed-contract method, as opposed to the percentage-of-completion method, would not usually produce a material difference in net income or financial position for a small plumbing contractor that performs primarily relatively short-term contracts during an accounting period; performance covers such a short span of time that the work is somewhat analogous to the manufacture of shelf production items for sale. An entity using the completed-contract method as its basic accounting policy should depart from that policy for a single contract or a group of contracts not having the features described in paragraph 31 and use the percentage-of-completion method on one of the bases described in paragraph 25. Such a departure should be disclosed.

32.   The completed-contract method is preferable in circumstances in which estimates cannot meet the criteria for reasonable dependability discussed in the section on the percentage-of-completion method or in which there are inherent hazards of the nature of those discussed in that section. An entity using the percentage-of-completion method as its basic accounting policy should depart from that policy and use the completed-contract method for a single contract or a group of contracts only in the circumstances described in paragraph 25.

33.   The use of the completed-contract method is recommended for the circumstances described in paragraphs 31 and 32. However, for circumstances in which there is an assurance that no loss will be incurred on a contract (for example, when the scope of the contract is ill-defined but the contractor is protected by a cost-plus contract or other contractual terms), the percentage-of-completion method based on a zero profit margin, rather than the completed-contract method, is recommended until more precise estimates can be made. The significant difference between the percentage-of-completion method applied on the basis of a zero profit margin and the completed-contract method relates to the effects on the income statement. Under the zero profit margin approach to applying the percentage-of-completion method, equal amounts of revenue and cost, measured on the basis of performance during the period, are presented in the income statement;

whereas, under the completed-contract method, performance for a period is not reflected in the income statement, and no amount is presented in the income statement until the contract is completed. The zero profit margin approach to applying the percentage-of-completion method gives users of general purpose financial statements an indication of the volume of a company's business and of the application of its economic resources.

Determining the Profit Center

34.   The basic presumption should be that each contract is the profit center for revenue recognition, cost accumulation, and income measurement. That presumption may be overcome only if a contract or a series of contracts meets the conditions described for combining or segmenting contracts. A group of contracts (combining), and a phase or segment of a single contract or of a group of contracts (segmenting) may be used as a profit center in some circumstances. Since there are numerous practical implications of combining and segmenting contracts, evaluation of the circumstances, contract terms, and management intent are essential in determining contracts that may be accounted for on those bases.

Combining Contracts

35.   A group of contracts may be so closely related that they are, in effect, parts of a single project with an overall profit margin, and accounting for the contracts individually may not be feasible or appropriate. Under those circumstances, consideration should be given to combining such contracts for profit recognition purposes. The presumption in combining contracts is that revenue and profit are earned, and should be reported, uniformly over the performance of the combined contracts. For example, a group of construction-type contracts may be negotiated as a package with the objective of achieving an overall profit margin, although the profit margins on the individual contracts may vary. In those circumstances, if the individual contracts are performed and reported in different periods and accounted for separately, the reported profit margins in those periods will differ from the profit margin contemplated in the negotiations for reasons other than differences in performance.

36.   Contracts may be combined for accounting purposes only if they meet the criteria in paragraphs 37 and 38.

37.   A group of contracts may be combined for accounting purposes if the contracts

a. Are negotiated as a package in the same economic environment
with an overall profit margin objective. Contracts not executed
at the same time may be considered to have been negotiated as a
package in the same economic environment only if the time
period between the commitments of the parties to the individual
contracts is reasonably short. The longer the period between the
commitments of the parties to the contracts, the more likely it is
that the economic circumstances affecting the negotiations have
changed.

b. Constitute in essence an agreement to do a single project. A
project for this purpose consists of construction, or related service activity with different elements, phases, or units of output
that are closely interrelated or interdependent in terms of their
design, technology, and function or their ultimate purpose or
use.

c. Require closely interrelated construction activities with substantial common costs that cannot be separately identified with, or
reasonably allocated to, the elements, phases, or units of output.

d. Are performed concurrently or in a continuous sequence under
the same project management at the same location or at different
locations in the same general vicinity.

e. Constitute in substance an agreement with a single customer. In
assessing whether the contracts meet this criterion, the facts and
circumstances relating to the other criteria should be considered. In some circumstances different divisions of the same entity would not constitute a single customer if, for example, the
negotiations are conducted independently with the different divisions. On the other hand, two or more parties may constitute
in substance a single customer if, for example, the negotiations
are conducted jointly with the parties to do what in essence is a
single project.

Contracts that meet all of these criteria may be combined for profit
recognition and for determining the need for a provision for losses
in accordance with ARB 45, paragraph 6. The criteria should be
applied consistently to contracts with similar characteristics in similar circumstances.

38.   Production-type contracts that do not meet the criteria in paragraph 37 or segments of such contracts may be combined into groupings such as production lots or releases for the purpose of accumulating and allocating production costs to units produced or delivered on the basis of average unit costs in the following circumstances:[6]

a. The contracts are with one or more customers for the production of substantially identical units of a basic item produced concurrently or sequentially.

b. Revenue on the contracts is recognized on the units-of-delivery basis of applying the percentage-of-completion method.

Segmenting a Contract

39.   A single contract or a group of contracts that otherwise meet the test for combining may include several elements or phases, each of which the contractor negotiated separately with the same customer and agreed to perform without regard to the performance of the others. If those activities are accounted for as a single profit center, the reported income may differ from that contemplated in the negotiations for reasons other than differences in performance. If the project is segmented, revenues can be assigned to the different elements or phases to achieve different rates of profitability based on the relative value of each element or phase to the estimated total contract revenue. A project, which may consist of a single contract or a group of contracts, with segments that have different rates of profitability may be segmented if it meets the criteria in paragraph 40, paragraph 41, or paragraph 42. The criteria for segmenting should be applied consistently to contracts with similar characteristics and in similar circumstances.

40.   A project may be segmented if all the following steps were taken and are documented and verifiable:

a. The contractor submitted bona fide proposals on the separate components of the project and on the entire project.

b. The customer had the right to accept the proposals on either basis.

---

[6]The division is preparing a separate statement of position on program accounting, which will provide guidance on the circumstances in which existing and anticipated production-type contracts may be combined for the purpose of accumulating and allocating production costs.

c. The aggregate amount of the proposals on the separate components approximated the amount of the proposal on the entire project.

41.   A project that does not meet the criteria in paragraph 40 may be segmented only if it meets all the following criteria:

a. The terms and scope of the contract or project clearly call for separable phases or elements.

b. The separable phases or elements of the project are often bid or negotiated separately.

c. The market assigns different gross profit rates to the segments because of factors such as different levels of risk or differences in the relationship of the supply and demand for the services provided in different segments.

d. The contractor has a significant history of providing similar services to other customers under separate contracts for each significant segment to which a profit margin higher than the overall profit margin on the project is ascribed.[7]

e. The significant history with customers who have contracted for services separately is one that is relatively stable in terms of pricing policy rather than one unduly weighted by erratic pricing decisions (responding, for example, to extraordinary economic circumstances or to unique customer-contractor relationships).

f. The excess of the sum of the prices of the separate elements over the price of the total project is clearly attributable to cost savings incident to combined performance of the contract obligations (for example, cost savings in supervision, overhead, or equipment mobilization). Unless this condition is met, segmenting a contract with a price substantially less than the sum of the prices of the separate phases or elements would be inappropriate even if the other conditions are met. Acceptable price variations should be allocated to the separate phases or elements in proportion to the prices ascribed to each. In all other situations a substantial difference in price (whether more or less) between

---

[7]In applying the criterion in paragraph 41(d), values assignable to the segments should be on the basis of the contractor's normal historical prices and terms of such services to other customers. The division considered but rejected the concept of allowing a contractor to segment on the basis of prices charged by other contractors, since it does not follow that those prices could have been obtained by a contractor who has no history in the market.

the separate elements and the price of the total project is evidence that the contractor has accepted different profit margins. Accordingly, segmenting is not appropriate, and the contracts should be the profit centers.

g. The similarity of services and prices in the contract segments and services and the prices of such services to other customers contracted separately should be documented and verifiable.

42. A production-type contract that does not meet the criteria in paragraphs 40 or 41 may also be segmented and included in groupings such as production lots or releases for the purpose of accumulating and allocating production costs to units produced or delivered on the basis of average unit cost under the conditions specified in paragraph 38.

Measuring Progress on Contracts

43. This section describes methods of measuring the extent of progress toward completion under the percentage-of-completion method and sets forth criteria for selecting those methods and for determining when a contract is substantially completed. Meaningful measurement of the extent of progress toward completion is essential since this factor is used in determining the amounts of estimated contract revenue and estimated gross profit that will be recognized as earned in any given period.

Methods of Measuring Extent of Progress Toward Completion

44. In practice, a number of methods are used to measure the extent of progress toward completion. They include the cost-to-cost method, variations of the cost-to-cost method, efforts-expended methods, the units-of-delivery method, and the units-of-work-performed method. Those practices are intended to conform to ARB 45, paragraph 4.[8] Some of the measures are sometimes made and certified by engineers or architects, but manage-

---

[8]ARB 45, paragraph 4, states:

The committee recommends that the recognized income [under the percentage-of-completion method] be that percentage of estimated total income, either:

(a) that incurred costs to date bear to estimated total costs after giving effect to estimates of costs to complete based upon most recent information, or

(b) that may be indicated by such other measure of progress toward completion as may be appropriate having due regard to work performed.

Costs as here used might exclude, especially during the early stages of a contract, all or a portion of the cost of such items as materials and subcontracts if it appears that such an exclusion would result in a more meaningful periodic allocation of income.

ment should review and understand the procedures used by those professionals.

45. Some methods used in practice measure progress toward completion in terms of costs, some in terms of units of work, and some in terms of values added (the contract value of total work performed to date). All three of these measures of progress are acceptable in appropriate circumstances. The division concluded that other methods that achieve the objective of measuring extent of progress toward completion in terms of costs, units, or value added are also acceptable in appropriate circumstances. However, the method or methods selected should be applied consistently to all contracts having similar characteristics. The method or methods of measuring extent of progress toward completion should be disclosed in the notes to the financial statements. Examples of circumstances not appropriate to some methods are given within the discussion of input and output measures.

Input and Output Measures

46. The several approaches to measuring progress on a contract can be grouped into input and output measures. Input measures are made in terms of efforts devoted to a contract. They include the methods based on costs and on efforts expended. Output measures are made in terms of results achieved. They include methods based on units produced, units delivered, contract milestones, and value added. For contracts under which separate units of output are produced, progress can be measured on the basis of units of work completed. In other circumstances, progress may be measured, for example, on the basis of cubic yards of excavation for foundation contracts or on the basis of cubic yards of pavement laid for highway contracts.

47. Both input and output measures have drawbacks in some circumstances. Input is used to measure progress toward completion indirectly, based on an established or assumed relationship between a unit of input and productivity. A significant drawback of input measures is that the relationship of the measures to productivity may not hold, because of inefficiencies or other factors. Output is used to measure results directly and is generally the best measure of progress toward completion in circumstances in which a reliable measure of output can be established. However, output

measures often cannot be established, and input measures must then be used. The use of either type of measure requires the exercise of judgment and the careful tailoring of the measure to the circumstances.

48.   The efforts-expended method is an input method based on a measure of the work, such as labor hours, labor dollars, machine hours, or material quantities. Under the labor-hours method, for example, extent of progress is measured by the ratio of hours performed to date to estimated total hours at completion. Estimated total labor hours should include (a) the estimated labor hours of the contractor and (b) the estimated labor hours of subcontractors engaged to perform work for the project, if labor hours of subcontractors are a significant element in the performance of the contract. A labor-hours method can measure the extent of progress in terms of efforts expended only if substantial efforts of subcontractors are included in the computation. If the contractor is unable to obtain reasonably dependable estimates of subcontractors' labor hours at the beginning of the project and as work progresses, he should not use the labor-hours method.

49.   The various forms of the efforts-expended method generally are based on the assumption that profits on contracts are derived from the contractor's efforts in all phases of operations, such as designing, procurement, and management. Profit is not assumed to accrue merely as a result of the acquisition of material or other tangible items used in the performance of the contract or the awarding of subcontracts. As previously noted, a significant drawback of efforts-expended methods is that the efforts included in the measure may not all be productive.

50.   Measuring progress toward completion based on the ratio of costs incurred to total estimated costs is also an input method. Some of the costs incurred, particularly in the early stages of the contract, should be disregarded in applying this method because they do not relate to contract performance. These include the costs of items such as uninstalled materials not specifically produced or fabricated for the project or of subcontracts that have not been performed. For example, for construction projects, the cost of materials not unique to the project that have been purchased or ac-

cumulated at job sites but that have not been physically installed do not relate to performance.[9] The costs of such materials should be excluded from costs incurred for the purpose of measuring the extent of progress toward completion. Also, the cost of equipment purchased for use on a contract should be allocated over the period of its expected use unless title to the equipment is transferred to the customer by terms of the contract. For production-type contracts, the complement of expensive components (for example, computers, engines, radars, and complex "black boxes") to be installed into the deliverable items may aggregate a significant portion of the total cost of the contract. In some circumstances, the costs incurred for such components, even though the components were specifically purchased for the project, should not be included in the measurement before the components are installed if inclusion would tend to overstate the percentage of completion otherwise determinable.

51. The acceptability of the results of input or output measures deemed to be appropriate to the circumstances should be periodically reviewed and confirmed by alternative measures that involve observation and inspection. For example, the results provided by the measure used to determine the extent of progress may be compared to the results of calculations based on physical observations by engineers, architects, or similarly qualified personnel. That type of review provides assurance somewhat similar to that provided for perpetual inventory records by periodic physical inventory counts.

Completion Criteria Under the Completed-Contract Method

52. As a general rule, a contract may be regarded as substantially completed if remaining costs and potential risks are insignificant in amount. The overriding objectives are to maintain consistency in determining when contracts are substantially completed and to avoid arbitrary acceleration or deferral of income. The specific criteria used to determine when a contract is substantially completed should be followed consistently and should be disclosed in the note to the financial statements on accounting policies. Circumstances to be considered in determining when a project is

---

[9]The cost of uninstalled materials specifically produced, fabricated, or constructed for a project should be included in the costs used to measure extent of progress. Such materials consist of items unique to a project that a manufacturer or supplier does not carry in inventory and that must be produced or altered to meet the specifications of the project.

substantially completed include, for example, delivery of the product, acceptance by the customer, departure from the site, and compliance with performance specifications.

Income Determination—Revenue Elements

53.   Estimating the revenue on a contract is an involved process, which is affected by a variety of uncertainties that depend on the outcome of a series of future events. The estimates must be periodically revised throughout the life of the contract as events occur and as uncertainties are resolved.

54.   The major factors that must be considered in determining total estimated revenue include the basic contract price, contract options, change orders, claims, and contract provisions for penalties and incentive payments, including award fees and performance incentives. All those factors and other special contract provisions must be evaluated throughout the life of a contract in estimating total contract revenue to recognize revenues in the periods in which they are earned under the percentage-of-completion method of accounting.

Basic Contract Price—General

55.   The estimated revenue from a contract is the total amount that a contractor expects to realize from the contract. It is determined primarily by the terms of the contract and the basic contract price. Contract price may be relatively fixed or highly variable and subject to a great deal of uncertainty, depending on the type of contract involved. Appendix B describes basic contract types and major variations in the basic types. The total amount of revenue that ultimately will be realized on a contract is often subject to a variety of changing circumstances and accordingly may not be known with certainty until the parties to the contract have fully performed their obligations. Thus, the determination of total estimated revenue requires careful consideration and the exercise of judgment in assessing the probabilities of future outcomes.

56.   Although fixed-price contracts usually provide for a stated contract price, a specified scope of the work to be performed, and a specified performance schedule, they sometimes have adjustment schedules based on application of economic price adjustment (esca-

lation), price redetermination, incentive, penalty, and other pricing provisions. Determining contract revenue under unit-price contracts generally involves the same factors as under fixed-price contracts. Determining contract revenue from a time-and-material contract requires a careful analysis of the contract, particularly if the contract includes guaranteed maximums or assigns markups to both labor and materials; and the determination involves consideration of some of the factors discussed below in regard to cost-type contracts.

Basic Contract Price—Cost-Type Contracts

57.   Cost-type contracts have a variety of forms (see Appendix B). The various forms have differing contract terms that affect accounting, such as provisions for reimbursable costs (which are generally spelled out in the contract), overhead recovery percentages, and fees. A fee may be a fixed amount or a percentage of reimbursable costs or an amount based on performance criteria.[10] Generally, percentage fees may be accrued as the related costs are incurred, since they are a percentage of costs incurred, and profits should therefore be recognized as costs are incurred. Cost-type contracts often include provisions for guaranteed maximum total reimbursable costs or target penalties and rewards relating to underruns and overruns of predetermined target prices, completion dates, plant capacity on completion of the project, or other criteria.

58.   One problem peculiar to cost-type contracts involves the determination of the amounts of reimbursable costs that should be reflected as revenue. Under some contracts, particularly service-type contracts, a contractor acts solely in the capacity of an agent (construction manager) and has no risks associated with costs managed. This relationship may arise, for example, if an owner awards a construction management contract to one entity and a construction contract to another. If the contractor, serving as the construction manager, acts solely as an agent, his revenue should include only the fee and should exclude subcontracts negotiated or managed on behalf of the owner and materials purchased on behalf of the owner.

59.   In other circumstances, a contractor acts as an ordinary

---

[10]Cost-type government contracts with fees based on a percentage of cost are no longer granted under government regulations.

principal under a cost-type contract. For example, the contractor may be responsible to employees for salaries and wages and to subcontractors and other creditors for materials and services, and he may have the discretionary responsibility to procure and manage the resources in performing the contract. The contractor should include in revenue all reimbursable costs for which he has risk or on which his fee was based at the time of bid or negotiation. In addition, revenue from overhead percentage recoveries and the earned fee should be included in revenue.

Customer-Furnished Materials

60. Another concern associated with measuring revenue relates to materials furnished by a customer or purchased by the contractor as an agent for the customer. Often, particularly for large, complex projects, customers may be more capable of carrying out the procurement function or may have more leverage with suppliers than the contractor. In those circumstances, the contractor generally informs the customer of the nature, type, and characteristics or specifications of the materials required and may even purchase the required materials and pay for them, using customer purchase orders and checks drawn against the customer's bank account. If the contractor is responsible for the nature, type, characteristics, or specifications of material that the customer furnishes or that the contractor purchases as an agent of the customer, or if the contractor is responsible for the ultimate acceptability of performance of the project based on such material, the value of those items should be included as contract price and reflected as revenue and costs in periodic reporting of operations. As a general rule, revenues and costs should include all items for which the contractor has an associated risk, including items on which his contractual fee was based.

Change Orders

61. Change orders are modifications of an original contract that effectively change the provisions of the contract without adding new provisions. They may be initiated by either the contractor or the customer, and they include changes in specifications or design, method or manner of performance, facilities, equipment, materials, site, and period for completion of the work. Many change orders are unpriced; that is, the work to be performed is defined, but the adjustment to the contract price is to be negotiated later. For some change orders, both scope and price may be unapproved

or in dispute. Accounting for change orders depends on the underlying circumstances, which may differ for each change order depending on the customer, the contract, and the nature of the change. Change orders should therefore be evaluated according to their characteristics and the circumstances in which they occur. In some circumstances, change orders as a normal element of a contract may be numerous, and separate identification may be impractical. Such change orders may be evaluated statistically on a composite basis using historical results as modified by current conditions. If such change orders are considered by the parties to be a normal element within the original scope of the contract, no change in the contract price is required. Otherwise, the adjustment to the contract price may be routinely negotiated. Contract revenue and costs should be adjusted to reflect change orders approved by the customer and the contractor regarding both scope and price.

62. Accounting for unpriced change orders depends on their characteristics and the circumstances in which they occur. Under the completed-contract method, costs attributable to unpriced change orders should be deferred as contract costs if it is probable that aggregate contract costs, including costs attributable to change orders, will be recovered from contract revenues. For all unpriced change orders, recovery should be deemed probable if the future event or events necessary for recovery are likely to occur. Some of the factors to consider in evaluating whether recovery is probable are the customer's written approval of the scope of the change order, separate documentation for change order costs that are identifiable and reasonable, and the entity's favorable experience in negotiating change orders, especially as it relates to the specific type of contract and change order being evaluated. The following guidelines should be followed in accounting for unpriced change orders under the percentage-of-completion method.

a. Costs attributable to unpriced change orders should be treated as costs of contract performance in the period in which the costs are incurred if it is *not* probable that the costs will be recovered through a change in the contract price.

b. If it is probable that the costs will be recovered through a change in the contract price, the costs should be deferred (excluded from the cost of contract performance) until the parties have agreed on the change in contract price, or, alternatively, they should be treated as costs of contract performance in the period in which they are incurred, and contract revenue should be recognized to the extent of the costs incurred.

c. If it is probable that the contract price will be adjusted by an amount that exceeds the costs attributable to the change order and the amount of the excess can be reliably estimated, the original contract price should also be adjusted for that amount when the costs are recognized as costs of contract performance if its realization is probable. However, since the substantiation of the amount of future revenue is difficult, revenue in excess of the costs attributable to unpriced change orders should only be recorded in circumstances in which realization is assured beyond a reasonable doubt, such as circumstances in which an entity's historical experience provides such assurance or in which an entity has received a bona fide pricing offer from a customer and records only the amount of the offer as revenue.

63.   If change orders are in dispute or are unapproved in regard to both scope and price, they should be evaluated as claims (see paragraphs 65 to 67).

Contract Options and Additions

64.   An option or an addition to an existing contract should be treated as a separate contract in any of the following circumstances:

a. The product or service to be provided differs significantly from the product or service provided under the original contract.

b. The price of the new product or service is negotiated without regard to the original contract and involves different economic judgments.

c. The products or services to be provided under the exercised option or amendment are similar to those under the original contract, but the contract price and anticipated contract cost relationship are significantly different.

However, even if the separate contract does not meet any of these conditions, it may be combined with the original contract if the contracts meet the criteria in paragraph 37 or 38. Exercised options or additions that do not meet the criteria for treatment as separate contracts or as separate contracts combined with the original contracts should be treated as change orders on the original contracts.

Claims

65.   Claims are amounts in excess of the agreed contract price (or amounts not included in the original contract price) that a contractor seeks to collect from customers or others for customer-caused delays, errors in specifications and designs, contract terminations, change orders in dispute or unapproved as to both scope and price, or other causes of unanticipated additional costs. Recognition of amounts of additional contract revenue relating to claims is appropriate only if it is probable that the claim will result in additional contract revenue and if the amount can be reliably estimated. Those two requirements are satisfied by the existence of all the following conditions:

a. The contract or other evidence provides a legal basis for the claim; or a legal opinion has been obtained, stating that under the circumstances there is a reasonable basis to support the claim.

b. Additional costs are caused by circumstances that were unforeseen at the contract date and are not the result of deficiencies in the contractor's performance.

c. Costs associated with the claim are identifiable or otherwise determinable and are reasonable in view of the work performed.

d. The evidence supporting the claim is objective and verifiable, not based on management's "feel" for the situation or on unsupported representations.

If the foregoing requirements are met, revenue from a claim should be recorded only to the extent that contract costs relating to the claim have been incurred. The amounts recorded, if material, should be disclosed in the notes to the financial statements. Costs attributable to claims should be treated as costs of contract performance as incurred.

66.   However, a practice such as recording revenues from claims only when the amounts have been received or awarded may be used. If that practice is followed, the amounts should be disclosed in the notes to the financial statements.

67.   If the requirements in paragraph 65 are not met or if those requirements are met but the claim exceeds the recorded contract costs, a contingent asset should be disclosed in accordance with FASB Statement no. 5, paragraph 17.

Income Determination—Cost Elements

68. Contract costs must be identified, estimated, and accumulated with a reasonable degree of accuracy in determining income earned. At any time during the life of a contract, total estimated contract cost consists of two components: costs incurred to date and estimated cost to complete the contract. A company should be able to determine costs incurred on a contract with a relatively high degree of precision, depending on the adequacy and effectiveness of its cost accounting system. The procedures or systems used in accounting for costs vary from relatively simple, manual procedures that produce relatively modest amounts of detailed analysis to sophisticated, computer-based systems that produce a great deal of detailed analysis. Despite the diversity of systems and procedures, however, an objective of each system or of each set of procedures should be to accumulate costs properly and consistently by contract with a sufficient degree of accuracy to assure a basis for the satisfactory measurement of earnings.

Contract Costs

69. Contract costs are accumulated in the same manner as inventory costs and are charged to operations as the related revenue from contracts is recognized. Contract costs generally include all direct costs, such as materials, direct labor, and subcontracts, and indirect costs identifiable with or allocable to the contracts. However, practice varies for certain types of indirect costs considered allocable to contracts, for example, support costs (such as central preparation and processing of job payrolls, billing and collection costs, and bidding and estimating costs).

70. Authoritative accounting pronouncements require costs to be considered period costs if they cannot be clearly related to production, either directly or by an allocation based on their discernible future benefits.

71. Income is recognized over the term of the contract under the percentage-of-completion method or is recognized as units are delivered under the units-of-delivery modification and is deferred until performance is substantially complete under the completed-contract method. None of the characteristics peculiar to those methods, however, require accounting for contract costs to deviate in principle from the basic framework established in existing authoritative literature applicable to inventories or business enterprises in general.

72.   A contracting entity should apply the following general principles in accounting for costs of construction-type and those production-type contracts covered by this statement. The principles are consistent with generally accepted accounting principles for inventory and production costs in other areas, and their application requires the exercise of judgment.

a. All direct costs, such as material, labor, and subcontracting costs, should be included in contract costs.

b. Indirect costs allocable to contracts include the costs of indirect labor, contract supervision, tools and equipment, supplies, quality control and inspection, insurance, repairs and maintenance, depreciation and amortization, and, in some circumstances, support costs, such as central preparation and processing of payrolls. For government contractors, other types of costs that are allowable or allocable under pertinent government contract regulations may be allocated to contracts as indirect costs if otherwise allowable under GAAP.[11] Methods of allocating indirect costs should be systematic and rational. They include, for example, allocations based on direct labor costs, direct labor hours, or a combination of direct labor and material costs. The appropriateness of allocations of indirect costs and of the methods of allocation depend on the circumstances and involve judgment.

c. General and administrative costs ordinarily should be charged to expense as incurred but may be accounted for as contract costs under the completed-contract method of accounting[12] or, in some circumstances, as indirect contract costs by government contractors.[13]

---

[11]The AICPA industry audit guide, *Audits of Government Contractors*, states, "Practice varies among government contractors as to the extent to which costs are included in inventory. Some contractors include all direct costs and only certain indirect costs. . . . Other contractors record in inventory accounts all costs identified with the contract including allocated general and administrative . . . expenses." The guide points out that many accountants believe that the practice of allocating general and administrative expenses to contract costs, which is permitted under the completed-contract method by ARB 45, paragraph 10, may appropriately be extended to government contracts because they believe that "all costs under the contract are directly associated with the contract revenue, and both should be recognized in the same period."

[12]Paragraph 10 of ARB 45, *Long-Term Construction-Type Contracts*, states
When the completed-contract method is used, it may be appropriate to allocate general and administrative expenses to contract costs rather than to periodic income. This may result in a better matching of costs and revenues than would result from treating such expenses as period cost, particularly in years when no contracts were completed.

[13]See the discussion of the AICPA industry audit guide, *Audits of Government Contractors*, in footnote 11.

d. Selling costs should be excluded from contract costs and charged to expense as incurred unless they meet the criteria for precontract costs in paragraph 75.

e. Costs under cost-type contracts should be charged to contract costs in conformity with generally accepted accounting principles in the same manner as costs under other types of contracts because unrealistic profit margins may result in circumstances in which reimbursable cost accumulations omit substantial contract costs (with a resulting larger fee) or include substantial unallocable general and administrative costs (with a resulting smaller fee).

f. In computing estimated gross profit or providing for losses on contracts, estimates of cost to complete should reflect all of the types of costs included in contract costs.

g. Inventoriable costs should not be carried at amounts that when added to the estimated cost to complete are greater than the estimated realizable value of the related contracts.

Interest costs should be accounted for in accordance with FASB Statement no. 34, *Capitalization of Interest Cost.*

Precontract Costs

73.   In practice, costs are deferred in anticipation of future contract sales in a variety of circumstances. The costs may consist of (*a*) costs incurred in anticipation of a specific contract that will result in no future benefit unless the contract is obtained (such as the costs of mobilization, engineering, architectural, or other services incurred on the basis of commitments or other indications of interest in negotiating a contract), (*b*) costs incurred for assets to be used in connection with specific anticipated contracts (for example, costs for the purchase of production equipment, materials, or supplies), (*c*) costs incurred to acquire or produce goods in excess of the amounts required under a contract in anticipation of future orders for the same item, and (*d*) learning, start-up, or mobilization costs incurred for anticipated but unidentified contracts.

74.   Learning or start-up costs are sometimes incurred in connection with the performance of a contract or a group of contracts. In some circumstances, follow-on or future contracts for the same goods or services are anticipated. Such costs usually consist of labor, overhead, rework, or other special costs that must be in-

curred to complete the existing contract or contracts in progress and are distinguished from research and development costs.[14] A direct relationship between such costs and the anticipated future contracts is often difficult to establish, and the receipt of future contracts often cannot reasonably be anticipated.

75.   The division recommends the following accounting for pre-contract costs:

a. Costs that are incurred for a specific anticipated contract and that will result in no future benefits unless the contract is obtained should not be included in contract costs or inventory before the receipt of the contract. However, such costs may be otherwise deferred, subject to evaluation of their probable recoverability, but only if the costs can be directly associated with a specific anticipated contract and if their recoverability from that contract is probable.

b. Costs incurred for assets, such as costs for the purchase of materials, production equipment, or supplies, that are expected to be used in connection with anticipated contracts may be deferred outside the contract cost or inventory classification if their recovery from future contract revenue or from other dispositions of the assets is probable.

c. Costs incurred to acquire or produce goods in excess of the amounts required for an existing contract in anticipation of future orders for the same items may be treated as inventory if their recovery is probable.

d. Learning or start-up costs incurred in connection with existing contracts and in anticipation of follow-on or future contracts for the same goods or services should be charged to existing contracts.[15]

e. Costs appropriately deferred in anticipation of a contract should be included in contract costs on the receipt of the anticipated contract.

f. Costs related to anticipated contracts that are charged to expenses as incurred because their recovery is not considered

---

[14]Statement of Financial Accounting Standards no. 2, *Accounting for Research and Development Costs*, requires that research and development costs be charged to expense when incurred.

[15]See footnote 3, which indicates that the division is preparing a statement of position on program accounting for consideration by the FASB.

probable should not be reinstated by a credit to income on the subsequent receipt of the contract.

Cost Adjustments Arising from Back Charges

76.   Back charges are billings for work performed or costs incurred by one party that, in accordance with the agreement, should have been performed or incurred by the party to whom billed. These frequently are disputed items. For example, owners bill back charges to general contractors, and general contractors bill back charges to subcontractors. Examples of back charges include charges for cleanup work and charges for a subcontractor's use of a general contractor's equipment.

77.   A common practice is to net back charges in the estimating process. The division recommends the following procedures in accounting for back charges:

• Back charges to others should be recorded as receivables and, to the extent considered collectible, should be applied to reduce contract costs. However, if the billed party disputes the propriety or amount of the charge, the back charge is in effect a claim, and the criteria for recording claims apply.

• Back charges from others should be recorded as payables and as additional contract costs to the extent that it is probable that the amounts will be paid.

Estimated Cost to Complete

78.   The estimated cost to complete, the other component of total estimated contract cost, is a significant variable in the process of determining income earned and is thus a significant factor in accounting for contracts. The latest estimate may be determined in a variety of ways and may be the same as the original estimate. Practices in estimating total contract costs vary, and guidance is needed in this area because of the impact of those practices on accounting. The following practices should be followed:

a. Systematic and consistent procedures that are correlated with the cost accounting system should be used to provide a basis for periodically comparing actual and estimated costs.

b. In estimating total contract costs, the quantities and prices of all significant elements of cost should be identified.

c. The estimating procedures should provide that estimated cost to complete includes the same elements of cost that are included in actual accumulated costs; also, those elements should reflect expected price increases.

d. The effects of future wage and price escalations should be taken into account in cost estimates, especially when the contract performance will be carried out over a significant period of time. Escalation provisions should not be blanket overall provisions but should cover labor, materials, and indirect costs based on percentages or amounts that take into consideration experience and other pertinent data.

e. Estimates of cost to complete should be reviewed periodically and revised as appropriate to reflect new information.

Computation of Income Earned for a Period Under the Percentage-of-Completion Method

79. Total estimated gross profit on a contract, the difference between total estimated contract revenue and total estimated contract cost, must be determined before the amount earned on the contract for a period can be determined. The portion of total revenue earned or the total amount of gross profit earned to date is determined by the measurement of the extent of progress toward completion using one of the methods discussed in paragraphs 44 to 51 of this statement. The computation of income earned for a period involves a determination of the portion of total estimated contract revenue that has been earned to date (earned revenue) and the portion of total estimated contract cost related to that revenue (cost of earned revenue). Two different approaches to determining earned revenue and cost of earned revenue are widely used in practice. Either of the alternative approaches may be used on a consistent basis.[16]

Alternative A

80. The advocates of this method believe that the portion of total estimated contract revenue earned to date should be determined by the measurement of the extent of progress toward completion and that, in accordance with the matching concept, the

---

[16]The use of Alternative A in the discussion and in the presentation of some of the provisions of this statement is for convenience and consistency and is not intended to imply that Alternative A is the preferred approach.

measurement of extent of progress toward completion should also be used to allocate a portion of total estimated contract cost to the revenue recognized for the period. They believe that this procedure results in reporting earned revenue, cost of earned revenue, and gross profit consistent with the measurement of contract performance. Moreover, they believe that, if there are no changes in estimates during the performance of a contract, the procedure also results in a consistent gross profit percentage from period to period. However, they recognize that a consistent gross profit percentage is rarely obtained in practice because of the need to be responsive in the accounting process to changes in estimates of contract revenues, costs, earned revenue, and gross profits. In accordance with this procedure, earned revenue, cost of earned revenue, and gross profit should be determined as follows:

a. *Earned Revenue* to date should be computed by multiplying total estimated contract revenue by the percentage of completion (as determined by one of the acceptable methods of measuring the extent of progress toward completion). The excess of the amount over the earned revenue reported in prior periods is the earned revenue that should be recognized in the income statement for the current period.

b. *Cost of Earned Revenue* for the period should be computed in a similar manner. Cost of earned revenue to date should be computed by multiplying total estimated contract cost by the percentage of completion on the contract. The excess of that amount over the cost of earned revenue reported in prior periods is the cost of earned revenue that should be recognized in the income statement for the current period. The difference between total cost incurred to date and cost of earned revenue to date should be reported on the balance sheet.

c. *Gross Profit* on a contract for a period is the excess of earned revenue over the cost of earned revenue.

Alternative B

81.   The advocates of this method believe that the measurement of the extent of progress toward completion should be used to determine the amount of gross profit earned to date and that the earned revenue to date is the sum of the total cost incurred on the contract and the amount of gross profit earned. They believe that the cost of work performed on a contract for a period, including

materials, labor, subcontractors, and other costs, should be the cost of earned revenue for the period. They believe that the amount of costs incurred can be objectively determined, does not depend on estimates, and should be the amount that enters into the accounting determination of income earned. They recognize that, under the procedure that they advocate, gross profit percentages will vary from period to period unless the cost-to-cost method is used to measure the extent of progress toward completion. However, they believe that varying profit percentages are consistent with the existing authoritative literature when costs incurred do not provide an appropriate measure of the extent of progress toward completion. In accordance with Alternative B, earned revenue, cost of earned revenue, and gross profit are determined as follows:

a. *Earned Revenue* is the amount of gross profit earned on a contract for a period plus the costs incurred on the contract during the period.

b. *Cost of Earned Revenue* is the cost incurred during the period, excluding the cost of materials not unique to a contract that have not been used for the contract and costs incurred for subcontracted work that is still to be performed.

c. *Gross Profit* earned on a contract should be computed by multiplying the total estimated gross profit on the contract by the percentage of completion (as determined by one of the acceptable methods of measuring extent of progress toward completion). The excess of that amount over the amount of gross profit reported in prior periods is the earned gross profit that should be recognized in the income statement for the current period.

Revised Estimates

82.   Adjustments to the original estimates of the total contract revenue, total contract cost, or extent of progress toward completion are often required as work progresses under the contract and as experience is gained, even though the scope of the work required under the contract may not change. The nature of accounting for contracts is such that refinements of the estimating process for changing conditions and new developments are continuous and characteristic of the process. Additional information that enhances and refines the estimating process is often obtained after the balance sheet date but before the issuance of the financial statements;

such information should result in an adjustment of the unissued financial statements. Events occurring after the date of the financial statements that are outside the normal exposure and risk aspects of the contract should not be considered refinements of the estimating process of the prior year but should be disclosed as subsequent events.

83.   Revisions in revenue, cost, and profit estimates or in measurements of the extent of progress toward completion are changes in accounting estimates as defined in APB Opinion 20, *Accounting Changes*.[17] That opinion has been interpreted to permit the following two alternative methods of accounting for changes in accounting estimates:

● *Cumulative Catch-up*. Account for the change in estimate in the period of change so that the balance sheet at the end of the period of change and the accounting in subsequent periods are as they would have been if the revised estimate had been the original estimate.

● *Reallocation*. Account for the effect of the change ratably over the period of change in estimate and subsequent periods.

Although both methods are used in practice to account for changes in estimates of total revenue, total costs, or extent of progress under the percentage-of-completion method, the cumulative catch-up method is more widely used. Accordingly, to narrow the areas of differences in practice, such changes should be accounted for by the cumulative catch-up method.

84.   Although estimating is a continuous and normal process for contractors, the second sentence of APB Opinion 20, paragraph 33, recommends disclosure of the effect of significant revisions if the effect is material.[18]

---

[17]Paragraph 31 of APB Opinion 20, *Accounting Changes*, requires that "the effect of a change in accounting estimate should be accounted for in (*a*) the period of change if the change affects that period only or (*b*) the period of change and future periods if the change affects both."

[18]APB Opinion 20, paragraph 33, states,

The effect on income before extraordinary items, net income and related per share amounts of the current period should be disclosed for a change in estimate that affects several future periods, such as a change in service lives of depreciable assets or actuarial assumptions affecting pension costs. Disclosure of the effect on those income statement amounts is not necessary for estimates made each period in the ordinary course of accounting for items such as uncollectible accounts or inventory obsolescence; however, disclosure is recommended if the effect of a change in the estimate is material.

Provisions for Anticipated Losses on Contracts

85.   When the current estimates of total contract revenue and contract cost indicate a loss, a provision for the entire loss on the contract should be made. Provisions for losses should be made in the period in which they become evident under either the percentage-of-completion method or the completed-contract method. If a group of contracts are combined based on the criteria in paragraph 37 or 38, they should be treated as a unit in determining the necessity for a provision for a loss. If contracts are segmented based on the criteria in paragraph 40, 41, or 42 of this statement, the individual segments should be considered separately in determining the need for a provision for a loss.

86.   Losses on cost-type contracts, although less frequent, may arise if, for example, a contract provides for guaranteed maximum reimbursable costs or target penalties. In recognizing losses for accounting purposes, the contractor's normal cost accounting methods should be used in determining the total cost overrun on the contract, and losses should include provisions for performance penalties.

87.   The costs used in arriving at the estimated loss on a contract should include all costs of the type allocable to contracts under paragraph 72 of this statement. Other factors that should be considered in arriving at the projected loss on a contract include target penalties and rewards, nonreimbursable costs on cost-plus contracts, change orders, and potential price redeterminations. In circumstances in which general and administrative expenses are treated as contract costs under the completed-contract method of accounting, the estimated loss should include the same types of general and administrative expenses.

88.   The provision for loss arises because estimated cost for the contract exceeds estimated revenue. Consequently, the provision for loss should be accounted for in the income statement as an additional contract cost rather than as a reduction of contract revenue, which is a function of contract price, not cost. Unless the provision is material in amount or unusual or infrequent in nature, the provision should be included in contract cost and need not be shown separately in the income statement. If it is shown separately, it should be shown as a component of the cost included in the computation of gross profit.

89.   Provisions for losses on contracts should be shown separately as liabilities on the balance sheet, if significant, except in circumstances in which related costs are accumulated on the balance sheet, in which case the provisions may be deducted from the related accumulated costs. In a classified balance sheet, a provision shown as a liability should be shown as a current liability.

Transition

90.   An accounting change from the completed-contract method or from the percentage-of-completion method to conform to the recommendations of this statement of position should be made retroactively by restating the financial statements of prior periods. The restatement should be made on the basis of current information if historical information is not available. If the information for restatement of prior periods is not available on either a historical or current basis, financial statements and summaries should be restated for as many consecutive prior periods preceding the transition date of this statement as is practicable, and the cumulative effect on the retained earnings at the beginning of the earliest period restated (or at the beginning of the period in which the statement is first applied if it is not practicable to restate any prior periods) should be included in determining net income for that period (see paragraph 20 of APB Opinion 20, *Accounting Changes*).

91.   Accounting changes to conform to the recommendations of this statement of position, other than those stated in paragraph 90, should be made prospectively for contracting transactions, new contracts, and contract revisions entered into on or after the effective date of this statement. The division recommends the application of the provisions of this statement for fiscal years, and interim periods in such fiscal years, beginning after June 30, 1981. The division encourages earlier application of this statement, including retroactive application to all contracts regardless of when they were entered into. Disclosures should be made in the financial statements in the period of change in accordance with APB Opinion 20, paragraph 28.

Schematic Chart of SOP Organization

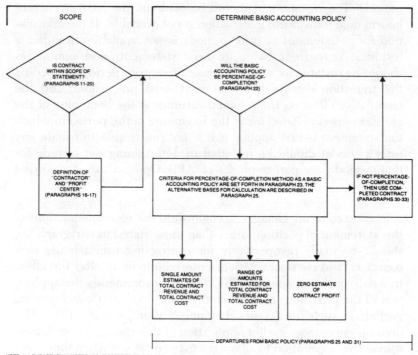

NOTE: ALL PARAGRAPH NUMBERS ABOVE REFER TO TEXT OF SOP.

*If computation results in a loss, see paragraphs 85-89

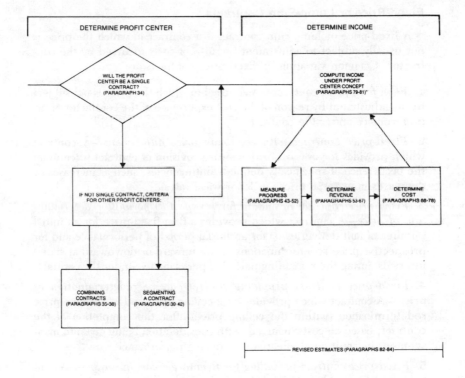

Types of Contracts

Four basic types of contracts are distinguished on the basis of their pricing arrangements in paragraph 15 of this statement: (a) fixed-price or lump-sum contracts, (b) time-and-material contracts, (c) cost-type (including cost-plus) contracts, and (d) unit-price contracts. This appendix describes the basic types of contracts in greater detail and briefly describes common variations of each basic type.

Fixed-Price or Lump-Sum Contracts

A fixed-price or lump-sum contract is a contract in which the price is not usually subject to adjustment because of costs incurred by the contractor. Common variations of fixed-price contracts are

1. *Firm fixed-price contract*—A contract in which the price is not subject to any adjustment by reason of the cost experience of the contractor or his performance under the contract.

2. *Fixed-price contract with economic price adjustment*—A contract which provides for upward or downward revision of contract price upon the occurrence of specifically defined contingencies, such as increases or decreases in material prices or labor wage rates.

3. *Fixed-price contract providing for prospective periodic redetermination of price*—A contract which provides a firm fixed-price for an initial number of unit deliveries or for an initial period of performance and for prospective price redeterminations either upward or downward at stated intervals during the remaining period of performance under the contract.

4. *Fixed-price contract providing for retroactive redetermination of price*—A contract which provides for a ceiling price and retroactive price redetermination (within the ceiling price) after the completion of the contract, based on costs incurred, with consideration being given to management ingenuity and effectiveness during performance.

5. *Fixed-price contract providing for firm target cost incentives*—A contract which provides at the outset for a firm target cost, a firm target profit, a price ceiling (but not a profit ceiling or floor), and a formula (based on the relationship which final negotiated total cost bears to total target cost) for establishing final profit and price.

6. *Fixed-price contract providing for successive target cost incentives*—A contract which provides at the outset for an initial target cost, an initial target profit, a price ceiling, a formula for subsequently fixing the firm

target profit (within a ceiling and a floor established along with the formula, at the outset), and a production point at which the formula will be applied.

7. *Fixed-price contract providing for performance incentives*—A contract which incorporates an incentive to the contractor to surpass stated performance targets by providing for increases in the profit to the extent that such targets are surpassed and for decreases to the extent that such targets are not met.

8. *Fixed-price level-of-effort term contract*—A contract which usually calls for investigation or study in a specific research and development area. It obligates the contractor to devote a specified level of effort over a stated period of time for a fixed dollar amount.[1]

Time-and-Material Contracts

Time-and-material contracts are contracts that generally provide for payments to the contractor on the basis of direct labor hours at fixed hourly rates (that cover the cost of direct labor and indirect expenses and profit) and cost of materials or other specified costs. Common variations of time and material contracts are

1. Time at marked-up rate.

2. Time at marked-up rate, material at cost.

3. Time and material at marked-up rates.

4. Guaranteed maximum cost—labor only or labor and material.

Cost-Type Contracts

Cost-type contracts provide for reimbursement of allowable or otherwise defined costs incurred plus a fee that represents profit. Cost-type contracts usually only require that the contractor use his best efforts to accomplish the scope of the work within some specified time and some stated dollar limitation. Common variations of cost-plus contracts are

1. *Cost-sharing contract*—A contract under which the contractor is reimbursed only for an agreed portion of costs and under which no provision is made for a fee.

2. *Cost-without-fee contract*—A contract under which the contractor is reimbursed for costs with no provision for a fee.

---

[1]AICPA Industry Audit Guide, *Audits of Government Contractors* (New York: American Institute of Certified Public Accountants, 1975), pp. 3–4.

3. *Cost-plus-fixed-fee contract*—A contract under which the contractor is reimbursed for costs plus the provision for a fixed fee.

4. *Cost-plus-award-fee contract*—A contract under which the contractor is reimbursed for costs plus a fee consisting of two parts: (*a*) a fixed amount which does not vary with performance and (*b*) an award amount based on performance in areas such as quality, timeliness, ingenuity, and cost-effectiveness. The amount of award fee is based upon a subjective evaluation by the government of the contractor's performance judged in light of criteria set forth in the contract.

5. *Cost-plus-incentive-fee contract (Incentive based on cost)*—A contract under which the contractor is reimbursed for costs plus a fee which is adjusted by formula in accordance with the relationship which total allowable costs bear to target cost. At the outset there is negotiated a target cost, a target fee, a minimum and maximum fee, and the adjustment formula.

6. *Cost-plus-incentive-fee contract (Incentive based on performance)*—A contract under which a contractor is reimbursed for costs plus an incentive to surpass stated performance targets by providing for increases in the fee to the extent that such targets are surpassed and for decreases to the extent that such targets are not met.[2]

Unit-Price Contracts

Unit-price contracts are contracts under which the contractor is paid a specified amount for every unit of work performed. A unit-price contract is essentially a fixed-price contract with the only variable being units of work performed. Variations in unit-price contracts include the same type of variations as fixed-price contracts. A unit-price contract is normally awarded on the basis of a total price that is the sum of the product of the specified units and unit prices. The method of determining total contract price may give rise to unbalanced unit prices because units to be delivered early in the contract may be assigned higher unit prices than those to be delivered as the work under the contract progresses.

[2]AICPA Industry Audit Guide, *Audits of Government Contractors*, pp. 4–6.

Summary of Disclosure Recommendations
in Statement of Position

SOP Par.	Nature of Disclosure
21	Accounting policy—methods of reporting revenue
45	Method or methods of measuring extent of progress toward completion
52	Criteria for determining substantial completion
65–67	Information on revenue and costs arising from claims
84	Effects of changes in estimates on contracts
90–91	Effects of accounting changes to conform to SOP

Glossary of Construction Accounting Terms

accelerated cost recovery system (ACRS)   A system of asset cost recovery for income tax purposes whereby all eligible property is assigned a prescribed recovery period of 3, 5, 10, 15, or 18 years, depending on the type of property and the time acquired, and is subject to prescribed accelerated recovery rates. Certain optional recovery periods can be elected as can the straight-line method of cost recovery.

accelerated performance   Usually occurs where the contractor accelerates his work either voluntarily or as a result of pressure from the owner in order to make up for delays. Increased costs almost always are the result of accelerated performance.

accrual method (for income tax reporting)   A method of accounting, whereby items of income are included in taxable income when all events have occurred that fix the right to receive such income and the amount of such income can be determined with reasonable accuracy; and items of cost or expense are deducted when all events have occurred that establish the liability, the amount of which can be determined with reasonable accuracy.

ACRS   *See* accelerated cost recovery system.

AGCA   Associated General Contractors of America.

aggregating contracts (for income tax purposes)   Accounting for two or more contracts as a single contract for income tax purposes. Usually, individual agreements will not be aggregated unless they are customarily treated as one contract as a commercial practice of the industry or unless there is no business purpose for the separate agreements.

AIA   American Institute of Architects.

AICPA   American Institute of Certified Public Accountants.

allocation base   This term denotes an accumulation of costs, units, or other measures of contracting activity upon which allocation of overhead or, under limited circumstances, general and administrative costs can be made to contracts, to tasks, or to bid items within contracts. Examples of allocation bases are: direct materials costs; direct labor costs or hours; subcontract costs or hours; total direct costs; machine hours; and cubic yards of concrete poured.

alternative A (approach to computation of income earned)   The estimate of progress toward completion is used to recognize contract revenues and contract costs in the statement of income. For example, if a contract is 50 percent complete, 50 percent of the revenues and 50 percent of the costs are recognized in income.

alternative B (approach to computation of income earned)   Provision for accounting for contract performance by recognizing all contract costs, other than costs of certain stored materials, as costs of contract revenue. Revenue is then recognized in whatever amount is necessary to achieve the amount of gross profit based on the percentage of completion of the contract.

anticipated losses on contracts   *See* provisions for anticipated losses on contracts.

APB   Accounting Principles Board of the AICPA.

apportionment (for income tax purposes)   The assignment of part of the taxable income of an entity to a particular state. Apportionment is normally done on the basis of a three-factor formula of property, payroll, and sales. Since definitions and formulas differ between states, it is not uncommon for duplications or omissions of taxable income to result when apportioning an entity's taxable income to the states within which it operates.

automated price book   Term used to describe price listing on magnetic media provided by a supplier, which contains the most current prices for materials and installation of labor.

back charges   Billings for work performed or costs incurred by one part that, in accordance with the agreement, should have been performed or incurred by the party to whom billed. Owners bill back charges to general contractors, and general contractors bill back charges to subcontractors. Examples of back charges include charges for cleanup work and charges for a subcontractor's use of a general contractor's equipment. (The AICPA's *Audit and Accounting Guide for Construction Contractors*.)

backlog   The amount of revenue that a contractor expects to be realized from work to be performed on uncompleted contracts, including new contractual agreements on which work has not begun.

basic review   An audit review performed by or under the direct supervision of the engagement partner.

bid   A formal offer by a contractor, in accordance with specifications for a project, to do all, or a phase of, the work at a certain price in accordance with the terms and conditions stated in the offer.

bid bond   A bond issued by a surety on behalf of a contractor, which provides assurance to the recipient of the contractor's bid that, if the bid is accepted, the contractor will execute a contract and provide a performance bond. Under the bond, the surety is obligated to pay the recipient of the bid the difference between the contractor's bid and the bid of the next lowest responsible bidder if the bid is accepted and the contractor fails to execute a contract or to provide a performance bond.

bid-day assistance program A computer program to assist the contractor in determining his bid amount for a particular project. The program accepts the contractor's bid by component and permits adjustments for differing overhead, profit and other "what if" assumptions.

bidding requirements The procedures and conditions for the submission of bids. The requirements are included in documents such as the notice to bidders, advertisement for bids, instructions to bidders, invitations to bid, and sample bid forms.

bid item *See* task.

bid (security) bond A bond that provides a project owner the protection that bidders will enter into a construction contract should they receive a contract award. Bid bonds serve to provide some assurance that only qualified, responsible contractors will submit bids.

bid shopping A practice by which contractors, both before and after their bids are submitted, attempt to obtain prices from potential subcontractors and material suppliers that are lower than the contractor's original estimates on which their bids are based or, after a contract is awarded, seek to induce subcontractors to reduce the subcontract price included in the bid.

billing schedule A contract schedule referencing the contractor's billing requirements to contract performance milestones.

billings in excess of costs and estimated earnings on uncompleted contracts A contractor's liability account, in which amounts billed on contracts that have not yet been earned (sometimes referred to as excess billings) are accumulated. This account includes only amounts pertaining to contracts on which billings exceed costs and estimated earnings.

bills of materials This term refers to "work packages" of similar items, common components, or assemblies, which are used as estimating tools. An example is the layout of the materials required for all the restrooms in a high-rise facility.

bonding A three-party arrangement whereby the surety and contractor join together to provide protection to the project owner. The surety underwrites the construction risk at all stages of construction, including preconstruction, and, in return, is paid a premium for assuming this risk. *See also* bid (security) bond; performance bond; payment bond.

bonding capacity The total dollar value of construction bonds that a surety will underwrite for a contractor, based on the surety's predetermination of the overall volume of work that the contractor can handle.

bonding company A company authorized to issue bid bonds, performance bonds, labor and materials bonds, or other types of surety bonds.

bonus clause A provision in a construction contract that provides for payments to the contractor in excess of the basic contract price as a reward for meeting or exceeding various contract stipulations, such as the contract completion date or the capacity, quality, or cost of the project.

break-even estimates   *See* zero estimate of profit/zero-profit margin.

broker   A party that obtains and accepts responsibility as a general contractor for the overall performance of a contract but enters into subcontracts with others for the performance of virtually all construction work required under the contract.

builders' risk insurance   Insurance coverage on a construction project during construction, including extended coverage that may be added for the contractor's protection or required by the contract for the customer's protection.

building codes   The regulations of governmental bodies specifying the construction standards that buildings in a jurisdiction must meet.

building permit   An official document issued by a governing body for the construction of a specified project in accordance with drawings and specifications approved by the governing body.

buying back the bond   A surety fulfilling his obligations upon default of a contractor by allowing the owner/obligee to complete the project, then reimbursing the owner.

buy-sell agreements   An agreement generally providing for the sale of a business interest at the occurrence of an event—usually death, disability, or retirement. Such agreements provide a ready market for an asset that might otherwise not be readily marketable.

capital expenditure authorization   A process used by the contractor whereby the need for additional equipment is justified on a financial basis prior to commitment to the expenditure. Capital expenditure authorizations will typically project expected equipment usage and return on investment.

cardinal change   A change order of a magnitude so great that the contractor cannot be forced to do the work. The contractor's further performance is excused as long as he negotiates in good faith for a price increase.

CASB   *See* Cost Accounting Standards Board.

cash receipts and disbursements method (for income tax purposes)   A method of accounting, whereby all items of income are included in taxable income when actually or constructively received and all items of cost or expense are deducted in the taxable year in which they are paid.

ceiling (or upset) price   Contract provision for a maximum price, thereby protecting the owner against costly cost overruns. Such contracts may also have provisions that call for any cost savings (actual vs. standard) to be divided between the parties.

change audit   Review of job progress by contract management, focusing on changes that may have been directed orally to field personnel.

change in accounting estimates APB Opinion 20 requires that a change in accounting estimate be accounted for in: (1) the period of change if the change affects that period only and (2) the period of the change and future periods if the change affects both. *See also* revisions in estimates/revised estimates.

change orders Modifications of an original contract that effectively change the provisions of the contract without adding new provisions. They include changes in specifications or design, method or manner of performance, facilities, equipment, materials, site, and period for completion of work.

claims Amounts in excess of the agreed-upon/contract price that a contractor seeks to collect from customers or others for customer-caused delays, errors in specifications and designs, unapproved change orders, or other causes of unanticipated costs.

claims made A type of professional liability insurance coverage protecting the contractor for claims brought against it during the policy period, irrespective of the time of occurrence of the alleged liability.

claims management A contractor's underlying system for identifying, documenting, and asserting claims against owners and others for owner-caused delays, errors in specifications and design, unapproved change orders, or other causes of anticipated costs.

clarifications *See* drawing revisions and clarifications.

COBOL *See* Common Business Oriented Language.

combined financial statements Financial statements presenting the combined financial position and results of operations of a group of affiliated companies to which consolidated financial statements are not applicable. Combined financial statements are preferable when such companies' operations are closely related and are economically interdependent; therefore, they should be viewed as a single economic unit.

combining contracts Grouping two or more contracts into a single profit center for accounting purposes. Appropriate under prescribed circumstances in which a group of contracts are so closely related that they are, in substance, parts of a single project with an overall profit margin. *See also* profit center.

committed costs In contract accounting, committed costs are those costs represented by purchase orders on materials not yet shipped and billed and similar costs to which the contractor is committed but for which he has not yet been billed. For managerial control purposes, some contractors charge costs to contracts when committed rather than when the liability is incurred.

Common Business Oriented Language (COBOL) A computer language that is generally accepted in the business community.

compensible delays Delays resulting, through the fault of the owner or his agents, in an increase in the contractor's costs or risks.

completed-contract method A method of accounting for contract profits that delays recognition until completion or substantial completion of the contract.

completion bond   A surety bond guaranteeing completion of the job as specified.

compliance testing   A process to evaluate the degree to which the internal accounting control procedures described by management are in use and are operating as planned.

consolidation method (joint ventures)   A method of accounting where the venture's accounts are fully consolidated into the venturer's financial statements, with the other venturers' interests shown as minority interests.

construction loan   *See* owner generated construction loan.

construction manager   An agency arrangement between a contractor and a project owner wherein the contractor agrees to supervise and coordinate the construction activity on the project. The role of the construction manager differs from that of the general contractor in the degree of responsibility taken for contract performance.

contract bond   An approved form of security executed by a contractor and a surety for the execution of the contract and all supplemental agreements, including the payment of all debts relating to the construction of the project.

contract changes   *See* change orders.

contract cost breakdown   An itemized schedule prepared by a contractor after the receipt of a contract, showing in detail the elements and phases of the project and the cost of each element and phase.

contract guarantees   Guarantees made by a contractor to an owner, such as a guarantee that a power plant, when completed, will generate a specified number of kilowatt hours.

contract item (pay item)   An element of work, specifically described in a contract, for which the contract provides either a unit or lump-sum price.

contract milestones   Important events in contract completion, such as completion of excavation or erection of facilities.

contract options   Provision for the performance of work beyond the basic contract scope or for the addition of new items under the same contract.

contract overrun/underrun   The amount by which the original contract price, as adjusted by change orders, differs from the total cost of a project at completion.

contract performance bond   The security furnished by the contractor to guarantee the completion of the work on a project in accordance with the terms of the contract.

corpnerships   A partnership in which all of the members are corporations. This form avoids the potentially adverse income tax consequences of the earnings of the partnership being attributed to the partners as individuals.

Cost Accounting Standards Board (CASB)   A governmental board charged with the responsibility of promulgating cost accounting standards for negotiated federal contracts.

cost method of accounting (joint ventures)   A method of accounting for a venturer's interest in a joint venture wherein the investment is recorded at cost and income is recognized as distributions are received from earnings accumulated by the venture since the date of acquisition by the venturer.

cost of earned revenue   The portion of estimated total contract cost that relates to the portion of total estimated contract revenue earned to date. *See also* alternative A (approach to computation of income earned); alternative B (approach to computation of income earned).

cost-plus award fee   A cost-type contract in which the contractor's fee is composed of a fixed part and a variable part. The variable part is an award based on performance, including timeliness and cost efficiency.

cost-plus fixed fee   A cost-type contract providing reimbursement to the contractor for his costs plus compensation of a predetermined fixed amount as the contract profit.

cost-plus/time-and-material contract   Contract types that address the contractor's performance and compensation from an efforts-expended point of view. The means of compensation is set forth in terms of specific costs incurred by the contractor or the time and materials he consumes.

costs and estimated earnings in excess of billings on uncompleted contracts   A contractor's asset account in which amounts earned on contracts that have not yet been billed (sometimes referred to as "underbillings") are accumulated. This account includes only amounts pertaining to contracts on which costs and estimated earnings exceed billings.

cost-sharing limitations   Ceilings or limitations on the amount of contract revenue otherwise available from the owner under cost-sharing provisions.

cost to complete   *See* estimated cost to complete.

cost-to-cost method   A method of measuring progress toward contract completion by the ratio of costs incurred to estimate total costs. Costs incurred would exclude costs of certain uninstalled materials.

cost-type contracts   Contracts providing for reimbursement of allowable costs, plus a fee that represents the contractor's profit. The "best efforts" of the contractor are the prime requirement in the cost-type contract. The contractor is required to put forth a standard of effort in performance of the contract that is based upon ordinary industry commercial standards.

cost without fee   A cost-type contract variation that is similar to a cost-sharing contract, except that all costs are reimbursed.

CPM   *See* critical path method.

CPU   Central processing unit of a computer system.

critical path method (CPM)   A network scheduling method that shows the sequences and interdependencies of activities. The critical path is the sequence of activities that shows the shortest time path for completion of the project.

custom-designed applications Software designed specifically to a user's needs as opposed to a pre-existing package of general application.

customer-furnished materials Construction materials that are purchased by the owner for the contractor's use on the construction project. Under such circumstances, the contractor normally informs the owner of the specifications of materials required and the owner purchases them or has the contractor act as his agent in purchasing them, using the owner's purchase orders and, sometimes, checks drawn against the owner's bank account.

customer's extras *See* extras.

default and termination Contract provisions setting forth the conditions under which the parties can be considered in default. The specific conditions under which the contractor may close the job down and other specific procedures for terminating the contract prior to completion may be given.

delay Failure to meet time requirements of a construction contract. *See also* compensible delays; excusable delays; no damage for delays.

demobilization A contractor's activities in shutting down a job-site after project completion.

design/build contractor A contractor who provides engineering or architectural design services in addition to construction work on a project.

direct costing The process of assigning only direct costs and variable indirect costs to contracts, thus accounting for the entity's contribution to fixed costs and profits.

direct costs Costs that are consumed and directly identifiable with the performance of the contract. They include direct labor, direct materials, and equipment.

direct labor variances Variances resulting from use of standard costing for direct labor. Direct labor variances can be broken down into rate and efficiency variances.

direct material variances Variances resulting from use of standard costing for direct materials. Direct material variances can be broken down into price and usage variances.

direct write-off (or specific charge-off) method of deduction for bad debts An income tax accounting method, whereby worthless accounts are deducted at the time they become worthless as contrasted with the reserve method of accounting for bad debts. *See also* reserve method.

draw The amount of progress billings on a contract that are currently available to a contractor under a contract with a fixed payment schedule.

drawing revisions and clarifications The practice of an architect issuing revised drawings as "clarifications," which are, in substance, changes to the scope of the contract.

economic price adjustments Contract provisions allowing price adjustments based on specified contingencies, such as material-price or labor-rate changes. The adjustments can either increase or decrease the contract price. A contract with a cost-of-living labor index adjustment clause is an example.

efficiency variance For labor costs, the efficiency variance is the difference between the two products obtained by multiplying the actual and standard amounts of labor by the standard labor rates. For variable overhead costs, the efficiency variance is the difference between the two products of the actual base units incurred versus the standard base units at the standard rates.

efforts-expended methods Methods of measuring progress of a contract toward completion based on a measure of the work, such as labor hours, labor dollars, or machine hours.

Eichleay Formula A formula that develops and documents a figure for home office overhead to be used in assertion of claims. Currently, the Eichleay Formula is not considered to be a generally acceptable approach.

equipment log A log detailing all of the equipment dedicated to a particular job, as well as the time that the equipment is planned to be used on the job. The equipment log is important to equipment cost recovery in the event of owner-caused delays or other claims involving equipment cost overruns.

equipment pools A cost center used to account for a contractor's equipment costs, including repairs and maintenance, depreciation, insurance, storage, and other equipment costs. These costs are accumulated in equipment pools and charged to contracts using rental rates, either calculated to recover costs or to return a profit on the equipment investment.

equity method of accounting (joint ventures) The equity method is the traditional one-line method prescribed by APB Opinion 18 for investments in corporate joint ventures and for investments in common stock that represent less than a majority interest but that evidence an ability to exercise significant influence.

ERISA Employee Retirement Income Security Act of 1974.

escalation clause A contract provision that provides for adjustments of the price of specific items as conditions change (e.g., a provision that requires wage rates to be determined on the basis of wage levels established in agreements with labor unions).

estimate (bid function) The amount of labor, materials, and other costs that a contractor anticipates for a project, as summarized in the contractor's bid proposal for the project.

estimated cost to complete The anticipated additional cost of materials, labor, and other items required to complete a project at a scheduled time.

excusable delays Delays caused by factors not the fault of the party seeking them.

excusable vs. inexcusable Contract change provisions that define liability as to changes, that is, who, if anyone, may be held liable for the impact of a change or changes to the project. This type of change generally sets forth what changes are or are not contractually allowed.

expanded equity method of accounting (joint ventures)   A method of accounting wherein the venturer presents its proportionate share of the venture's assets and liabilities in capsule form, segregated between current and noncurrent. The elements of the investment are presented separately by including the venturer's equity in the venture's corresponding items under current assets, current liabilities, noncurrent assets, noncurrent liabilities, revenues and expenses, using a caption such as "investor's share of net current assets of joint venture."

extended period long-term contract   A contract that the taxpayer estimates will not be completed within two years (long-term contracts other than construction) or within three years (long-term construction contracts) of the contract commencement date. However, a construction contract will not be considered an extended period long-term contract if the taxpayer's average annual gross receipts over the three taxable years ending before the taxable year in which the construction contract was signed do not exceed $25 million. An extended period long-term construction contract is subject to more extensive cost allocation rules than are other construction rules.

extras (customer's extras)   Additional work, not included in the original plan, requested of a contractor, that is billed separately and does not alter the original contract amount.

face checking   Frequent unexpected visits to jobs by timekeepers or supervisory personnel who are expected to know by sight each person working to keep control over actual hours worked.

FASB   The Financial Accounting Standards Board.

fast-tracking   Accelerated performance requiring performance of portions of interdependent phases of project work simultaneously, such as having the design phase continue after the excavation phase has begun.

favorable and unfavorable variances   Differences between actual and standard costs. Favorable variances are those in which actual is less than standard. Unfavorable are those in which actual exceeds standard.

FCPA   *See* Foreign Corrupt Practices Act.

field funds   Imprest accounts used for the purpose of either satisfying payroll expenditures or for the payment of incidental expenditures at the job-site.

FIFO   *See* first-in, first-out method.

final acceptance   The customer's acceptance of the project from the contractor on certification by an architect or engineer that the project is completed in accordance with contract requirements. The customer confirms final acceptance by making final payment under the contract, unless the time for making the final payment is otherwise stipulated.

final inspection   The final review of the project by an architect or engineer before issuance of the final certificate for payment.

firm target cost incentives   An establishment of target profits and costs at the outset of the contract, along with a price ceiling. A formula is then devised, based on the final negotiated cost-to-target cost ratio, for establishing the final profit and the total contract price.

first-in, first-out method (FIFO)   Flow of cost assumption used to cost inventories. It is assumed that the first items purchased are the first sold or consumed.

fixed-price contract   A contract providing for a single price for the total work to be performed on a construction project.

flow-down clauses   Clauses in a contractor's contracts with subcontractors providing that certain terms of the prime contract "flow down" to, and bind, the subcontractors. Such a clause could incorporate time limits for submission of change orders, special methods of dispute resolution, and a requirement for complete release prior to final payment.

forbearance   *See* loan or forbearance.

force majeure   Acts of God (e.g., floods, earthquakes, or other types of natural disasters) that are excuses for delay.

Foreign Corrupt Practices Act (FCPA)   Enacted into law in 1977 as an amendment to the Securities Exchange Act of 1934, the FCPA requires companies that are required to file periodic reports to comply with certain accounting standards. The FCPA also makes it unlawful to engage in certain corrupt practices involving foreign officials.

fringe benefit pools   Cost centers used for the accumulation of fringe benefit costs. Such costs are normally accumulated in fringe benefit pools and charged to jobs and other cost objects using appropriate fringe benefit rates.

fringe benefit rates   Rates developed and used to charge fringe benefit costs to contracts and other cost objects and to contract tasks or bid items.

front-end loading   A procedure under which progress billings are accelerated in relation to costs incurred by assigning higher values to contract portions that will be completed in the early stages of a contract than to those portions that will be completed in the later stages, so that cash receipts from the project during the early stages will be higher than they otherwise would be.

full absorption (to contracts)   The process of assigning all overhead costs and, where permitted, general and administrative costs to contracts.

full absorption (to contract tasks)   The process of assigning all overhead costs and, where permitted, general and administrative costs to contract tasks or bid items.

functional audits   Audits that review the duties of a specific type of department within the company. The department will typically be one whose function affects all the projects on which the contractor is working.

GAO The U.S. General Accounting Office.

general and administrative cost pools Cost centers used for the accumulation of general and administrative costs. Where permitted, such costs are accumulated in a general and administrative cost pool and charged to jobs using an appropriate allocation base.

general contractor A contractor who enters into a contract with the owner of a project for the purpose of construction and who takes full responsibility for its completion, although the contractor may enter into subcontracts with others for the performance of specific parts or phases of the project.

governmental prequalification reporting Reports required to be filed by contractors with agencies of federal, state, or county governments in order to qualify for bidding on or performing work for such agencies.

gross-profit approach *See* alternative B (approach to computation of income earned).

HVAC Heating, venting, and air conditioning task of a construction project.

incentive based on cost A cost-type contract variation where the determination of the contractor's fee is by a formula (a type of flexible revenue budget) that contains target, minimum and maximum fees, and the method for changing the application of the formula.

incentive based on performance A cost-type contract variation in which increases and decreases to a stated fee are determined by reference to the contractor's performance measured against specific targets provided in the contract.

incentives Contract clauses that seek to reward or penalize a contractor for favorable or unfavorable contract performance. *See also* bonus clause; penalty clause.

independent review Audit review performed by a specialist or firm partner independent of direct supervision of the engagement.

inherent hazards (that make estimates doubtful) Hazards involving events and conditions that would not be considered in the ordinary preparation of contract estimates and would not be expected to recur frequently, given the contractor's normal business environment. Inherent hazards that may cause contract estimates to be doubtful usually differ from inherent business risks.

input measure Those measures of percentage of completion that depend on measures of effort and resources applied to a project. The labor-hours method is an example of an input measure.

interactive time sharing Provides continuous access to the computer system by the user. Data is entered and processed immediately. The user has virtually complete control over when jobs are processed. In most respects, interactive time sharing provides most of the benefits of an in-house computer system.

interest-on-default clauses   Contract clauses that provide for payment of interest in the event of default.

internal control   The system in any organization that assures that management's plans and intentions are being carried out.

investment carrying costs (contractor's equipment investment)   Insurance, taxes, storage, other equipment division costs, and interest, less equipment division income.

investment credit   A credit against income taxes for investment in eligible business property. The credit is a specified percentage of eligible investment expenditures, subject to limitations.

investment maintenance costs (contractor's equipment investment)   Repairs and maintenance, repair parts, and depreciation (a type of maintenance cost in that it measures the failure to keep the equipment in its original state).

IRC   Internal Revenue Code.

IRS   Internal Revenue Service.

job cost system   A system whereby a contractor accounts for his costs separately by job and by task or bid item within a job.

job diary   A daily, updated "picture" of the job and its progress. The level of recovery from job changes, delays, or other events is usually tied closely to the quality of the job diary. The job diary is the cornerstone of the whole job documentation system.

job overhead   Costs that are specifically identifiable to contracts but not specifically identifiable to tasks or bid items within contracts. Examples are mobilization and demobilization costs, job support services (e.g., field payroll and supervision), and costs of physical security.

joint venture   An entity owned, operated, and jointly controlled by a small group of participants as a separate and specific business or project for the mutual benefit of the participants, including arrangements for pooling equipment, bonding, financing, and sharing skills (e.g., engineering, design, and construction).

last-in, first-out method (LIFO)   Flow of goods assumption used to cost inventories. It is assumed that the last items purchased are the first sold or consumed.

letter agreement (letter of agreement)   A letter stating the terms of an agreement between addressor and addressee, usually prepared for signature by the addressee as indication of acceptance of those terms as legally binding.

letter of intent   A letter signifying an intention to enter into a formal agreement and usually setting forth the general terms of such an agreement.

level-of-effort term contract   Provision applied, usually to research and development projects, wherein the contractor must devote a specified level of effort for a stated period of time to earn a fixed amount of revenue.

lien/lien rights An encumbrance that usually makes real or personal property the security for payment of a debt or discharge of an obligation.

LIFO *See* last-in, first-out method.

liquidated damages Construction contract clauses obligating the contractor to pay specified daily amounts to the project owner as compensation for damages suffered by the owner because of the contractor's failure to complete the work within a stated period of time.

loan or forbearance A legal characterization of amounts withheld from payment that could subject such transactions to limits of usury laws.

loss contract A contract on which the estimated cost to complete exceeds the contract price.

lump-sum contract *See* fixed-price contract.

maintenance bond A document, given by the contractor to the owner, guaranteeing rectification of defects in workmanship or materials for a specified time following completion of the project. A one-year bond is normally included in the performance bond.

management information systems (MIS) The systems of information upon which a contractor manages his business. An effective management information system produces timely, accurate, and relevant information upon which to base management decisions.

manufacturing contractor A contractor who produces for order under production-type contracts, generally involving a portable product.

mechanics lien bond A mechanics lien bond guarantees that a project of private work for unpaid laborers and materialmen will be free of mechanics liens.

mechanics lien law A lien law intended to provide a source of recovery for unpaid laborers and materialmen on private work. A mechanics lien law affords security to the claimants, since the owner must discharge all liens before he can obtain a clear title to the property.

microcomputer An inexpensive computer system consisting of a processing unit, or CPU, a CRT-type terminal input data display device, and a mass storage unit storing files or data bases—typically hard disks or diskettes ("floppy disks").

MIS *See* management information systems

mobilization A contractor's activities in starting up a job-site after contract award.

negotiated contract A contract for construction developed through negotiation of plans, specifications, terms, and conditions without competitive bidding.

negotiated insurance placements   A process of negotiating insurance coverage and related costs as contrasted with obtaining coverage through competitive bid.

no damage clauses   Contract clauses that exculpate the owner for owner-caused delays by compensating the contractor with additional time to complete the contract.

no damage for delays   Delays an owner is liable for, caused by his acts or omissions that are not forgiven by a specific contract provision.

noninsurance transfer of risk   Transfer of risk to other than professional risk takers (insurance companies). Examples of noninsurance transfers of risk are hold-harmless agreements and/or waivers of subrogation. A waiver of subrogation in the contractor's subcontract agreement can prevent the subcontractor's insurer from pursuing a recovery for damage to its insured's equipment, even if the contractor's employee caused the damage.

nonqualified plan   A deferred compensation arrangement that can either be funded or unfunded. Generally, if the arrangement is funded and nonforfeitable, the employee recognizes income and the employer takes a current deduction. Otherwise, both are deferred until either cash payments are made to the employee or his rights to such payments become nonforfeitable.

off-balance-sheet financing   Financing a project without increasing the debt on the investor's balance sheet, such as by having a joint venture incur the debt and accounting for the joint venture by the equity method.

office automation   The automation of the information input, processing, storage, and output functions of an office.

OMB   U.S. Office of Management and Budget.

oral instructions or modifications   Informal directions given on the job-site or elsewhere by inspectors, owners, and others with apparent authority, which may have the effect of changing the scope of work under the contract.

other direct costs   Contract direct costs other than direct materials, direct labor, and subcontract costs.

output measure   A method of measuring progress of a contract toward completion in terms of results achieved (e.g., units produced, units delivered, contract milestones, and value added).

overhead costs   Those costs, other than general and administrative costs, that cannot be specifically identified to contract profit centers. Examples of overhead costs are general supervision, supplies, quality control, and inspection, and support services such as procurement, payroll preparation, job billing, and accounting.

overhead pools   Cost centers used for accumulation of overhead costs. Such costs are normally accumulated in an overhead pool and charged to jobs using an appropriate allocation base.

owner generated construction loan   Construction financing obtained by the project owner against which the construction project is pledged as collateral.

partial or proportionate consolidation (joint ventures)   A method of accounting whereby the venturer records its proportionate interest in the venture's assets, liabilities, revenues, and expenses on a line-by-line basis and combines the amounts directly with its own assets, liabilities, revenues, and expenses without distinguishing between the amounts related to the venture and those held directly by the venturer.

payment bond   A bond executed by a contractor to protect suppliers of labor, materials, and supplies to a construction project.

payment schedule   Provisions in a contract for payments to a contractor of agreed-upon sums upon attainment of contract milestones.

pay request   A form of contractors' billings, made by a general contractor to the owner or by a subcontractor to the general contractor.

PDL (price determined later) letter   A letter from an architect/owner directing a change in scope and providing that the price will be determined later.

penalty clause   A provision in a construction contract that provides for a reduction in the amount otherwise payable under a contract to a contractor as a penalty for failure to meet targets or schedules specified in the contract or for failure of the project to meet contract specifications.

percentage-of-completion method   A method of accounting in which contract profit is recognized as contract performance and that occurs using a measure of percentage-of-completion appropriate to the circumstances.

performance bond   A bond issued by a surety and executed by a contractor to provide protection against the contractor's failure to perform a contract in accordance with its terms.

performance incentives   Provisions made in the contract whereby the contractor is rewarded for surpassing, and penalized for failing to meet, specific performance targets.

performance pay plan   A bonus plan, usually for employees other than executive officers, based on specific performance, such as a supervisor completing a job budgeted for 1,000 hours in 900 hours.

PERT   *See* Program Evaluation and Review Technique.

point estimates   A contractor's estimates of total contract revenue and total contract costs in terms of single amounts.

precontract costs   Costs incurred in anticipation of future contracts, including: (1) costs of mobilization, engineering, architectural, or other services incurred on the basis of commitments or other indications of interest in negotiating a contract; (2) costs incurred for assets to be used in connection with specific anticipated contracts (e.g., costs for the purchase of production equipment, materials, or supplies); (3) costs incurred to acquire or produce goods in excess of the amounts required under a contract in anticipation of future orders for the same item; and (4) learning, start-up, or mobilization costs incurred for anticipated but unidentified contracts.

prequalification The written approval of an agency seeking bids on a project that authorizes a contractor to submit a bid in circumstances in which bidders are required to meet certain standards.

price variance A direct materials variance of actual direct materials costs from standard direct materials costs. The difference between the actual cost of materials and what the cost would have been had the same quantities been purchased at standard cost.

prime contract A contract between an owner of a project and a contractor for the completion of all or a portion of a project, under which the contractor takes full responsibility for the completion of the work.

Private Companies Practice Section One of the two sections of the Division of Firms of the AICPA, which contains both voluntary and self-regulatory sections.

profit center A unit for the accumulation of revenues and costs for the measurement of income. For business enterprises engaged in the performance of contracts, the profit center for accounting purposes is usually a single contract; but under some specified circumstances, it may be a combination of two or more contracts, a segment of a contract, or a group of combined contracts.

Program Evaluation and Review Technique (PERT) A network-type scheduling system for contract performance.

progress (advance) billings Amounts billed, in accordance with the provisions of a contract, on the basis of progress to date under the contract.

prospective periodic redetermination of price A revision of the contract price, subject to a ceiling price, after the passage of an initial period of performance (or a delivery of a defined base number of units). This type of pricing is used in some government contracting.

provisions for anticipated losses on contracts Accounting recognition that a loss is expected on a contract. Provisions for the entire estimated loss on a contract should be made in the period that such loss becomes evident.

punch list A list made near the completion of work indicating items to be furnished or work to be performed by the contractor or subcontractor in order to complete the work as specified in the contract.

qualified plan A deferred compensation plan where current funding is permitted and the employer gets a current income tax deduction while the employee is not taxed until he actually receives the funds.

quantity takeoffs   An itemized list of the quantities of materials and labor required for a project, with each item priced and extended, which is used in preparing a bid on the project.

quantum meruit   That which is merited or deserved.

quid pro quo   Fair or equitable compensation for services or work performed.

range estimates   A contractor's estimates of total contract revenue and total contract cost in ranges of amounts. If the contractor can determine the amounts in the range that are most likely to occur, these amounts should be used in accounting for the contract under the percentage-of-completion method. If the most likely amounts cannot be determined, the lowest probable level of profit in the range should be used in accounting for the contract until the results can be estimated more precisely.

rate variance   A direct labor variance of actual direct labor costs from standard direct labor costs. The rate variance is the difference between the two products obtained by multiplying the actual labor hours by the actual rates and the standard rates.

reasonably dependable estimates   The ability of a contractor to estimate the extent of contract progress toward completion, contract revenues, and contract costs, at least to the degree that the minimum total revenue and the maximum total cost can be estimated with sufficient confidence to justify the contractor's bids on contracts.

Regulation S-X   A regulation of the SEC governing the form and content of financial statements.

remote-batch processing technique   A technique whereby data that is entered through the terminal is accumulated either at the terminal or at the computer site. Processing is performed at the convenience of the vendor, usually during off hours. The processed data is then transmitted back to the user's terminal or printer.

Report Program Generator (RPG)   A computer language of general acceptance.

resale certificate   A certification from a purchaser of goods or equipment that he has purchased for resale and therefore will be liable for the sales tax liability upon ultimate sale.

reserve method   An income tax accounting method, whereby provision is made for losses on accounts receivable based on estimates as contrasted with the direct write-off method. *See also* direct write-off (or specific charge-off) method of deduction for bad debts.

retainage   *See* retentions.

retentions   Amounts withheld from progress billings until final and satisfactory project completion.

retrospective rating plan A workers' compensation plan that contracts for a specified premium at the beginning of the term with the understanding between the parties that the rating used in computing the original premium will be retroactively adjusted for the actual loss experience of the contractor.

revenue-cost approach *See* alternative A (approach to computation of income earned).

revisions in estimates/revised estimates Adjustments to the original estimates of the total contract revenue, total contract cost, or extent of progress toward completion. As experience is gained, revisions in estimates are usually required, even though the scope of the work required under the contract may not change.

ripple effect The impact of a great number of changes to a contract that, taken individually are small but, in the aggregate, have a substantial effect on the overall performance plan.

risk management A management approach to controlling the costs associated with the possibility of fortuitous loss. The risk management approach recognizes that insurance is the most expensive technique to use and seeks to use the insurance mechanism only as a last resort.

risk retention The planned acceptance of losses through the use of insurance policy deductibles, the deliberate use of noninsurance, or the implementation of a formal self-insurance program.

RPG *See* Report Program Generator.

SAP Statements on Auditing Procedures issued by the Committee on Auditing Procedure of the AICPA.

SAS Statements on Auditing Standards of the Auditing Standards Board (previously the Auditing Standards Executive Committee) of the AICPA.

scheduling Planning and marshalling of manpower, materials, subcontractors, and other resources to requirements of contract performance.

SEC The Securities and Exchange Commission.

SEC Practice Section One of the two Divisions of Firms of the AICPA into voluntary, self-regulatory sections.

segmenting contracts Dividing a single contract or group of contracts into two or more profit centers for accounting purposes. This is appropriate under certain circumstances where the contract or contracts include several elements or phases, each of which has been negotiated separately with the same customer without regard to the others. Revenues are assigned to the different segments to achieve different rates of profitability based on the relative value of each segment to the total. *See also* profit center.

segregated cost approach An approach to documenting claims, which segregates and identifies each matter of delay and defines the related costs incurred as a result of the delay.

separating contracts (for income tax purposes)   The division of a single contract into several segments for income tax reporting purposes. Contracts usually subject to such division and acceleration of income are those that separate delivery and acceptance of contract phases.

severing contracts (for income tax purposes)   *See* separating contracts.

SFAS   Statements of Financial Accounting Standards issued by the FASB.

software directory subscription service   Services offered by publishers containing information on available software packages, including construction industry software.

SOP   Statements of Position of the Accounting Standards Division of the AICPA.

specialty contractor   *See* trade or specialty contractor.

spending variance   A variance of actual variable overhead from standard variable overhead. The difference between actual variable overhead and the amount obtained by multiplying the base units (i.e., direct labor hours) by the standard rate.

split dollar life insurance   A type of life insurance policy whereby the company owns the policy and has the rights to its cash value while the employee names the beneficiary and has the insurance protection. The employer gets no tax deduction for premium payments. The employee is taxed on an amount equal to the cost of a one-year nonrenewable term policy.

standard costing   Accounting for costs comparing predetermined standards for materials, labor, and variable overhead with actual costs. An accounting for variances from standard can be used to identify variances of direct materials price and usage, direct labor rate and efficiency, and variable overhead spending and efficiency.

stop order   A formal notification to a contractor to discontinue some or all work on a project for reasons such as safety violations, defective materials or workmanship, or cancellation of the contract.

stored materials   *See* uninstalled materials.

subcontractor   A contractor who enters into a contract with the general contractor of a construction project and who is responsible for completion of a specific phase of the construction project. Subcontractors are usually specialists within the industry, such as electrical or mechanical subcontractors.

subcontractor bond   A bond executed by a subcontractor and given to the prime contractor to assure the subcontractor's performance on the subcontract, including the payment for all labor and materials required for the subcontract.

substantial completion   The point at which the major work on a contract is completed and only insignificant costs and potential risks remain. Revenue from a contract is recognized under the completed-contract method when the contract is substantially completed.

substantive tests   All of the analytical review procedures and details of the particular classes of transactions and balances that the auditor deems necessary under the circumstances.

successive target cost incentives   A contract provision for initial target costs and profits, a price ceiling, and a formula by which to fix the target profit. This variation further provides a point in production for the application of the formula.

surety   A three-party relationship between the bonding company, the contractor, and the owner (or the bonding company, the subcontractor, and the contractor), whereby the surety guarantees the performance of the contractor or subcontractor (can also refer to a bonding company).

system conversion   The implementation of a new system by conversion of an existing manual or automated one.

takeout commitment   A commitment for permanent financing obtained by a project owner at, or prior to, the time of obtaining the construction loan. The takeout commitment provides assurance to the construction lender of a source of repayment of the construction loan.

task   A separate phase of a construction contract, such as excavation or erection. A contractor's bid price is normally built up by estimating costs for each task or bid item separately.

time-and-material contract   An agreement under which the contractor says he will complete a project of general or specific scope in exchange for rates for the time incurred and materials consumed. The rates would normally be calculated to recover indirect labor, overhead, and profit.

time sharing   A system of making the resources of a large computer system available to many users simultaneously. Generally, the computer is accessed via telephone lines from terminals in the users' offices.

total cost claim   A concept in contract claims theory that a contractor who has bid the job at a certain price, and the job costs more than the bid price, is therefore entitled to the difference. The assumption is that anything that happened that caused the job to run over is someone else's fault. Today, the concept of the total cost claim may suggest that the contractor does not have the necessary information and documentation to present a detailed and well-supported claim.

trade or specialty contractor   Contractors specializing in particular phases of construction, such as steel erection or roofing. Trade or specialty contractors normally function as subcontractors on construction projects.

trade stacking   Crowding on the job-site by the stacking of various trades that are not normally together on the job at the same time. Trade stacking could result from owner-caused delays.

turnkey project A project for which a contractor undertakes to deliver under contract a fully operational and tested facility before being entitled to payment.

unbalanced bids A proposal under which the contract price is allocated to phases or items in the contract on a basis other than that of cost-plus overhead and profit for each bid item or phase. A common practice is to front-end load a bid proposal to obtain working capital to finance the project. Another form of unbalanced bid on unit-price contracts assigns higher profits to types of work for which the quantities are most likely to be increased during the performance of the contract.

unbilled receivables Receivables that arise when revenues have been recorded but the amount cannot be billed under the terms of the contract until a later date. Such balances may represent unbilled amounts arising from the use of the percentage-of-completion method of accounting, incurred costs to be billed under cost-reimbursement-type contracts, or amounts arising from routine lags in billing.

undivided interests in joint ventures Each investor/venturer owns an undivided pro rata share of each of the assets and owes an undivided pro rata share of each of the liabilities of the joint venture; thus, he has an undivided interest in the net equities of the venture.

uninstalled materials Materials acquired for a specific contract and usually stored at the job-site awaiting installation.

unitary jurisdiction The practice of some state taxing jurisdictions to group the financial activities of related entities for income tax reporting purposes. Apportionment of the combined entities' tax results may then have to be made to two or more states. *See also* apportionment (for income tax purposes).

unit-price contract A type of construction contract providing for compensation to the contractor based on a specific price for a specific unit of production.

units-of-delivery method An output method of measuring progress of a contract toward completion in terms of results achieved (e.g., tonnage of linear steel erected).

units-of-work performed An output method of measuring progress of a contract toward completion in terms of results achieved (e.g., cubic yards of concrete poured).

unpriced change orders Change orders for which the work to be performed has been defined and agreed to, but for which the adjustment to the contract price is to be negotiated later.

upset price *See* ceiling (or upset) price.

usage variance A variance of actual direct materials cost from standard direct materials cost. The difference between the two products obtained by multiplying the actual and standard amounts of materials used by the standard price of materials.

use rate theory A method of establishing operating unit costs for construction equipment. The following factors are considered: (1) the cost of the equipment, less estimates of its salvage value or rental if it is leased; (2) the probable life of the equipment; (3) the average idle time during the life or period of hire of the equipment; and (4) the costs of operating the equipment (e.g., repairs, storage, insurance, and taxes). A rate may be arrived at that, based on the reported use of the equipment, will serve as a basis for charging the contracts on which the equipment is used.

usury Unlawfully high interest rates.

waiver of lien An instrument by which the holder of a mechanics or materials lien against property formally relinquishes that right.

warranty (maintenance) period A specified period, which is normally specified in the contract, after the completion and acceptance of a project, during which a contractor is required to provide maintenance construction and for which the contractor is required to post a maintenance bond.

work-in-progress Construction contracts in an uncompleted or undelivered state.

wrap-up programs Owner-controlled insurance programs, whereby the owner decides to furnish the insurance for the project and requests that the contractor reduce his bid accordingly. In some instances, the programs are "contractor-controlled," where the contractor controls the placement of casualty insurance and possibly surety bonds for subcontractors.

zero estimate of profit/zero profit margin A contractor's estimate of total contract revenue and total contract costs of equal amounts, so that no profits are assumed. This is appropriate in circumstances where estimating the final outcome may be impractical, either in terms of single amounts or ranges of amounts, except to assure that no loss will be incurred. A change to a more precise estimate should be made where results can be estimated more precisely.

Index

[Chapter numbers are boldface and are followed by a colon; lightface numbers after the colon refer to pages within the chapter.]

A

[Chapter numbers are boldface and are followed by a colon; lightface numbers after the colon refer to pages within the chapter.]

[Chapter numbers are boldface and are followed by a colon; lightface numbers after the colon refer to pages within the chapter.]

[*Chapter numbers are boldface and are followed by a colon; lightface numbers after the colon refer to pages within the chapter.*]

[Chapter numbers are boldface and are followed by a colon; lightface numbers after the colon refer to pages within the chapter.]

[*Chapter numbers are boldface and are followed by a colon; lightface numbers after the colon refer to pages within the chapter.*]

[Chapter numbers are boldface and are followed by a colon; lightface numbers after the colon refer to pages within the chapter.]

[Chapter numbers are boldface and are followed by a colon; lightface numbers after the colon refer to pages within the chapter.]

[*Chapter numbers are boldface and are followed by a colon; lightface numbers after the colon refer to pages within the chapter.*]

[Chapter numbers are boldface and are followed by a colon; lightface numbers after the colon refer to pages within the chapter.]

[Chapter numbers are boldface and are followed by a colon; lightface numbers after the colon refer to pages within the chapter.]

[Chapter numbers are boldface and are followed by a colon; lightface numbers after the colon refer to pages within the chapter.]

[Chapter numbers are boldface and are followed by a colon; lightface numbers after the colon refer to pages within the chapter.]

[Chapter numbers are boldface and are followed by a colon; lightface numbers after the colon refer to pages within the chapter.]